UNIVERSITY OF WINNIPEG
LIBRARY
515 Portage Avenue
Winnipeg, Manitoba R3B 2E9

Developments in Geotechnical Engineering 4B

COASTAL ENGINEERING, II

Further titles in this series:

1. *G. SANGLERAT*
THE PENETROMETER AND SOIL EXPLORATION

2. *Q. ZARUBA AND V. MENCL*
LANDSLIDES AND THEIR CONTROL

3. *E.E. WAHLSTROM*
TUNNELING IN ROCK

4A. *R. SILVESTER*
COASTAL ENGINEERING, I

Developments in Geotechnical Engineering 4B

COASTAL ENGINEERING, II
Sedimentation, estuaries, tides, effluents, and modelling

by

RICHARD SILVESTER

Department of Civil Engineering, University of Western Australia, Nedlands, W.A., Australia

ELSEVIER SCIENTIFIC PUBLISHING COMPANY
Amsterdam   London   New York   1974

ELSEVIER SCIENTIFIC PUBLISHING COMPANY
335 Jan van Galenstraat
P.O. Box 211, Amsterdam, The Netherlands

AMERICAN ELSEVIER PUBLISHING COMPANY, INC.
52 Vanderbilt Avenue
New York, New York 10017

Library of Congress Card Number: 72-97435

ISBN 0-444-41102-X

With 140 illustrations and 17 tables.

Copyright © 1974 by Elsevier Scientific Publishing Company, Amsterdam

All rights reserved. No part of this publication may be reproduced, stored in a retrieval system, or transmitted in any form or by any means, electronic, mechanical, photocopying, recording, or otherwise, without the prior written permission of the publisher,
Elsevier Scientific Publishing Company, Jan van Galenstraat 335, Amsterdam

Printed in The Netherlands

PREFACE

This book and its companion volume have evolved from courses developed at the Asian Institute of Technology (A.I.T.) in Bangkok, Thailand. These consisted of Coastal Engineering, Wave Hydrodynamics, Coastal Sedimentation, Tidal and Estuarine Hydraulics, and Coastal Modelling. It is envisaged, therefore, that similar courses at graduate or higher undergraduate level could be served by these tomes. Examples and problems have been included to aid in any such teaching programme.

However, another main purpose of the works has been to collect the latest information available in this fast developing field and to present it in a form for ready application. They should therefore serve as hand-books for the practising engineer who is concerned with coastal problems, either periodically or perennially. Even if a maritime project is considered too involved for him or his organisation, he can become aware of the complexities before calling in a consultant.

A third group of workers who could benefit from these books, particularly Volume II, are geologists and geographers who are interested in coastal geomorphology. Wave action can produce many submarine and shoreline features, which have here-to-fore been "explained" by tidal and other oceanic currents. To the author's knowledge, there is no aspect of submarine geology on which unanimous agreement has been reached respecting its origin. This may be due to the omission of this forceful factor of wave generation in the storm areas of the oceans and the concomitant spread of energy to their distant margins. A better understanding, therefore, of the principles of wave characteristics and beach processes should aid the descriptive scientist in his observations of nature and subsequent reporting of same.

Unlike many other treatises on hydraulics and wave mechanics, these omit the differential equations and manipulations leading to the final formulae. Where a researcher wishes to further the topic, and must go to the original works, ample references are supplied as a lead into the literature. It was felt that the practising engineer, always a busy person, wants to have a handy solution, preferably in the form of graphs or tables, which is suggested to him from the two or more generally available. The author has tried to digest the available information in order to recommend a course of action. This is a great responsibility, of which he is fully conscious, but is happy in the thought that his suggestions should be better (if only a little) than the apparently undisciplined guesses made by some engineers in the past and even the present.

The minimization of mathematics is made without remorse. The engineer's role, in any case, is to proceed where angles, scientists, and mathematicians alike, fear to tread. There are far too many practical problems for him to be overly concerned with deriving, or redeveloping or checking theoretical analyses. He must, of course, be aware of the premises upon which they are based, and know the simplifications used in obtaining a solution. There are occasions when he must substitute constants into a relationship which are not proven beyond reasonable doubt, but therein lies the "art" of engineering.

The necessity to divide this treatise into two volumes has the one advantage that readers interested mainly in shoreline processes do not have to invest money, and perhaps time, in the more mathematical topics of wave forecasting, propagation, and influence of structures. However, one major disadvantage is the unavailability of discussion on wave generation and general wave characteristics. To obviate this disability somewhat the characteristics of ocean waves are reiterated in chapter 1, especially as they relate to short- and long-term sedimentary processes. It will be assumed that the reader is aware of wave and meteorological terminology, either from reference to Volume I (the term to be used for the companion tome to this book), or from the other references listed therein, which supply a glossary of marine expressions.

Acknowledgements

The author's interest in coastal engineering was initiated in 1954 by Professor K.L. Cooper, then Head of the Department of Civil Engineering at the University of Western Australia, who subsequently encouraged the author in his researches. During sabbatical leave from the University in 1956-7 the author spent six months at the National Institute of Oceanography in the United Kingdom, where the complexities of wave generation and other coastal processes were imbibed from the comprehensive N.I.O. library and discussions with oceanographers. Thanks are expressed to Dr. G.E.L. Deacon F.R.S. for his warm support. It was during this period that the importance of the crenulate shape of shoreline first struck the author, which led later to model studies at his own campus. During a second study leave (1963-4) the author spent twelve months at the University of California, Berkeley, whilst in receipt of a National Science Foundation fellowship. There he studied many geological phenomena which could be related to the action of ocean waves. His contact with such workers as Professors J.W. Johnson, R.L. Wiegel and H.A. Einstein helped his understanding of many facets of wave forces and sediment transportation.

In 1967 the University of Western Australia granted the author leave of absence to fulfil a SEATO appointment as Professor of Coastal Engineering at the SEATO

Graduate School of Engineering (now an autonomous body called the Asian Institute of Technology). During his four years in Thailand the author was supported by the Australian Department of Foreign Affairs. It was here that a treatise was conceived, which ended up as two volumes. The thesis projects supervised by the author helped clear up some outstanding problems. Special mention should be made of Dr. Suphat Vongvisessomjai, who devoted much spare time with the author in modifying the latest wave forecasting formulae to the spectral form found in this volume. Dr. M.E. Bender, President of A.I.T., was extremely encouraging in the introduction of this new specialty, and provided funds for the conduct of a three-week "Institute" for practicing coastal engineers from a dozen Southeast-Asian countries. This helped the author learn of the problems facing men in the professional front line and has, it is hoped, improved the presentation. After returning to Western Australia in 1971 the author was granted sabbatical leave on campus in order, among other things, to complete this work. To these various institutions and personnel in authority the author expresses his deep sense of gratitude.

Besides the many oceanographers and engineers with whom the author has been in direct contact, there are others to whom he has written for personal explanations of points not understood, or for data not contained in publications. Thanks are expressed to Professors N.H. Brooks of Caltech, A.T. Ippen of M.I.T., T. Hayashi of Chuo University, J.B. Herbich of Texas A & M, K. Horikawa of Tokyo University, W.J. Pierson Jr. of New York University and W.H. Munk of University of California, San Diego, and to Doctors D.E. Cartwright of National Institute of Oceanography (U.K.), B.D. Dore of University of Reading and S. Senshu of Central Research Institute of Electrical Power Industry, Tokyo, and Messrs. P. Donnelly of Department of Public Works, Canada, H. Sakai of Ministry of Construction, Japan, H. Yamashito of Ministry of Transport, Japan, and T. Yano of Ministry of Agriculture and Forests. These and many others have borne the brunt of the author's copious correspondence. Special thanks are due to Mr. L. Draper of the National Institute of Oceanography (U.K.) who supplied copies of wave records for analysis.

Richard Silvester

CONTENTS

Preface V

Chapter 1. Shoreline processes 1
  Nature of ocean waves 1
    Storm waves 2
    Swell waves 3
    Angled wave trains 4
  Beach processes 6
    Surf zone 7
    Offshore zone 7
    Swell profile 10
    Storm profile 12
    Meteorological cycles 15
    Enclosed seas 17
    Tidal range 18
    Ocean currents 20
    Estuarine conditions 22
  Calculation of beach profiles 24
    Offshore profile 24
    Swell to storm profile 25
    Theoretical analysis 26
  Calculation of littoral drift 28
    Littoral current 29
    Littoral drift 29
  Offshore conditions 33
    Theory for motion of sediment 33
    Shear-stress approach 35
    Other shear-stress approaches 39
    Practical aspects of shear-stress approach 40
    The macroscopic approach 42
    Description of particle motion 43
    Effects of permeability 52
    Reach of waves 53
  Transport rates in surf and offshore zones 57
  Problems 62
  References 64

Chapter 2. Long-term longshore drift ... 71

- Equilibrium shapes of bay ... 72
  - Field observations ... 74
  - Model studies ... 75
  - Equation of beach plan ... 77
  - Test of stability ... 81
- Physiographic units ... 84
  - Contour spreading ... 85
  - Shoals ... 87
  - End conditions ... 89
- Continental patterns ... 90
- World-wide pattern of net sediment movement ... 94
- Geomorphological study of a coastline ... 98
  - Sediment supply ... 98
  - Wave climate ... 99
- Physiography and climate of Japan ... 100
  - Wave climate ... 101
  - Sediment movement ... 106
- Deltas ... 113
  - Pattern-forming forces ... 114
  - Regression analysis ... 119
  - Apex to sea length ... 120
  - Delta area ... 120
  - Maximum width of delta ... 121
  - Maximum width to number of mouths and of distributaries ... 122
  - Delta topography ... 122
  - Side topography ... 123
- Problems ... 124
- References ... 125

Chapter 3. Coastal defense ... 127

- Groins ... 128
- Headland control ... 131
- Inlet entrances ... 134
- By-passing ... 139
- Shoreline harbours ... 140
- Walls and revetments ... 143
- Beach nourishment ... 145
- Estuarine problems ... 146
- Dredging ... 148
- Problems ... 153
- References ... 155

Chapter 4. Long-period waves 159
- Tides 160
  - Influence of continental shelf 163
  - Transitions—no friction 163
  - Transitions—with friction—progressive wave 166
  - Transitions—with friction—total reflection 172
  - Typical tides 174
- Storm surge 178
  - Wind stress 178
  - Enclosed bodies of water 180
  - Continental shelf 183
  - Examples of Hong Kong surges 188
- Examples 190
- Problems 196
- References 199

Chapter 5. Estuarine problems 203
- Salt-water wedge 203
- Mixing processes 211
- Estuary pollution 218
  - Dispersion model 220
  - Tidal prism concept 222
- Estuary sedimentation 224
- Estuary control 228
- Examples 230
- Problems 233
- References 236

Chapter 6. Marine hydraulic works 239
- Sewage disposal 240
  - Circular jet uniform density medium 243
  - Circular jet in non-uniform density medium 246
  - Slotted jet in uniform density medium 249
  - Slotted jet in non-uniform density medium 253
  - Experimental verification 254
  - Mixing tubes 256
  - Surface spreading 259
  - Miscellaneous considerations 260
- Cooling water structures 262
  - Outfalls 262
  - Intakes 266

Examples	274
Problems	279
References	282
Chapter 7. Hydraulic modelling	285
Similitude	286
Waves	287
Tides and currents	292
Movable bed models	294
Onshore wave action	296
Offshore wave action	298
Examples	307
Problems	309
References	311
Appendix. Tables of functions of d/L_o	315
Index	329

Chapter 1

SHORELINE PROCESSES

After a summary of wave incidence, the action of waves on the beach and sea bed will be discussed in what will be termed a macroscopic outlook, which considers the coastal profile as a whole. The influence of meteorological cycles, geographical location, and tidal fluctuations will be outlined in turn. The state of the art in determining actual beach profiles is then treated, plus the special conditions introduced by ocean currents or estuarine locations. Finally, the importance of the continental shelf as a sediment-transporting feature is stressed. This entails a microscopic approach, in which the forces acting on individual sand grains must be considered.

The complexities of the sources of energy and of the boundary conditions should soon become apparent. Due to this wide spectrum of soil and active forces, it is very difficult to supply quantitative assessments of currents or of rates of sediment transport. However, a knowledge of the processes involved can help the engineer to reach a decision on ultimate tendencies regarding siltation or erosion. Because of the difficulties of verifying theoretical analyses, either in models (due to the necessary distortions) or in the field (due to the changeable energy input and magnitude of the measurements), some of the views expressed can be termed tentative. They should be tested against observations of nature and, it is hoped, indicate fruitful lines of research.

Shoreline processes have been described in many articles [1–8], as well as in many text books. Views expressed herein may differ either slightly or substantially from those expressed by others. The author believes that the area beyond the breakers is of paramount importance in beach processes, whereas other workers have emphasized the role of the surf zone. But what occurs near the beach is influenced greatly by what has occurred offshore previously. The lag time for this influence to be felt has caused engineers to seek other explanations for failures in maritime structures. There is a strong need for publication of the success or otherwise of coastal works some years after they are carried out.

NATURE OF OCEAN WAVES

It is worthwhile to summarize wave characteristics as they apply to movement of sediment on the sea bed. All water waves originate in zones of medium to high wind velocities, which are normally associated with storm centres. Some ocean areas experience such storm sequences almost continuously, for example the 40°–60° latitudes in both hemispheres, whereas most coastlines experience them for short durations throughout the year. When a storm is in close proximity to the coast, the waves are still being generated as they arrive at the continental shelf, and are therefore termed *storm waves*. The waves that have emerged from the fetch, and possibly travelled hundreds or even thousands of miles to the shore, are called *swell waves* or just *swell*. These latter waves are generally of much longer duration, or more persistent, than the storm waves. This is because storm waves last only as long as the winds producing them, whereas the swell consists of these waves spread out across large expanses of ocean. The long-period waves arrive first, with the shortest waves of engineering significance (say 5-sec period) reaching the coast some days later. Snodgrass et al. [9], in measuring swell waves traversing the Pacific Ocean, observed that a wave system generated over two days in the Antarctic lasted for one week on the coast of Alaska. Thus, storms occurring for about 30% of the time could provide continuous swell at some distant shore.

Storm waves

The storm wave will have the specific connotation that it is still under the influence of the wind, or is in the fetch. Thus, for storm waves to reach a shoreline, the wind field must be close to the coast and the wind itself directed at some angle towards the land.

Storm waves are characterized by their many wave lengths or wide spectrum. From the inception of the wind the small-period components grow quickly until they reach an unstable steepness, at which stage they break. This breaking process appears to add energy to the longer waves which grow more slowly. For any given wind velocity a steady wave condition can be reached when the energy being transferred from the wind to the sea is being dissipated at an equal rate by all components, except the longer ones, breaking. The distribution of wave energy within the spectrum has been detailed in *Coastal Engineering, I*. The important aspect to be noted here is the excessive steepness of the waves within the fetch. Each train contains a large volume of water above the mean sea level and so exerts a special influence when it is virtually poured onto the beach.

Whilst the waves experience the force of the wind, they are tilted forwards. This tends to instability and hence promotes breaking as the shallower depths are reached. This ready formation of surf has its unique influence in the beach cycle.

Another feature of storm waves is the many directions in which the various wave components are travelling. This arises from the mode of generation. The smaller-period waves present will, in general, be angled more obliquely to the average wind direction, whilst the orthogonals of the longer waves will be closely aligned to it.

An additional factor that amplifies the multi-directional nature of storm waves is the motion of the storm centre itself. The fetches bringing waves to a specific point on a coast are changing location and direction very swiftly. Thus, throughout a storm cycle waves can arrive from a wide fan of directions.

The random direction of approach of storm waves, together with their relatively short duration throughout the year (compared to the more persistent swell) results in them having little influence on the movement of sediment *along* the shore. However, as will be seen, they effect a prodigious transfer of beach material laterally to the offshore zone. Where a section of coast can only receive waves generated locally, or swell is minimal, a net longshore drift can ensue from the repeated oblique approach of the larger storm waves. In essence, this vector resultant would include the directional component of wave energy times duration.

Swell waves

Once storm waves move out of the fetch, and are therefore no longer receiving energy from the wind, they enter what is commonly termed the *decay area*. A better title is *dispersal area,* since negligible "decay" is taking place through viscous, turbulent or eddy processes. The energy of the waves in the fetch, in fact, is just dispersing across the ocean. This involves spreading of waves both radially and circumferentially, mainly to the downwind zone outside the area of generation. As noted by measurements in the Pacific Ocean [9], little or no energy is actually lost by swell waves as they propagate thousands of miles across the globe. They are, therefore, a very efficient distributor of wind energy, from the storm centres of the seas to the margins of the land masses.

The direction in which waves are first generated is maintained by each train in its passage from the fetch. The waves arriving at a point some distance from the fetch can only arrive within a directional fan which encompasses the width or diameter of the generation area. The further away the point, the narrower the band of directions in which waves can be received. This means that the crest alignments are closer together and the waves become more or less aligned with each other. Thus, at a reasonable distance from a storm, 1000 miles or so, swell waves may arrive at a point on the coast from essentially one direction. Even if the storm zone were to be displaced laterally a hundred miles or so, the approach angle to a coast may not alter more than a degree or two.

This angular dispersion reduces the height of swell waves significantly within two fetch widths from the fetch (down to 60% of the combination wave termed the significant height or $H_{1/3}$) as discussed under "Wave-generation processes" in *Coastal Engineering, I.* At very long distances the reduction is to one third of the $H_{1/3}$ value in the fetch itself, reduced further as noted below. The distribution of energy in the various components of the full spectrum is such that the bulk of the energy in the dispersal area is contained within a 30°-fan either side of the wind vector of the fetch. At any radius from the downwind end of the fetch (outside three fetch widths distance) the energy within this fan is fairly evenly distributed around the circumference.

The waves also suffer radial dispersion, with the longer-period components of the storm-wave spectrum racing ahead of the smaller ones. Since there are less wave trains to interact, the composite wave height ($H_{1/3}$) is reduced. It is probable that a single train of waves, or a very narrow band of periods, will be present at any one time. This occurs when the distance from the fetch is great or the duration of the storm is small.

It has already been noted that radial spreading results in a swell duration time much longer than the storm sequence. Even so, the wave conditions do not remain steady, since the longer waves arrive first, followed by components containing the optimum energy, and followed again by the waves with reduced height and period. This makes it difficult to assess average values for a coastline, even for the simple case of swell incidence alone.

In travelling through ocean areas of little or no wind, the swell waves suffer no pressure or shear stresses. This, together with their reduction of height due to the dispersion discussed above, causes the wave profile to be nearly sinusoidal. This is a very stable form for propagation and assists in the conservation of energy over thousands of miles of ocean. Even where local breezes generate short choppy waves on top of the longer-period swell, there is negligible attenuation of the latter. This was proven in the Pacific measurements [9]. It is understandable when it is realized that these short-period waves may produce water-particle oscillations down to very limited depths (say 6 fathoms for 4-sec waves), whereas swell of any significance penetrates much deeper (for example 60 fathoms for 12-sec waves).

Angled wave trains

The mode of bed disturbance by two wave trains travelling at an angle to each other has not been investigated very fully by researchers to date. It alters the problem of progressive waves from two dimensions to three dimensions. The mathematics becomes much more difficult, even when only linear theory is employed. As noted in the section on wave theory in Volume I of *Coastal Engineering*, such angled wave trains produce peaks in the water surface where the crests of the trains intersect, and so give the appearance of a short-crested system. It is believed that the water-particle motions near the bed, which have the greatest influence on the sedimentary material, can contain eddies and net motions which are of paramount importance in the transport mechanism. Since such combinations of waves are the rule rather than the exception in nature, there appears to be a dire need for research into this phenomenon.

Although not quite in the context of the above description, it should be recounted that waves inside a fetch will be travelling in many directions, producing a short-crested system of extreme complexity. The water particles at all depths and lateral locations are moving in random orbits and producing a turbulent structure which is conducive to suspension of sediment or any other material, even though there may be little capacity for horizontal transport. This production by storm waves of macro-turbulence can be hazardous for power stations and refineries, when sand and seaweed are inducted into their cooling water intakes, even when these are set up some metres from the sea bed.

Within one fetch width distance of the storm area the waves will still contain a large variety of trains, which will be travelling in a wide range of directions. Thus storm-type waves can approach a beach even though the wind there may be negligible. The only difference in the sea surface from the normal fully-arisen state will be the absence of breaking and the more stable shape of all the component trains. The water-particle motions near the bed, however, will be as complex as those within the fetch. In fact, being shallower water near the beach, the oscillations will be magnified over those present at the bed in the deeper offshore zone.

Across the width of the continental shelf, running out to a depth of around 60 fathoms, swell waves are generally short-crested. This can arise from a storm not far distant from the shelf edge as noted above. But also, two trains aligned in deep water and approaching the shelf obliquely, will be refracted at different rates. This will cause them to become angled to each other at intermediate depths across the shelf, until they become almost aligned again near the beach prior to breaking. The angular difference will reach a maximum at some intermediate depth dependent upon the periods of the two trains concerned.

Another source of angled wave trains is diffraction behind an offshore breakwater or island. As waves propagate beyond each end of the obstruction they spread there energy into the shadow area, forming arcuate crests that intersect each other to form a short-crested wave system. The resultant net movement of sediment on the bed is to the centre of the leeward zone, which soon becomes shoaled. It will be seen later how this can be a menace or a useful measure, depending upon its application as a boat shelter or as a coast protection unit.

Another important source of angled wave trains is reflection of oblique waves from sea walls or cliffs. In essence, this is a double application of wave energy to the sea bed. This expedites the passage of sediment past this area, even though the water may be deeper than normal. Such rapid transport of material around headlands has been reported [10]. The use and abuse of reflected waves will be discussed in the section on coastal defense.

From the above remarks it is not difficult to conceive that much of the continental shelf is traversed by angled wave trains. The near-normal approach and parallel crests of waves arriving at the beach, all at approximately the same speed, can be quite deceptive as far as offshore conditions are concerned. They tend to conceal the complexity of the offshore conditions, both at the surface and at the bed. Improvements are needed in the replication of prototype water motions near the ocean bed. To more truly represent the boundary layer and sediment-particle dynamics, such modelling needs to be conducted full-scale. Also to cope with the problem of short-crestedness, the water particles need to be oscillated in two directions simultaneously, with a phase relationship of each to suit the angle of the trains and the position of the point being represented in the system: a detailed discussion of previous and current attempts at reproduction of sediment movement due to wave action is included in Chapter 7.

BEACH PROCESSES

As waves approach the shore they slow down and hence become shorter in length and greater in height. This increase in steepness reaches a limit of stability, after which the waves break. This process of forming a bore or surge of water to the beach, produces two distinct zones of action, as far as sediment movement is concerned. There is the *surf zone*, between the breaker-line and the beach, and the *offshore zone* seawards of the breaker-line.

The extent of the surf zone is generally well defined, although its width varies substantially with the height and period of the waves reaching the shore. As will be shown later, the width of the surf zone depends also on whether swell or storm waves are present, and the stage of transition from one beach profile to another. The slope of the bed near the beach is also a factor, which is determined in part by the size of the sand particles present, but this is relatively constant for any section of coast.

The offshore area that is active in sediment movement has yet to be fully ascertained by laboratory and field measurement. However, there are clear indications that the complete continental shelf, where this is sedimentary, is a highway for such transport. In the outer reaches the longer-period waves alone can generate sufficient water-particle motion near the bed to effect a shear stress and turbulence on the sediment particles. In the discussion on wave generation in *Coastal Engineering, I* the predominance of the swell around 13 sec was stressed. Such waves can readily be produced by 30-knot winds in a fully arisen sea and by higher velocities over shorter fetches or durations of wind. The deep-water wave lengths of a 13-sec wave is around 850 ft., so that it commences to be affected by the bed at the 70-fathom depth, or in turn commences to influence the bed by the production of water particle oscillations at this depth. As noted by Shepard [11, p.258], the more detailed hydrographic surveys have indicated the edge of continental shelves, where waves are prevalent, average 59.2 fathoms. Beyond this point the sea-bed steepens to form the continental slope, which runs down to the abyssal depths.

The implied massive distribution of terrigenous material, plus marine debris, out to these widths and along such stretches of shoreline, can readily be conceived when geologic time is considered. Since the genesis of the ocean, waves have been generated, and their energy has been distributed to the continental margins per medium of the swell, there to act on the material fed to them by the rivers. Over such eons of time, not only have continental shelves been constructed, but wave action has been instrumental in building the sedimentary portions of the land masses. These constitute 75% of their area. This aspect will be discussed more fully later.

Surf zone

Waves arriving obliquely to the coast are refracted as they proceed across the continental shelf. Their crests thus become more aligned with the underwater contours, or their paths of motion become more normal to the shore. However, they are bound to be still slightly angled to the beach when they break, so that the resultant surge will have a longshore component. The height and angle of breaking of the wave depends on wave steepness as well as the deep-water approach angle, both of which can change from hour to hour. This fact reduces confidence in computations of average wave energy.

The longshore component of the surge force induces a current which is termed the *littoral current*. Even though the energy input to this current may change continually, its direction and velocity can remain reasonably steady over days and weeks. The direction from which this water arrives at a point on the coast is termed *up-coast* and the area to which it is flowing is called *down-coast*. Due to the turbulence of the breaking and surging of the waves, large volumes of sediment are placed in suspension or rolled along the bed in the surf zone. The littoral current will act on these particles and so transport them along the shore. This material is termed *littoral drift*.

Offshore zone

The nature and width of the offshore zone varies greatly, it being influenced as much by tectonic movements as by sedimentation. In this respect, average characteristics of continental shelves are of little significance to the engineer, but some figures quoted by Shepard [11, p.257] may give a sense of proportion: (*a*) average width 40 nautical miles (N.M); (*b*) average edge depth 72 fathoms; (*c*) average of flattest portion 35 fathoms; (*d*) average slope 0.002 or $\tan^{-1} 0°07'$ (steeper near the shore).

Shelves with predominant wave action have an edge depth around 60 fathoms. Shepard suggests that another survey be made of these averages, using more up-to-date hydrographic data. It is submitted that classifications be made for which averages may be more relevant, such as: (a) oceanic margins — western and eastern; (b) enclosed sea margins; (c) latitudinal divisions $0°-25°$, $25°-40°$, $> 50°$; (d) sedimentary or base rock.

From these the influence of waves in constructing this submarine feature could then be verified. One method of determining the sedimentary nature of a section of shelf is to observe its smooth outer contours. If it is of orogenic origin these contours will be more indented. However, more information on the nature of shelf material is becoming available each year in the geologic literature.

It has been stressed in the discussion of wave incidence above that the waves are likely to be more angled to each other in the outer regions of the shelf. Thus, besides the two-dimensional aspects associated with ripple and dune formation on the bed (which could be minimal at these great depths), there is the possibility of vortex generation which can place material into suspension for both short and long periods of time. Details of these forces on particles in the offshore zone will be presented later; for the present, certain tendencies existing in this zone will be outlined.

One of the most important actions of the waves is to sweep the sediment shorewards. This is effected by the mass-transport, or net movement, of the water particles, which at the bed is in the direction of advance of the waves. This applies particularly to long swell waves. In the case of storm waves, where there is a multitude of trains, the net motion at the bed is zero, and in the reverse direction of wave propagation just above the bottom. Because of the extra turbulence present on these storm occasions, this seawards mass-transport could help in spreading material away from the coast. However, the short duration of these storm sequences must be compared to the greater persistence of the swell which forces the sediment shorewards. This constructional tendency for shelf and coastal plain alike is indeed fortunate for mankind, since the majority of the land masses would not otherwise exist. If a continuous seaward motion ensued the weathered rock material of the world would be spread as large submarine fans across the ocean floor. As it is, the continental shelves make up only 4% of the ocean areas of the world.

Another characteristic of the offshore zone is the size sorting of bed material. The oscillatory motion of the water particles at the bed incorporates the net force of mass-transport shorewards on the sand particles. But the slope of the beach profile causes gravity to exert itself in a seaward direction. With their successive suspension or rolling, the particles find a location on the bed profile where these two forces balance [12–14]. Since the slope increases as the beach is approached, so does the size of the sediment increase in this zone of the profile. The coarsest material reaches the breaker-line and is forced high up the beach, the finer fractions being activated to the deeper zones offshore. The very fine particles, which are readily suspended, can be easily transported seawards during storm sequences.

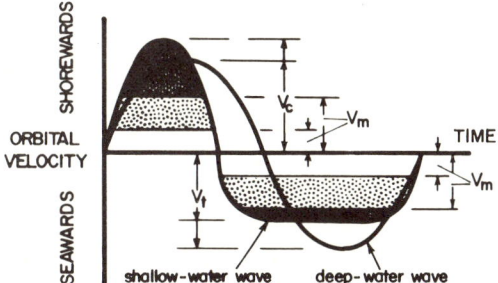

Fig. 1-1. Water-particle velocities at the ocean bed due to wave propagation.

The selective transport of coarser particles shorewards arises from the greater orbital velocities under the wave crests than under the troughs. These differential velocities are depicted in Fig. 1-1, where it is seen that motion under the crest (maximum value V_c) is shorewards and that under the trough (maximum value V_t) is seawards. If a minimum velocity (V_m) is required to move a sand particle of given size, the area formed by the intercept of this value with the variable velocity will be proportional to the distance the particle moves. As can be surmised from the figure, a minimum velocity V_m could exceed V_t but be less than V_c, so that a net shoreward motion would result. Again, a V_m value could be so small that the active area in the trough curve could exceed that under the crest zone. Thus small particles can migrate seawards whilst larger ones suffer a net motion shorewards.

The above discussion is an extreme simplification of a complex phenomenon, as will be gauged from discussion later, when the role of turbulence and of mass-transport of water due to boundary layer influences is discussed. It is the differential crest and trough orbital velocities that produce this mass-transport, the influence of sediment suspension on which is not fully understood. It should be realized also that a flat bed is implied in the simple picture above. A sloping or rippled bed will have differential effects on large and small sediment particles. However, there is no denying the general sorting tendencies as described.

The height and spacing of sand ripples and dunes, formed by the oscillatory motion of the water particles near the bed, is dictated both by the wave and sediment characteristics. For a long duration of a single wave train, with constant height, period and direction, bed undulations will reach an equilibrium form. This stable shape impedes lateral transport, as is the case for dunes in open channel flow. But when wave conditions change, either height, period or direction, a new dune system must be established. During the transition period from one system to another, there is an increase of material in pseudo-suspension, and therefore more opportunity for horizontal transport through mass-transport or any other currents existing near the bed. Changes in wave conditions, especially wave direction, would be more pronounced in the deeper regions of the shelf than nearer shore, making for shorter periods of stability. Thus tests on the ability of waves to move sediment in deep water, conducted with single wave trains on the one alignment (as in a flume), may give little appreciation of conditions existing in nature. The proof of this may soon be available through observations of the sea bed from research submersible vehicles.

Swell profile

As waves reach depths of around half their deep-water wave length they commence to disturb the bottom. The longer-period waves do so before the shorter ones. The mass-transport associated with this disturbance forces particles in the direction of wave advance. Although this may be minimal in magnitude at this outer limit, the fraction of it directed alongshore is greater than that for the same wave after refraction in the shoaler water. Even when gravity due to bed slope acts on the particle, together with mass-transport, the resultant vector can be directed downcoast. This is due to mass-transport being aligned with the wave orthogonal and gravity acting normal to the bed contour. When these are not co-linear sediment particles can oscillate in saw-tooth fashion along the shore, whilst still maintaining their equilibrium distance from the beach line. This is of great significance, since substantial longshore drift can take place well beyond any man-made structures meant to intercept it. This transport is also independent of the cyclic exchange of material from beach to offshore which, as seen later, is a feature of storm sequences.

Since the bed profile is determined in part by the wave climate, particularly the persistent swell, any change in this climate should result in changes of this profile. Changes of any significance would take a long time to effect [15] and hence, with a repetitive annual cycle of waves, the offshore region of the beach profile remains reasonably static over decades. This can be seen in successive hydrographic surveys made at intervals of 50 or more years.

As waves proceed into shallow water the mass-transport at the bed increases so that material is brought into the area where they finally break. The surging mass of water carries much material in suspension, the concentration increasing towards the bed. Each swell wave covers the beach face in an uprush, as depicted in Fig. 1-2A, most of which percolates through the sand down to the water table. Whilst there is a reasonable time between each wave and a fairly permeable beach face, there is not likely to be any saturation of the soil. With such disposal of the uprush the backwash from any wave will be minimal [16,17]. This reduced downflow velocity causes the sediment carried in suspension to the beach to be stranded on the beach face. The hydraulic jump associated with this return flow is also small. For these many reasons, swell waves cause accretion of the beach, so long as material is

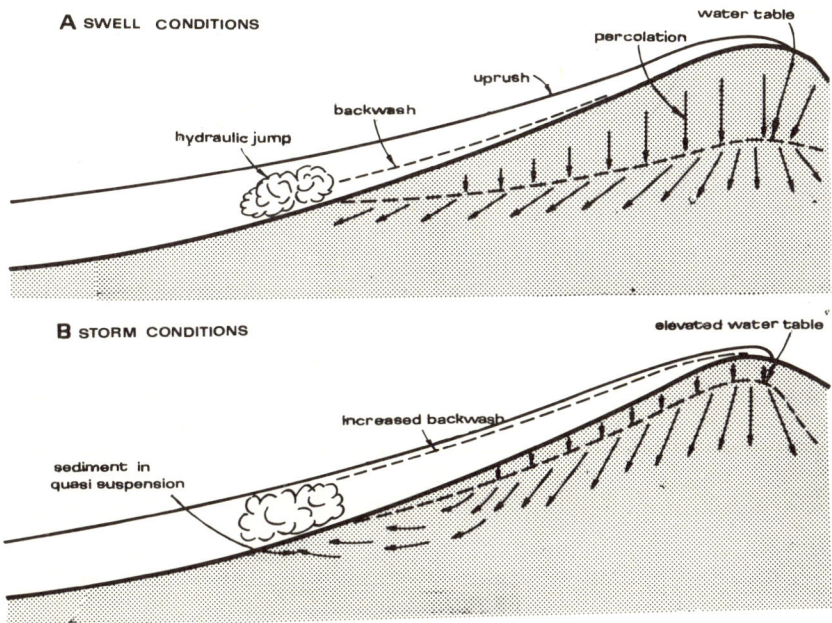

Fig. 1-2. Beach processes, A. under swell conditions; B. under storm conditions.

available from offshore to feed the incoming breakers. When the offshore zone deepens and steepens sufficiently the swell can no longer bring material in, and so the whole profile, from berm to offshore, becomes reasonably static.

Whilst the beach is aggrading at a rapid rate, the sand particles are lightly compacted, which aids in the percolation of the uprush. Walking along the beach face of such newly placed material results in ankle deep penetration. When the offshore supply has diminished the waves come in "empty-handed", but continue to pound the beach face and so compact it.

As noted earlier, a sorting process takes place, which leaves the coarsest particles high on the beach face and berm. The slope of this face is dictated by the size of this sediment [18], varying from 1:30 to 1:10 for fine and coarse sand, respectively, and about 1:4 for gravel. This slope, in turn, determines the degree of reflection of the waves [19]. Compensatory tendencies are active here: the coarser the material the steeper the beach face, making for greater reflection, thus increasing the capacity of the wave to suspend and roll sediment [20]. The partial standing waves, produced by reflection from steeply sloped beaches and cliff faces, are sufficient to lift large boulders from the sea floor [21]. The influence of angled wave trains, resulting from reflection at walls and cliffs, has been emphasized already, suffice it to say that it is very difficult to maintain beaches in front of such features[22].

Storm profile

The swell-built profile described above will exist for most of the year on oceanic margins, where swell waves are persistent. Consider now the action of storm waves arriving at this relatively steep beach. These waves will have characteristics of excessive steepness, many wave trains and a great variability in direction. A crest will be arriving at the beach almost every second. Large volumes of water, in essence, will be poured onto the beach face. The beach soon becomes saturated and the water table becomes coincident with the beach face. This situation is shown in Fig. 1-2B, where it is seen that the backwash is nearly equal to the uprush. This downflow drags material from the surface of the beach face, which also gets undermined by the large hydraulic jump at its foot. At this mean water line also the increased groundwater flow is returning to the sea. Its upflow tendency at this point provides a quick-sand effect, which sets the stage for rapid removal of material. It is little wonder that large slices of beach can disappear within hours of the inception of a storm.

The large volume of water reaching the beach must return to the sea, to maintain a balance. This return flow is sediment-laden and therefore denser than normal, so that it travels outwards near the bed. This is encouraged by the strong shoreward drift at the surface caused by the onshore wind. The outgoing water carries its load until its velocity is sufficiently reduced that it drops it. In this manner an *offshore bar* is constructed, as illustrated in Fig. 1-3, made up of material taken from the beach. This bar continues to grow until its crest is high enough to steepen and break the majority of the incoming storm waves. This action is assisted by the seaward counter-current, which is fed by the excess water entering the longshore trough between the bar and the beach. In this respect the asymmetry of the storm waves promotes premature breaking. Once this wave dissipation develops fully, erosion of the beach ceases, since only the broken and turbulent surge proceeds shorewards, attenuated somewhat by the seaward current beneath it.

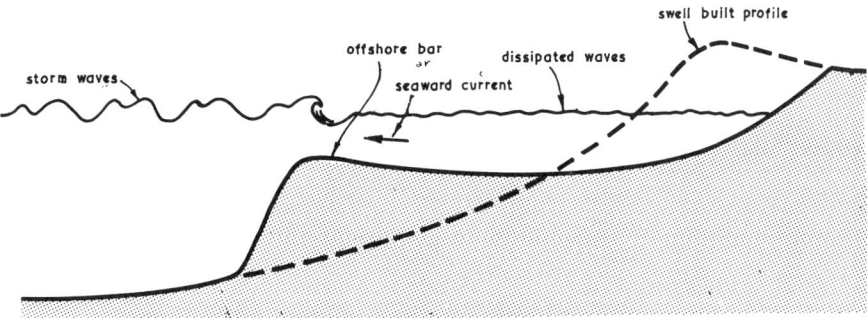

Fig. 1-3. Profile of beach under storm conditions.

Whilst waves are breaking at the bar, an inward flow is occurring at the surface and an outward flow near the bed of the storm-profile apron. This maintains a circulation of water from the beach to the bar and back again. This will be associated with suspension of material, but the net effect will be a consolidation of the bar at the expense of the beach. So that a prolongation of the storm sequence could result in a slightly increased erosion. This certainly is not commensurate with the initial onslaught during the bar's initial formation.

It should be obvious from the above discussion that, on an oceanic coastline which experiences persistent swell, the first storms of a winter season will cause the quickest and most spectacular beach erosion. It is generally on these occasions that the engineer concerned with coastal works is blamed for mis-management, notwithstanding that partial replenishment will be effected soon after the abatement of the storm. Erosion reaches its peak within hours, partial recuperation a day or two later, whilst complete restoration of the bar material to the beach takes weeks or even months. Any successive storm in the same season is unlikely to cause denudation of the beach to any further degree, since the offshore bar (perhaps partly returned) is soon reformed to carry out its role of wave dissipator. Should the succeeding storms be accompanied by extra high water levels, of tidal or storm-surge origin, the bar will have to be heightened in order to break the incoming waves, which will require more material from the beach.

The return of sand from the offshore bar and apron is accomplished by the swell waves. Those arriving from the same storm soon skin off the top of the bar and sweep it to the beach. Hence the swift recovery noted above. As normal swell conditions return, so more of the material is returned to the beach. It may take three or four weeks after the last storm of the season for the hump of the winter profile to be removed.

During this recovery period the water pouring into the trough by the breaking swell waves must return to the sea. This is generally by means of a *rip current*, formed at some low point of the bar. This swift seaward flowing stream is fed by longshore currents either side. These can be a hazard to bathers who are having their first swim after the storm season. Coastal engineers should publicize this danger annually and inform the public that the best way to stop being carried out to sea is to swim along the shore (that is transverse to the outgoing rip) and not attempt to swim against the current. Even if a person is unable to extricate himself completely from the rip, he is more likely to reach the edge of it, and so be contained in a whirlpool just beyond the breakers. Once such a depression has been established in the offshore bar it is likely to promote rips for some weeks, so should be clearly marked as a danger area. This applies also to beach zones either side, in which feeder currents exist.

The volume of beach material demanded by the sea for the construction of a bar depends upon the depth offshore that requires filling. The steeper the swell profile the greater the erosion. Thus, the more time swell waves have available to sweep offshore material to the beach, the more severe the beach recession in the following storm season. A calm winter, with its lack of erosion, may thus provide a false sense of security, and could even encourage further encroachment of the beach area by business and leisure promoters.

It should have become clear by now that nature has provided an effective mechanism for protecting its mobile sections of shoreline. If this were not so, mankind would not enjoy the massive accumulations of coast that have obviously taken place over geologic time. This sedimentary land is necessary for the growth of vegetation. The coastal engineer should appreciate this protective procedure and provide an adequate "bank" of material at the beach to "lend" to the "stormy borrowers". This requires that structures of economic value be kept sufficiently far from the water line for the beach to provide the offshore bar, without jeopardizing their safety. Of course, the greater the long-term stability exhibited by a section of coast the less the reserves required in the "bank".

Before leaving the topic of storm action on a beach it is worthwhile noting certain features, man-made and natural, which can promote erosion. The impor-

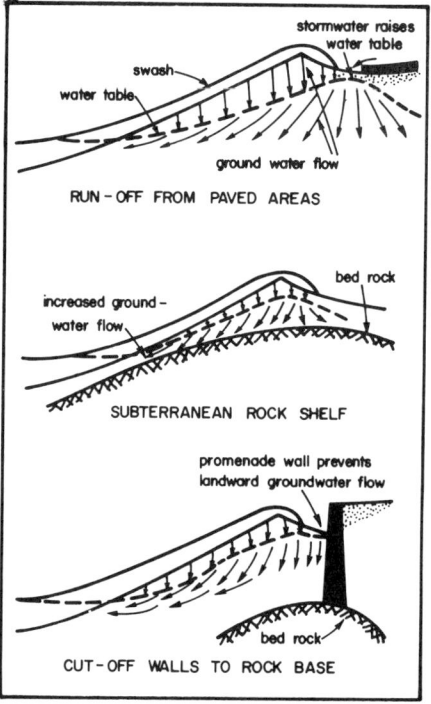

Fig. 1-4. Features that promote beach erosion.

tance of the water table in this mechanism is apparent, hence any condition that helps the saturation of the beach detracts from its stability. Three of these conditions are illustrated in Fig. 1-4, namely:

(*a*) the disposal of storm water onto the beach from large catchment areas, man-made or natural;

(*b*) the existence of a rock shelf just below the beach surface, which concentrates the flow of ground water to the sea;

(*c*) the construction of promenade walls (usually sited on base rock) which prevent the temporary flow of uprush water inland during storm sequences [17]. Once the water line reaches the vertical wall, the resulting standing waves accelerate the removal of sand and make re-establishment of the beach more difficult.

Notwithstanding the above tendencies, the existence or non-existence of the above features cannot influence greatly the disposal of thousands of tons of water being thrown onto the beach during the course of an hour's storm. Only the provision of sufficient width of sedimentary material for bar formation can provide the ultimate defense against erosion of coastlines. This point will be stressed again and again in this treatise.

Meteorological cycles

From the characteristics of storm waves, it can readily be conceived that there is generally little longshore drift during an erosion sequence. The multi-directional nature of these waves in any fetch, the swift changes in fetch direction during the passage of a cyclone, and the proportionately, small duration of storm conditions throughout the year; all combine to make the transfer of material offshore appear to be near-normal to the beachline. This is not to infer that there cannot be specific coastwise drift due to repetitive obliquity of storm waves in an area. Where swell waves are minimal, as on the coasts of inland seas and the eastern margins of continents, longshore drift due to storm waves is predominant, and important. This aspect will receive special treatment later, so that the following discussion will concentrate on coasts subjected to persistent swell. These constitute a large proportion of the coastlines of the world.

When the offshore bar is being returned to the beach under the action of the swell waves the littoral current will effect a longshore displacement. During the transition period the surf zone is wider than normal and the swell waves arrive more obliquely at the steep rise of the outer face of the bar as they break. If the swell arrives normal to the coast, the material is returned directly to shore. However, where littoral drift occurs, creating the coastal engineers' problems, the annual storm cycle results in a saw-tooth circuit, transversally offshore and then angled downcoast. By this process it is the swell that determines the net direction of motion of sediment along the shore, both within the surf zone, and beyond the breaker-line as noted earlier. Such cyclic "winter" removal of beach and its "summer" return downcoast would be of little concern if the longshore drift were in fact "in regime". By this is meant that material removed from a particular section of coast is replaced at the same rate from the adjacent upcoast area. This, however, is not often the case, either in the short term or in the long run.

In the short term many fluctuations in the rate of longshore drift occur. For example, the supply of sediment from rivers is in batches which pass along the coast in "humps" [23]. Again, the concentration of wave energy, particularly the longshore component, can vary significantly from place to place, from season to season, and from year to year. Also, the construction of a barrier to the longshore drift, either in the form of a structure, or of a dredged channel, will quickly exert its influence downcoast by intercepting material. The further a point is from this impediment the less the influence of any fluctuation and the longer the delay time before it is felt. This emphasizes the need for the coastal engineer to be conversant with developments upcoast, and to provide information on the status of his beach and his proposals to authorities downcoast. There is ample scope for a consultant to solve a problem at one point and create another further along the shore, either wittingly or unwittingly. There is certainly need for greater liaison between local councils and other bodies responsible for beach maintenance. This point has been promulgated for some decades[21] with apparently poor response.

In the long run, non-replenishment of the offshore zone in a particular stretch of coast will result in steepening of the beach profile, with a concomitant increase in erosion during subsequent winters. As stressed before, this offshore transport and removal is influenced more by the persistent swell, and hence it is the cause of beach "mal-de-mer". The storm waves only disclose the symptoms of the differential disease. If this were a problem of the medical profession it would probably be called "swellistence". In the discussion of mass-transport due to wave action, it will be remembered that the long swell of even minor height — and most of it is difficult to observe offshore — exerts a very strong sweeping action on the bed material. With any imbalance of supply and demand so the offshore bed is raised or lowered, without noticeable effect at the beach until the storm waves come to construct their protective bar. Since the speed of waves is determined by depth, any offshore deepening helps subsequent swell to approach the beach at greater velocity, obliquity and energy content.

The insidious influence of swell waves in eroding the offshore bed can only be observed by successive surveys of sections of coastline. Where the bed contours are seen to be approaching the coast, increased erosion at the beach line can be expected. Remedial measures should immediately be planned, since any continuation of the erosion can involve prodigious volumes of material. The annual or biennial hydrographic survey can be likened to an insurance premium which can save inestimable property damage if acted upon.

Enclosed seas

Mention has already been made of special conditions existing on margins of seas where oceanic swell is excluded or is otherwise minimal. These zones vary in size, shape and wind conditions, so that the wave heights and periods are quite dissimilar, but the common factor is their storm character. Such enclosed seas include the Mediterranean, Baltic, Black and Caspian Seas, the Sea of Japan and the Great Lakes of the United States of America. On these bodies of water waves are generated locally, or have little space in which to disperse before reaching a shoreline.

The result of this intermittent storm wave action is a more or less permanent offshore bar, because the swell is not available to sweep it back to shore. The longshore transport of sediment is almost completely confined to the surf zone, which necessarily is wider than the surf zone on an oceanic coast averaged throughout the year. The difference in the beach profile is not sufficiently emphasized in the literature, or in littoral-drift computations. Because of the extensive studies made of margins of enclosed or semi-enclosed seas, the picture is widely publicized of beaches generally having offshore bars and of most longshore drift occurring within the surf zone. A glance at the coastlines of the world would soon indicate the limited extent of these non-swell applications.

Although storm waves on enclosed seas will be generated in a reasonable variety of directions during the passage of a cyclonic centre, there could well be a resultant vector applicable to any specific point on the coast, due to the repetitive nature of storm paths. The duration and paths of low-pressure centres over a year, or decades, can be studied by the use of cyclonicity charts [24,25]. Knowledge of the general path of cyclones, their mean diameter and wind structure, should provide information on the likely approach direction of the bulk of the storm wave energy.

What is desired in the study of longshore drift is only the determination of the quadrant, either side of the normal to the coastline, in which the resultant occurs. This may involve an alternation throughout the year, from which an annual vector resultant might be computed. Where wind roses are available around the periphery of an enclosed body of water of moderate dimensions, proportionate durations of particular velocities in particular directions might be assessed. From this a resultant may be derived from vectoral addition of the wave energies over time. Since $E \propto H^2$ and $H \propto U^2$ then each vector should represent values of duration $t \times$ energy $E = tU^4$.

Another aspect of fully enclosed seas of major importance is the small or non-existent tidal oscillation. The water level remains sensibly the same throughout the year. This helps to retain the offshore bar at a fixed height above the bed and at a constant distance from the shore. In fact, all beach processes are carried out at the one level, such as construction of beach berms, sand spits and wave attack on structures. This may provide a false sense of security in the construction of harbour facilities, or protective dykes, at some modest level to cope with waves alone. However, the possibility should not be overlooked of a storm surge being generated by a strong wind field over a wide expanse of relatively shallow water. In November 1969, for example, such a surge was developed up the Adriatic Sea, which forced the water level 4 ft. above normal. This, with the upset level due to the storm waves, caused a catastrophe along the low-lands in the vicinity of Venice. The computation of storm surge is dealt with elsewhere in this treatise.

Partially enclosed seas receive both tides and swell with some restriction. Such sea zones include the North Sea, Gulf of Mexico and Caribbean, the East and South China Seas, the Sea of Okhotsk and the Bering Sea. Swell that enters through a limited entrance has a constant direction of travel within the sea. Around the periphery swell conditions will vary with locations in respect to this opening to the major ocean. Depending upon its duration, so will the offshore bar be completely or partially returned to shore. In respect to tides, some partially enclosed seas can be funnel-shaped and shallow, which conditions are conducive to large tidal ranges, which may have a paramount influence on beach processes.

Tidal range

It is proposed here to discuss only the effect of water-level fluctuations and not the currents associated therewith. The influence of tidal currents will be discussed under the heading of estuarine conditions. Currents, in fact, will be shown to be insignificant unless they are concentrated between islands or in narrow channels. For the majority of the open ocean coastlines of the world the mean spring tidal range is 2 m or less, as discussed in the section on tides. This variation of water level will have little influence on the beach processes previously discussed. In fact, larger fluctuations will have negligible effect in the offshore zone, even though they can produce significant change within the surf zone.

It should be stated at the outset that large tides tend to be reproduced by wide continental shelves, which in turn imply accretion of vast quantities of sediment adjacent to a shore. Thus, in the over-all picture, zones of large tidal range are associated with accretion rather than erosion. Nevertheless, problems of local beach degradation could occur where man-made structures, or variations in wave conditions or sediment supply, upset the material balance of a limited area. The problem for the coastal engineer in physiographic zones suffering large tides is generally siltation of harbours and channels.

As can readily be imagined, the main effect of large fluctuations in water level is the spreading of wave action over a greater width of the shore profile [26]. This involves a changing location of the surf zone, whilst the offshore width remains ostensibly constant in character, due to its vast width. The spreading of this energy also produces a flatter profile, which enlarges the surf zone, causing waves to break further from the water line. This fact is important during storm wave incidence, since the steep swell beach profile implied in the previous discussion of erosion does not now exist. It is more difficult for storm waves to devour a slice of beach, as is the case with an attack at essentially the one level.

Consider the case of storm waves arriving during a rising tide. Any demand made by them for construction of an offshore bar is spread over a greater horizontal width of beach. The seaward bed current carrying this material to sea is balanced somewhat by the flow of the incoming tide. This could well result in a steeper offshore bar which requires less volume of sand. Since most maritime structures in this situation would be sited beyond the high-water spring-tide level, they should be safe for most other stages of the tide. Should a structure be so located that during normal high tides it presents a vertical wall to the sea, care must be taken to see that the resulting reflection does not expedite undermining of the foundations. Even where facilities are positioned well above high tide, it should be remembered that the wide shelf producing the high tide can also produce storm surges, as discussed in the relevant section of this treatise.

Storm waves arriving during a falling tide will be readily dissipated on the bar built during the high-tide period. This is due to the decreasing depth over the bar as the water level drops. Hence, at this stage, the storm wave attack is prematurely prevented. This is so, in spite of the beach face being saturated from the excessive ground water discharge resulting from the "high-water" level. By the same token, storm waves on a rising tide will have fresh beach to saturate before exerting their full erosive influence.

The most active part of the beach profile in the short run is undoubtedly the surf zone. Hence the influence of tidal oscillations in changing the active level and location of this zone is of paramount importance. One instance where its influence can be crucial is in the maintenance of dredged channels across the inner portions of the shelf. Because of the large width of shelf, and its milder slope, concomitant with the large tidal range, approaches to harbours necessarily have to be excavated. The depths and lengths of channel must cope with ships arriving over certain periods of the tidal cycle. The investments are therefore vast and so knowledge of their possible siltation assumes great importance. The proper location of spoil is also associated with this problem.

Where wave breaking occurs there is a strong possibility of material accreting in a channel running to shore. In the case of a large tidal range, this zone of siltation can cover a fair length of channel, so that continuous surveys should be conducted over these critical zones. In geographic locations where there is persistent swell the above problem is magnified. In others, including many tropical areas ($0°-25°$ latitude), where the major wave climate consists of locally generated storm waves from tropical cyclones, the sediment movement problem is not so predominant.

Where significant changes in depth can occur throughout the tidal cycle, and hence throughout the month and year, cognisance should be taken of this when examining the path of incoming storm waves or swell. The offshore contours could be such that, during specific stages of the tide, waves from a predominant deepwater direction could be refracted towards a structure or harbour entrance, which could make navigation difficult, or create conditions where bed material is swiftly moved. In this context the extra high surge levels should not be forgotten.

Ocean currents

It has been widely contended by some geographers, geologists and engineers that ocean currents are the main mechanism for the distribution and transport of material along the coast. This view is disputed very strongly by the writer for several reasons. These pertain to the magnitude and direction of such water velocities near the bed and the fact that strong littoral drift occurs where no currents, tidal or otherwise dissociated with the wave action, are present.

Even where a strong surface flow is produced by tidal inflow and outflow, or alongshore, the overall motion of any batch of water is likely to be a closed or near-closed loop. The net movement in any direction over the tidal cycle is therefore likely to be small. Velocities can admittedly vary throughout such an orbit, so altering the capacity of the water to suspend or transport material, but this fluctuation is unlikely to produce a commensurate change in conditions near the bed.

As the bed is approached the velocities are attenuated. This means that the circuit followed by the water layers nearer the bottom will be far smaller than those near the surface (see Fig. 1-5). This differential magnitude, in both speed and amplitude of orbit, causes large eddies to form. These are more prominent in the vicinity of the bed where the velocity gradients are greatest, and bed undulations could exert an influence. This results in short-term fluctuations in velocity and

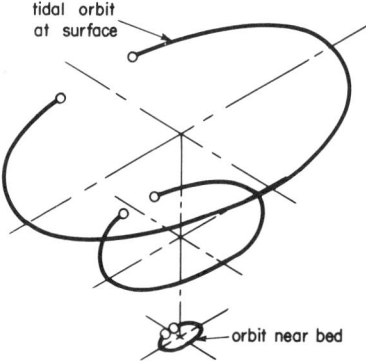

Fig. 1-5. Paths followed by water particles at different levels during a tidal cycle.

direction at any point near the bottom. The writer has observed tidal current records made within 3 ft. of the sea bed in 30 ft. of water in which several 180°-changes in direction have occurred within the 20-sec period required for the measurement. These eddies or whirl pools could be instrumental in drawing batches of sediment into suspension, but the net lateral transport could be negligible.

The majority of coastal margins of the world suffer very little longshore-current of tidal origin, even though they experience strong littoral drift through wave action. As noted previously, enclosed seas exclude any effects of tidal oscillations or currents and yet exhibit the features indicative of mass transfer of material along their shores. On some oceanic margins the direction of longshore drift, as gauged by physiological factors to be discussed later, is contra to surface tidal currents offshore as published in hydrographic charts.

This great influence exerted by waves on bed material is understandable when the basis for it is appreciated. The mass-transport of the water particles, resulting from their differential forward and backward oscillation, is maximum at the bed because of the viscous effects in that vicinity. In fact, the mass-transport for progressive swell waves is optimum within the boundary layer. For waves of around 10-sec period such a boundary layer is in the order of 1 or 2 cm thick. Thus maximum net motion is produced in the zone where the majority of the suspended or jumping sediment particles are present. Mass-transport distribution throughout a vertical plane, which takes much time to be established, contains a net drift in the direction of wave advance for nearly half the overall water depth. Thus, material placed temporally in suspension by the turbulent eddies of the tidal currents near the bed will be moved horizontally by swell waves. As noted earlier, storm waves tend to have a mass-transport counter to the waves in the lower half of the water body. Such seawards removal of sediment lasts only as long as the storm, so is negligible compared to the shorewards transport of the more persistent swell.

Currents can alter the characteristics of incoming waves and so control their capacity to shift sediment. To accomplish this they must have a reasonable velocity, as produced by concentration of flow in river or inlet mouths, or between land masses. A current flowing contra to waves will steepen them and so promote breaking. Where a spectrum of waves is present the shorter and steeper components will be dissipated. The remaining longer-period components will be shortened in length, which will reduce the amplitudes and hence velocities of water-particle orbits at the bed. This decreases the influence of the waves on the bed. Likewise, a current running with the waves will elongate them. This increases the orbital motion of the particles whilst also causing the wave crests to become peaked. The difference in the forward and reverse velocities becomes greater, so increasing the mass-transport at the bed. Thus, an ingoing or flood tide assists the waves in landward transport of material, whilst an outgoing or ebb tide substantially reduces this mechanism.

The effect of tidal streams on waves has been discussed by Unna [27] and proved in experiments by Yu [28]. Johnson [29] has analyzed the case of waves arriving obliquely to a uniform current in deep water. Hunt [30] has dealt with the problem of opposing and following currents in transitional depths where velocities are attenuated with depth in a specific relationship.

Estuarine conditions

An estuary is a zone in which fresh water from a river or lake system issues to the sea and results in stratified flow due to density differences. For this influence to be significant the mixing waters must generally be contained within limited side boundaries, such as a funnel-shaped indentation of the coast. This specific physiographic feature will be analyzed below; any different shape of river outlet can be examined by the reader in the light of the processes outlined. Many features and peculiar problems pertaining to estuaries are discussed in detail elsewhere in this treatise, but those concerning wave action in particular are presented here.

The discharge of the lighter fresh water is concentrated at the surface of the denser salt water, so forming an interface between the two. This layer slopes downwards into the estuary or river mouth and so forms what is termed a salt-water wedge. The shearing effect of the fresh water on the interface causes mixing to take place, with a consequent removal of salt water to the sea, accompanying the fresh-water discharge. This causes the induction of more salt water to replace that discharged, so producing an inflowing current near the bed (see Fig. 1-6).

This inward current will fluctuate over the tidal cycle. It is enhanced during the flood tide and is decreased, to even become seaward, during the ebb tide. On balance, it can be accepted that there is a stronger inward tendency at the bed and a stronger outward tendency at the surface. This should be verified by current measurements throughout the full depth in any particular estuary. Longer-term fluctuations will occur due to the annual cycle of fresh-water flow. As discussed elsewhere in this treatise, an increase in fresh-water discharge will demand more salt water at the interface, and so generate a stronger inflow near the bed. The wedge itself is

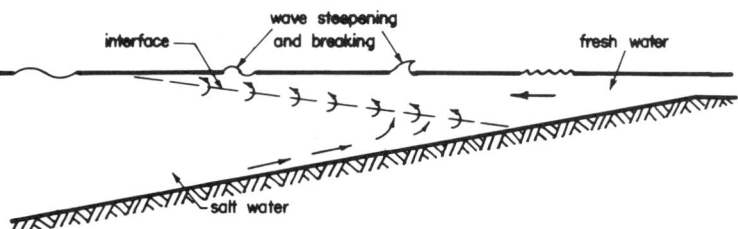

Fig. 1-6. Section through estuary mouth showing salt water wedge.

forced closer to, or even outside, the river mouth on such occasions, so that the zone of siltation changes throughout the year.

The discussion here will concentrate on the interaction of the currents involved in the saline wedge with the incoming waves. The contra current at the surface, which persists for the majority of the tidal cycle, will steepen and dissipate part of the wave spectrum and shorten the lengths of the remaining waves. Their influence on the bed is thus reduced during the period when an outflow at the bed might effect a seaward clearance of material being brought to the head of the estuary from any source. During a flood tide, waves are carried into the river mouth whilst being lengthened and distorted at the crest in a manner to enhance their mass-transport at the bed, just when the inflow at the bed is optimum. This differential influence on the bed by waves and currents concurrently results in a strong tendency for siltation of estuaries.

Besides coming to the mouth from upstream, sediment may be supplied from the sea and the boundaries of the estuary. That from the ocean entrance of the estuary could be supplied from along the coast and carried by mass-transport or littoral current into the indentation. The assistance of the flood and ebb tides in this mechanism has been discussed above. Other sources of marine sediment, as illustrated in Fig. 1-7, are shell debris and precipitation of salts from sea water. The bulk of the latter is calcium carbonate which is the main constituent of shell material. Estuaries are normally shallow zones of the ocean whose water is impeded from mixing with the adjacent oceanic mass. It can therefore become warm and so promote the precipitation noted above; either through the activity of marine fauna or plain chemical processes. The volume of sediment so produced depends greatly

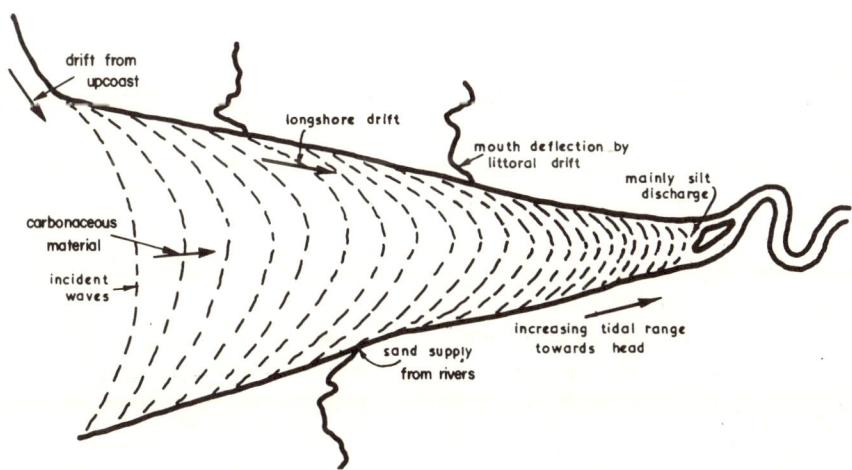

Fig. 1-7. Typical estuary showing sources of sediment.

upon the latitude of the estuary, or section of continental shelf. In some coastal zones of the tropics the carbonate content of shelf material has reached 97%. It is important, therefore, to analyze deposits in an estuary to find the probable source of the bulk of them. Should they be mainly of maritime origin, any control of terrigenous output from the river will not greatly affect the estuarine shoaling problem.

The other source of material in the river mouth is the littoral drift generated by the incoming waves along the estuary boundaries. Sediment may be fed to the funnel-shaped shoreline by rivers debouching on to it. These streams are secondary to the main river, whose outlet is the major geologic feature of the coastal indentation. The general physiographic nature of such large inlets would normally cause these side streams or rivers to be short and steep in character. They therefore tend to transport coarser material than the longer rivers, which of necessity must meander through lengthy plains of very modest slope. Tests would soon indicate the proportion of material coming from such sources. Means of reducing siltation through this longshore transporting mechanism are discussed in Chapter 3.

CALCULATION OF BEACH PROFILES

It has been stressed above that longshore drift occurs both inside and beyond the surf zone. The mode of transport is distinctly different in the near-shore and offshore areas, as described previously. To attempt an assessment of rates in the surf zone, it is necessary to have the beach profile specified, so that depths and widths over which waves are operating are known. Normal hydrographic charts, produced in a number of countries, can suffice for the deeper zones offshore where profile changes are imperceptible over decades. For an accurate picture of depths less than 10 fathoms the most up-to-date survey data should be used.

With such information available, it may be questioned why computation of beach profiles is necessary, together with correlations of them against sediment characteristics. The answer to this is concerned with long-term stability of the swell profile and short-term transposition from swell to storm beach profile.

Offshore profile

In regard to a long-term equilibrium profile of the offshore ocean bed, which is produced by a particular wave climate and type of coastal sediment, it may be possible to derive an equation which could predict shapes for new wave or sedimentary conditions.

A fresh wave climate may be created on a section of coast, due to a barrier impeding the approach of swell waves, for example. Reduction in wave height may be effected by refraction or diffraction. What is the long-term effect on a section of coast previously subjected to much larger waves? Or the shoreline might be reoriented, by the establishment of a littoral-drift barrier, to face the incoming swell more directly. This will increase the wave energy which was previously spread along a greater length of beach.

New sedimentary conditions might ensue due to interference of the normal supply from a river mouth. Dams upstream could intercept coarse or fine material or both. The shoreline previously receiving this will have to adapt itself to the new conditions, which could necessitate a long-term change in the offshore profile. Also the replenishment of beach material from offshore or from nearby land areas may introduce a change in the average sediment size or other characteristic. In this case it would be useful to know the probable equilibrium profile, so as to determine the volume of fill required to reach stability.

If a theoretical or empirical relationship could be developed between average wave conditions and average sediment characteristics for a *stable* underwater profile, a measuring stick would be available with which to gauge any shoreline. Should a beach section be significantly steeper than the "standard", reasons for erosion should be sought and remedial measures instituted. Should the profile be flatter than normal, transient accretion may be indicated, such as humps traversing the coast [23]. For oceanic margins the wave climate in this respect is closely associated with the persistent swell, but in enclosed seas storm waves are the sole source of wave energy. In the latter case the profile beyond the pseudo-permanent offshore bar must be related to a widely varying spectral width. Some relationship may be derivable that employs average energy per unit area of ocean, together with mean sediment characteristics [31]. In computing such an average, some statistical value may have to be used for the energy fluctuations throughout the year, such as the average of the highest third of the peak energies. It should be obvious from the previous discussion that a stable profile derived for a margin of an enclosed sea cannot serve as a measure for a section of an oceanic coastline, even where energy levels are found to be the same. A factor to be included in any such energy relationship is wave length or wave period, which will vary from storm to swell conditions.

Swell to storm profile

A major concern of the coastal engineer is the almost annual transition of the beach from a swell to a storm profile. He may desire to know the probable loss of beach in the next storm, or the depth to which the eroded surface may reach. In the former case he could advise on the safe limit of beach encroachment for commercial or other purposes. In the latter case he could recommend depths to which pipelines or cables be laid with safety beneath the beach zone. In any of these assessments the abnormal conditions of surge height on top of high-tide level

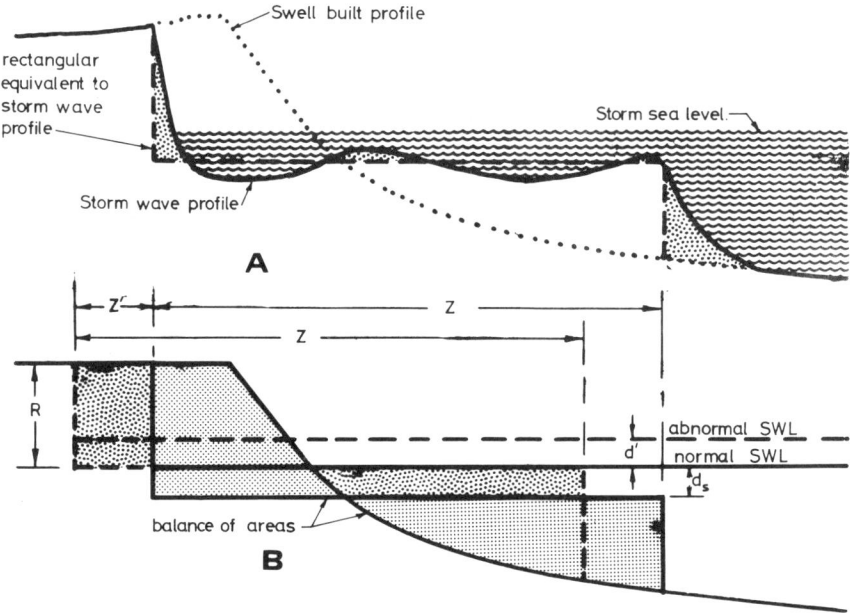

Fig. 1-8. A. Transposition from swell to storm-wave beach profile. B. Relevant variables for computing beach recession.

must be included. Whilst the present discussion assumes a specified swell-built profile, from which the storm beach evolves, the long-term erosion or accretion trends must also be considered, as outlined in later sections.

A typical transposition from swell to storm-wave profile is depicted in Fig. 1-8A, in which the secondary undulations of the bed can be smoothed to a horizontal plane as indicated, whilst preserving equal areas of "cut and fill". If the swell profile can be defined up to the average beach level R above normal high-tide level (marked *normal SWL* in Fig. 1-8B), plus the depth d_s down to the storm beach average level, and the overall width z, then manipulation of the profile can be carried out to give equal volumes of beach erosion and submarine accretion. As noted in Fig. 1-8B, a possible elevation d' to an abnormal surge level during a storm cycle could take the beach line back by an amount z'.

Theoretical analysis

Sitarz [32] has developed equations, based upon energy principles, for profiles beyond the breaker-line to the limit of active bed disturbance. He has determined a relationship for the width of the surf zone for swell and storm-wave profiles, plus the cross-sectional area of the water in this zone, from which its average depth can

CALCULATION OF BEACH PROFILES

be derived. The offshore profile is given by:

$$x = ky^2 \quad \text{where} \quad k = 0.95/[(s-1)^{1/2}DH^{3/2}] \tag{1-1}$$

where x is the distance seawards from the breaker line; y is the depth below SWL; s is the ratio (density of sediment)/(density of seawater) = 2.65; D is the median sediment diameter at some appropriate depth (mm); and H is the wave height just offshore from the breaker zone (m).

From the above equation it can be seen that the milder slopes occur with larger values of k, which in turn result from finer sediment or smaller wave heights. The usual value of s = 2.65 is applicable to most sand situations. It appears strange that wave period does not enter this relationship, since this determines in part the water-particle oscillation. Also, the mass-transport "velocity" at the bed is strongly dependent upon period or depth to wave length ratio. Taken overall it is period that dictates the refraction pattern and hence the proportion of the deep-water height occurring near the breaker zone.

To complete the swell-built profile it is necessary to continue the offshore bed slope or curve up the beach face to the berm. The beach face is best measured at the site, but some general values are graphed in Fig. 1-9 for three conditions of exposure [33]. The D (mm) in Fig. 1-9 represents median diameters of sediment particles in the swash zone. The berm height should be measured where possible, since calculations contain too many variables for practical purposes.

The width of the surf zone (Z) is given by:

$$Z = BH^{3/2}/[(s-1)^{1/2}D^{1/2}] \tag{1-2}$$

where H and D have dimensions similar to those in equation 1-1, although D repre-

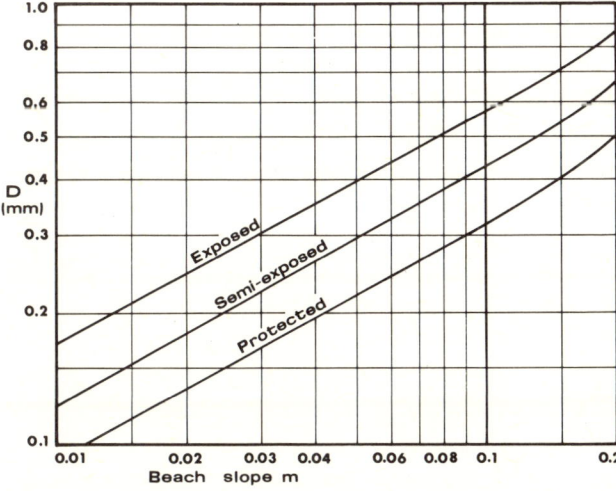

Fig. 1-9. Slopes of beach face for different exposures to waves. (From Ref.[33].)

sents in this case the median diameter of sediment in the beach face (near the SWL).

For the swell-built profile factor B above has the value 43.5, whilst for the storm-wave surf zone it is 75.0 [32]. In this latter case, a full spectrum of waves is implied, so that H must be representative of this. Brunn [34,35] has employed $H_{1/3}$ for such a purpose.

The cross-sectional area of water in the storm beach surf zone is given by Sitarz [32] as:

$$A = (3.4 g^{1/2} H_{1/3}^2 T_{1/3}) / [(s-1)^{1/2} D^{1/2}] \qquad (1\text{-}3)$$

where $T_{1/3}$ is the significant wave period; g is 9.81 m/sec^2; D is the median sediment diameter across the surf zone; and $H_{1/3}$ and s are defined as before.

The mean depth of this surf zone (d_s) in Fig.1-8B is then given by A/Z. This should be compared to the depth of $H_{1/3}$ suggested by Brunn[34].

Having established the swell profile existing at any time, either completely by survey or partly by survey and partly by computation above, the width Z and depth d_s of the probable storm profile can be overlain so as to predict the likely beach degradation. In this respect the slope of the storm beach face, which is very steep, is of little consequence compared to the other relevant dimensions (see Fig. 1-8B).

Other workers have concerned themselves in this field [15,36,37] but there is need for further analysis of the dynamics of beach material. It is salutary to consider the difficulties in arriving at relationships and subsequently in substituting values into them. As has been noted previously, swell waves vary in height and period from hour to hour and week to week. The recognition of a reasonable mean energy input and a stable profile for a given sediment size is therefore very difficult.

Added to the above problem are the seasonal storm and swell sequences. The offshore bar formed during the winter season may take some months to be swept completely back to the beach. Not until this is accomplished can the cross-section be considered a fully-fledged swell profile. Certainly the bed is not in equilibrium, whilst material is still actively being removed. Thus, to obtain a "standardized" profile for any given wave climate and sediment characteristic, measurements should be made at the end of a lengthy period of swell. The wave climate should express the average swell conditions prior to this survey. Relationships should be derived for oceanic and enclosed sea margins separately, because of the different nature of the predominant waves in these two situations.

CALCULATION OF LITTORAL DRIFT

The longshore transport of sediment has been seen to take place both in the surf zone and the offshore area. Discussion here will relate to the former only, in the next section the major factors influencing bed disturbance in the deeper zone will be considered.

Since the water momentum that generates the littoral current can be derived from values of wave energy and its longshore component, it is possible to relate deep-water wave characteristics to these, and more empirically to the volume of sediment carried. There are factors such as energy dissipation, bottom percolation and wave reflection which are difficult to allow for, but information is accumulating which indicates general values.

Littoral current

Sonu et al. [38] tested various formulae [39—46] but found they did not agree with measurements made over some six months. They concluded: "Under natural conditions, the near-shore topography participates in the longshore current mechanism as a dynamic variable, not only redistributing the breaker influx into different positions along the shore but also itself undergoing displacements and transformations due to the waves and the currents thus affected." A beach profile was illustrated which represented three of the above six months. Examination of average wind and swell data [47] for these three months indicates swell incidence from the northeast quadrant for 21% of the time and from the southeast quadrant for 18%, with an average of 14% of confused seas. It appears to the writer almost impossible under these conditions to correlate a littoral current with waves breaking on a coast, which was oriented almost north—south. The beach cross-section shown also exhibited a pronounced hump, which could have been a partly returned offshore bar, even though it was an oceanic margin. These points are made here, not to detract from the attempt to verify these formulae, but to stress the temporal variables involved.

Further formulae have become available, the theoretical aspects of which disagree amongst themselves and previous papers [48—57]. These differences arise from using either momentum or energy fluxes of the breaking wave, and from employing theory for sinusoidal, cnoidal or solitary waves in the transition and shallow-water zone. Most workers have based their computations on a single sinusoidal wave which could well represent the persistent swell situation on an oceanic margin, if its characteristics are suitably chosen. Brunn [34,35] has specifically dealt with the storm beach profile, but the transient nature of this respecting wave height, period and direction makes application in this case extremely subjective.

Littoral drift

The step from littoral current to littoral drift itself, introduces factors which must be derived empirically from model or field tests. The variables involved will be outlined by presenting a relationship published by Castanho[55]. Using the solitary-wave theory, he obtained:

$$G = E_r P / [(1 - 1/S) \tan \theta] \qquad (1\text{-}4)$$

where G = the weight of sediment moved per second across a plane normal to the beach;

$$E_r = \frac{\text{energy dissipated}}{\text{longshore energy component}} = \sin \alpha_b f\left(\frac{mH_b}{K' \tan \alpha_b L_b}\right)$$

where m = mean beach slope; $H_b/L_b = H_b/C_b T = H_b/\sqrt{1.78 g d_b} T = H_b/\sqrt{1.78 g \, 1.3 H_b} T$; K' = roughness factor; P = wave power per unit length of shore = $2.2 \, w H_b^2 \sin \alpha_b \cos \alpha_b / T = w H_o^2 L_o \sin \alpha_b \cos \alpha_o / 16 \, T$, where w is the specific weight of seawater; S = specific gravity of sediment; and θ = angle of internal friction of the sediment.

By substituting certain values suggested by Castanho for sandy shores, namely:

$$mH_b/K' L_b = 0.12 \, , \, \theta = 35° \, , \text{ and } (1 - 1/S) = 0.623 \, ,$$

eq. 1-4 can be reduced to:

$$7GT/wH_o^2 L_o = E_r \sin \alpha_b \cos \alpha_o \qquad (1\text{-}5)$$

The value of α_b depends upon α_o and the wave steepness (H_o/L_o) [58,59], so that both $E_r = \sin \alpha_b \, f(0.12/\tan \alpha_b)$ and $\sin \alpha_b \cos \alpha_o$ have been graphed against H_o/L_o in Fig. 1-10 and 1-11. Since eq. 1-5 is dimensionally homogeneous any appropriate units may be employed in solving for G.

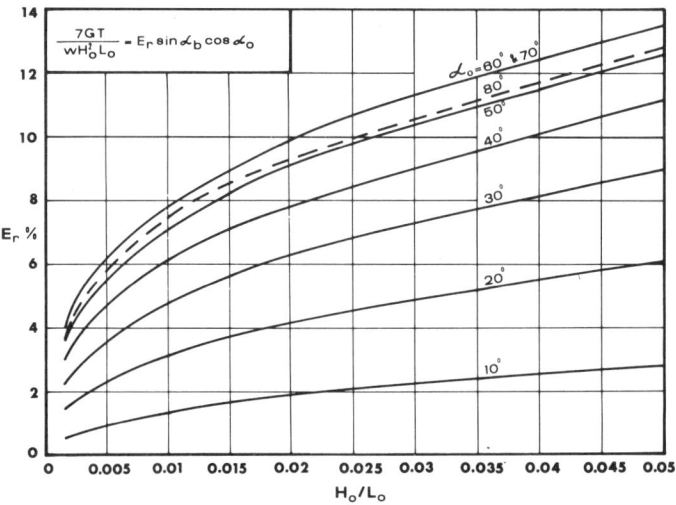

Fig. 1-10. Relationships for calculating littoral drift (eq. 1-5).

CALCULATIONS OF LITTORAL DRIFT

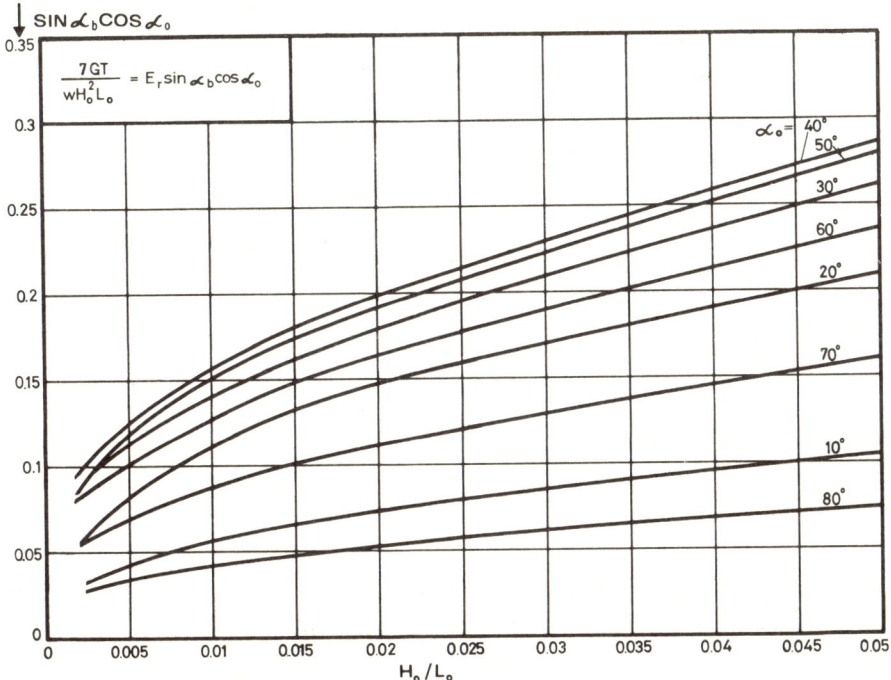

Fig. 1-11. Relationships for calculating littoral drift (eq. 1-5).

Example. Waves of 2.5 ft. height and 10 sec period arrive at various angles to a coast, the continental shelf of which can be considered uniform along the length of coast under study. Using the normal sediment and beach characteristics utilized in the text, determine the rate of littoral drift for deep-water approach angles of 20°, 40°, 60° and 80°.

$$H_o = 2.5 \text{ ft.}, L_o = 512 \text{ ft.}, H_o/L_o \doteq 0.005, \frac{wH_o^2 L_o}{7T} = \frac{64 \times 2.5^2 \times 512}{7 \times 10} = 2930.$$

α_o	E_r	$\sin \alpha_b \cos \alpha_b$	G (lb/sec)	V (yd^3/day)
20	2.3	0.08	5.40	180
40	4.7	0.125	17.20	570
60	6.2	0.10	18.20	605
80	5.8	0.03	5.10	170

The specific weight of dry sand used to determine V in the table is 100 lb./ft.^{3} = 1600 km/m^3. To assess the annual rate of transport it would be necessary to know the number of days during which waves of a given order actually arrive. Even

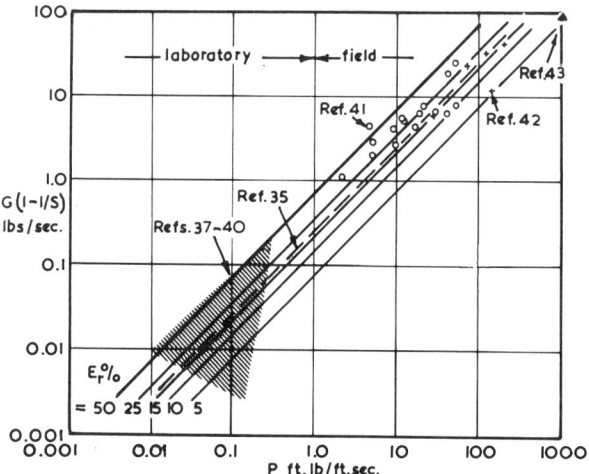

Fig. 1-12. Littoral drift relationship from Ref. [35] with data from other sources.

with the modest energy implied in this problem the optimum littoral drift at $\alpha_o = 50°$ is equivalent to a 6-yard truck passing along the coast about every 30 minutes.

Inman and Frautschy[60] have derived an empirical relationship between immersed weight per second passing any point and wave power per unit length of coast. In terms of eq. 1-4 this is:

$$G = P/4(1 - 1/S) \tag{1-6}$$

and is graphed in Fig. 1-12, together with data supplied from laboratory tests [61–64] and from field tests [65–67]. The hatched zone indicates the location of the bulk of the laboratory results.

Accepting the value of $\tan \bar{\theta} = 0.7$ previously employed, eq. 1-6 implies that $E_r = 17.5\%$. As seen in Fig. 1-10 this should apply to waves steeper than 1/20, which is the limit of the figure. Inman and Frautschy commented on the likely differences in laboratory and field investigations as follows: "The equations for wave power are usually written for the root-mean-square wave height, which is the wave parameter most commonly measured in the laboratory. However, the power computation for the field data appears to be based on significant wave height and thus may be too high by a factor of 2. On the other hand, the practice of representing the entire power spectrum in terms of a single significant wave may result in the omission of important energy contributions of waves of different frequency, causing the computations to be too low. A rigorous evaluation of this relation is essential to beach planning.'

In this respect the substantial difference between storm waves and the more

OFFSHORE CONDITIONS   33

prolonged swell condition should be kept in mind. With the former the concept of significant wave height or energy of the spectrum is important, whereas with swell an average height and period over a lengthy period could prove adequate. The overall wave climate of a coastal area is important to consider, in terms of the total wave energy or, more correctly, the longshore component of it. This includes both storm waves and swell throughout the year, which will be acting upon different beach profiles and suffering different losses (E_r).

Castanho [55] gave examples of two beaches in which E_r was 7% and 16%, respectively. In Fig. 1-12 lines are drawn which represent a range of E_r. It can be seen that the data from various sources are contained within the range of E_r from 5% to 50%. It would seem that eq. 1-4 and 1-6 could lead to a general solution of the problem. The variables of beach slope (m) and friction factor K' might be related in some manner to sediment size.

OFFSHORE CONDITIONS

Beyond the breaker-line the water-particle motion near the bed for a single progressive wave train is an oscillatory back and forth orbit. It is on such a basis that shear velocities are computed in order to assess forces on sediment particles, which purport to lift them from the floor. As soon as this commences ripples and dunes form on the bed material, which in turn affect the fluid movements. Existing and proposed methods for computing the net transport across the bed will be outlined below.

In the light of the previous discussion on angled wave trains and the presence of large eddies or macro-turbulence the role of vortex generation will receive attention. Treatment of the problem of sediment transport as a stochastic process, in which random pressure and velocity-fluctuations occur in the vicinity of the bed, is presently being developed for streams. Its application to oscillatory flow appears paramount for future advance in this field.

Theory for motion of sediment

The forces exerted on sediment particles by a uni-directional flow have been presented over a number of decades, but there is a great need for rationalisation of the many relationships developed. Theory and experiments involving wave action are of more recent date, with the bulk of it being accomplished in the past twenty years.

In comparing water-particle motion near the bed of a river or channel to those at the ocean bed (subjected to wave action), it becomes immediately obvious that significant differences exist. The nature of the movement is dissimilar, oscillatory as against continuous flow, attended by fluctuations in boundary layer thickness as against steady turbulent conditions. The reasonably uniform distribution of energy over the floor of a long stretch of channel must be compared to the variable amplitudes and speeds of oscillation produced by continually changing wave conditions. Again, the depths in the two realms are significantly different, waterways are shallow but of relatively constant depth, coastal zones vary in depth from the breaker-line to the edge of the continental shelf. Thus offshore zones must be specified in terms of depth and wave characteristics when sediment transport is being discussed.

The oscillatory motion, produced by waves, involves accelerations and decelerations of water particles. These exert forces on sand grains in addition to those of drag and lift produced by peak velocities. These inertial forces are somewhat out of phase with the velocity components and have been shown to be in the order of 5% of the maximum force on the particle [68]. However, the computation of either group of forces still depends upon the use of coefficients of drag, lift and inertia, as computed from uni-directional flow conditions. The degree of scatter in their values, as will be presented later, tends to undermine any confidence that may be placed in them.

The vortices produced by waves will differ from those in steady flow. They are generated in alternate directions by water motions of short duration. This motion is relatively uniform along the length of a wave crest, which differs from the more random velocity fluctuations in uni-directional flow. Whilst the vortices from waves are produced during the acceleration and high-velocity portions of the water-particle orbits, their influence is felt in a decelerating, zero, or even reverse flow situation.

The nature of the fluid flow results in differing ripple and dune formations on the bed. In uni-directional flow dunes are asymmetrical, with steep downstream and mild upstream slopes. These help the streamlines to separate at the crest and re-attach partway up the slope of the adjacent downstream dune. The spatial randomness of the fluid vortices also makes for a random dune distribution, although the dune dimensions have a specific peak in their frequency distribution commensurate with the mean flow velocity and other boundary conditions. Oscillatory flow, on the other hand, initially produces parallel ripples and dunes. These later turn into a brick pattern and finally a relatively random distribution. The crests of the bed undulations are symmetrical and reach a steepness (ratio of height to distance between crests) similar to those in uni-directional flow, namely that dictated by the angle of repose of the sediment. Separation of the streamline, from either side, can thus occur, but the eddies so generated are more cylindrical than spherical in shape.

The comparison to be made is between transport down a river channel and that along the coast over the width of the continental shelf. Depths down a channel can be considered sensibly constant whilst those across the shore profile vary continuously. For a given wave train arriving obliquely to the coast, refraction results in a change of its direction as it traverses the shelf. This is accompanied by a change in wave length and wave height as the depth varies, so that the amplitudes and velocities of water particles at the bed vary over the whole width of "highway". This means that any calculation of sediment transport along the shore is applicable, at best, to a particular bed contour offshore and this only for a straight length of coast with a uniform profile lengthwise. If the coast is curved in plan, or has an undulatory floor, such a computation can only refer to one small zone on the shelf.

The suspension of bed material will not only depend upon the fluid motions mentioned above, but also upon size, density, and shape of these grains. As discussed previously, sorting takes place across the shoreline profile. Local distributions can appear anomalous due to changes in coastal alignment and modes of sediment supply from rivers. In any particular zone the grains are likely to be uniform in diameter, more so than in the river situation where a wide range of sizes exist together. Also, the particle shape of coastal sand is likely to be more spherical than its terrigenous counterpart, due to its longer period of abrasion in the sea. Either wash or bed load in a river has had a relatively short life in contact with flowing water, so that its particle shapes are dictated more or less by the weathering process that produced them.

Although the swell waves arriving from the deep ocean may be from a predominant direction, they may have to penetrate embayments, after suffering refraction and diffraction to different degrees. The near-shore profiles will thus vary along the shore and so the rate of transport, even for a steady and uniform wave climate beyond the shelf. Such variations make the assessment of longshore drift for any reasonable length of coastline a formidable and most-times an unrewarding task. Added to this, the changeability in wave energy over time and space, even over periods of hours, and the difficulty in recording this information, make the annual fluctuations of river discharge appear extremely simple.

Depths in the ocean, where sediment movement is of engineering interest, vary from 10 to 200 ft. and may extend to the shelf edge at around 400 ft. Depths of channels and rivers, on the other hand, vary from 5 to 50 ft. Thus the roughness factor (roughness dimension)/(depth) varies significantly between the two. In this context either the grain size or dune height in the marine situation will be much less than in a river, whilst the depths are much greater. The maximum bed velocities may be similar in the two cases, but the boundary layer development and macro-turbulence will be significantly different.

Shear-stress approach

This approach involves the calculation of the velocity and acceleration of water particles at some level within the boundary layer at which they are purported to exert lift, drag and inertial forces on the sand grains. A flat bed is inferred and the computation is carried over successive intervals of the wave cycle. For the purposes of outlining the procedure, the most recent method presented to date will be discussed, namely that by Kamphuis [69,70], to which the reader is referred for the detailed analysis. Only the various steps will be described below, with comments on the relevant assumptions and simplifications.

In summary, the suggested procedure is to assess a suitable water-particle velocity which initiates sediment motion, after which the proportion of the surface so influenced is computed. There is an acceleration of sand grains to some terminal velocity at the end of the first interval, which must be integrated over this interval to provide sediment discharge. During the second time interval a further acceleration takes place, of particles already in motion, besides more material being rolled across the bed. Further time intervals are considered until saturation of sediment movement is reached, which was found to occur after the third time interval. The above procedure for forward velocities under the crest was repeated for reverse velocities under the trough.

Variables included in the analysis were:

waves: height, period, depth, time;

sediment: density, diameter, percentage in motion, angle of internal friction;

coefficients: drag, lift, inertia, bed roughness.

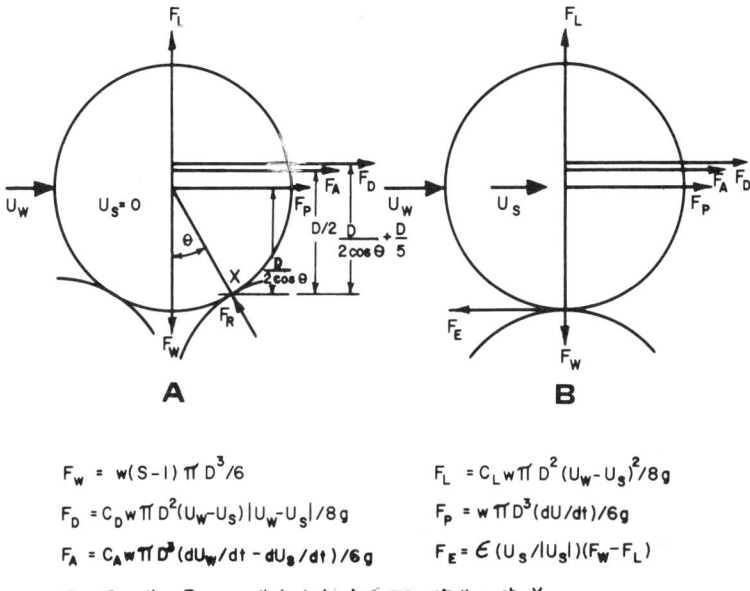

$F_W = w(S-1) \pi D^3/6$

$F_D = C_D w \pi D^2 (U_W - U_S) |U_W - U_S|/8g$

$F_A = C_A w \pi D^3 (dU_W/dt - dU_s/dt)/6g$

F_R = Reaction Force — eliminated when moments through X.

$F_L = C_L w \pi D^2 (U_W - U_s)^2/8g$

$F_P = w \pi D^3 (dU/dt)/6g$

$F_E = \mathcal{E}(U_s/|U_s|)(F_W - F_L)$

Fig. 1-13. Forces on grains due to shear velocities and acceleration for: A. at rest; B. in motion. (Ref.[69].)

OFFSHORE CONDITIONS

From the free velocity, some representative value in the boundary layer was determined. Since this layer was normally turbulent [71], an empirical approach to the velocity distribution was involved. From the level of assumed zero velocity, dependent upon packing conditions of the grains on the surface and the number of grains in motion, a value of velocity at the centreline of the stationary grains was computed by an integration process.

The forces accepted as acting on a particle of sediment at rest, and later in motion, are depicted in Fig. 1-13, with the various relationships listed. It is seen that the angle of internal friction θ enters into the computation of moments about point X for the majority of forces. Kamphuis was not confident in values of θ he obtained for his sediments.

The drag force F_D was located 0.30 from the top of the particle as suggested by Chepil [72], who noted that this could change with the exposure of the grain. For established motion, C_D was used with a Reynolds number influenced by sheltering, wake and boundary layer proximity. Graf [73] has summarized data from a number of workers [74–77] on C_D for accelerated flow and found that for spheres it could exceed 10 and generally exceeded steady-state values by an order of magnitude, as exhibited in Fig. 1-14, which has been reproduced from data presented by Graf.

The added mass force F_A was located $D/2$ above the point of contact "the average value during the actual process of initiation of motion". Kamphuis cited literature on C_A where values had been found to vary from 0.5 to 1.59. It was

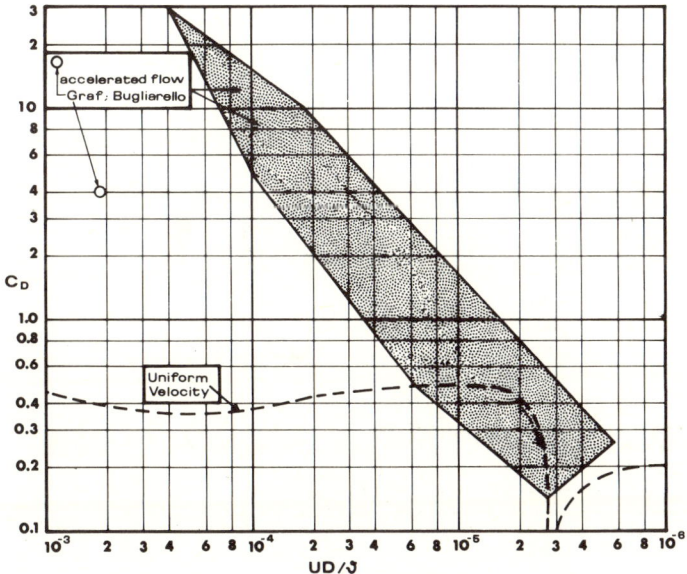

Fig. 1-14. C_D for spheres in steady and accelerated flow. (Ref.[73].)

assumed constant throughout the wave cycle. Graf had found variations of 0.5—40 for a range of the acceleration parameter AD/U^2 from 0.01 to 1.0, as illustrated in Fig. 1-15 [76—79]. The moment arm of the lift force F_L depends upon θ. The value of C_L was taken by Kamphuis as 0.85 C_D as obtained by Chepil [72,80] for particles of several mm diameter, even though it was suggested by Chepil that increases would ensue for thinner boundary layers. Errors in evaluating C_D are also incorporated into C_L. The weight F_w varies with D^3 and the moment arm varies with θ and D. Errors in the measurement of D are therefore greatly magnified in F_w.

The equation for the balance of forces permitted the determination of diameter for the particle about to be moved. Probability of motion of this critical diameter was assessed from the Rayleigh distribution of random motion, caused by fluctuation in angle of repose, packing and sheltering. Restrictions in the calculations as noted by Kamphuis were: the omission of turbulent fluctuations of velocities; the omission of vertical velocities that could be present through percolation; the sensitivity of the equation to errors in θ, C_A and (sediment/fluid) density ratio.

The proportion of the critical to the mean particle diameter gave the number of particles set in motion per unit area during the first interval of time, based upon knowledge of the above probability function and the number of particles per unit area of bed surface. During this commencement interval the first particles are set in motion at some velocity, after which the number increases. Once a particle is rolling, the friction force F_E (see Fig. 1-13) must be assessed (involving a knowl-

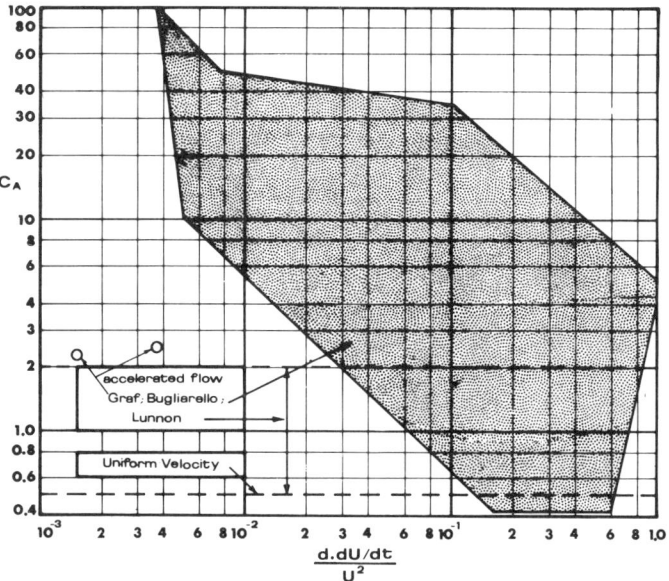

Fig. 1-15. C_A for spheres in steady and accelerated flow. (Ref.[73].)

edge of F_L), as well as new values of F_A and F_D. These vary with distance of the particle from the bed, but allowance was nor made for this, nor for the Magnus effect that could keep a particle suspended longer. An iterative procedure was required to turn mean water and particle velocities into incremental values for the time interval. From this velocity, the proportion of particles in motion, and the mean mass of each particle, the incremental discharge of material per unit width was finally assessed for the first time interval.

The above was repeated for the second interval, when the velocity increased further. A new batch of particles commenced movement and was accelerated, so adding to those already in motion and increasing in velocity. Further time intervals were considered until all particles in motion reached sensibly the same speed. After this the water decelerates and the sand grains come to rest, within two time intervals of each other. The above was then repeated for the reverse flow under the trough, the whole operation being computerized.

Other shear-stress approaches

Einstein [81] at the University of California, Berkeley campus, has headed a group of researchers who have employed a similar microscopic approach [82–84]. All tests conducted to study the type of boundary layer, or the motions and discharge of sediment, were conducted with an oscillating horizontal bed in still water. The proportion of particles lifted from the bed was assessed from the volume of material caught in trays inserted in the bed. From this "transport" and velocity of oscillation, the concentration of sediment raised from the floor was calculated. The proportion of the bed so involved was then derived by empirical coefficients based upon previous results of uni-directional flow [85]. Only the boundary layer area was considered, no allowance being made for turbulent suspension. The overall transport was obtained by multiplying the sediment concentration by the mass-transport velocity (for a laminar boundary layer condition) and integrating over a thickness of two grain diameters. The resultant equations, as those of Kamphuis, are not in a readily applicable form. Discussion on the experimental procedures is reserved for the section on modelling of sediment movement.

Workers at the Massachusettes Institute of Technology have concentrated on the equilibrium beach profile for waves arriving normal to the coast and the resulting sorting of sediments [86–88]. The theoretical profiles did not compare favourably with those measured in flume tests, probably through the assumptions of laminar boundary layer conditions for the mass-transport velocity.

Other workers who have contributed to this field are listed in the references [89–95]. It is not proposed to examine their hypotheses or results here. It is believed a completely new concept is required, involving the spectrum of turbulence, which is related to sediment concentration. If such a spectrum, either of velocity or pressure fluctuations, could be correlated with suspension of material of given characteristics, it could be applied to uni-directional or oscillatory flow alike. In modelling, similarity could be obtained by reproducing a prototype spectrum, or the bulk of it, in a model flume or wave basin, possibly with oscillating rods or screens, as has been used in some salt water diffusion studies.

Practical aspects of shear-stress approach

In spite of the mathematical sophistication of the analyses outlined above, the results apply only to a single sinusoidal wave train, and even then omit consideration of turbulence. Before too much effort is exerted in putting any relevant equations into more usable form, certain complications in the natural conditions need to be considered.

Eddies are bound to be present in the body of and at the bottom boundary of the sea. These provide pressure and velocity fluctuations at the bed, which must exert an overriding force on the bed particles, certainly more than that produced by some mean shear stress. The observations of Singamsetti [96] would seem to suggest that even correcting the shear velocity for r.m.s. values, such turbulent variations would not provide the desired solution of the sediment problem. He states: "These fluctuations are only a result of circular motion in the eddies. A sediment particle in water, subjected to this circulatory motion, will tend to be thrown out of the eddies because of the failure to accomplish the required normal acceleration, and this tendency increases as the relative inertia of the particle increases."

Gyr [97] and Muller [98] have suggested that suspended material enlarges the size of vortices, and Rouse [99] has shown that this decreases their rate of dissipation. Thus, any analysis which omits the influence of turbulence is unlikely to produce a comprehensive relationship for sediment transport.

The complexities of water-particle motions resulting from the passage of storm waves has been alluded to previously. But even two swell waves of differing period which are travelling along the same path, will generate velocities and accelerations near the bed that any simple sinusoidal assumption cannot predict readily. A drag force exerted on a particle beneath one crest may coincide with the peak lift force produced by the other wave. Such interactions will also produce more macro-turbulence than even a single sine wave of equal energy to the two.

In the discussion of wave climate the importance of angled wave trains was stressed. As noted in Ch. 4 *Coastal Engineering, I*, the orbital motions resulting from them is extremely complex. At the bed either linear or circular orbits are followed, depending upon the transverse location of the observation point across the path of the combination crests. The rotational direction of these orbits is constant throughout time, so that the vortices could build up a strong capacity for drawing sediment upwards like tornadoes in the equivalent atmospheric situation.

These effects are multiplied many times when the multi-directional wave system exists inside a fetch. Hom-ma and Horikawa [100] have measured accumulations of sand in bamboo samplers over three days in an offshore zone suffering storm waves from a nearby typhoon. They found a variable concentration that extended some three metres from the bed, increasing towards the bottom as might be expected.

Another practical aspect of the shear-stress approach to sediment movement is the interaction of a uni-directional current and the water oscillations produced by waves at the bed. Bijker [101] has examined such combined effects, mainly to establish laws for movable bed coastal models. This involved the assessment of velocities within a "viscous sublayer". Commencing with a current only, mean velocity v, the shear stress τ_c at a height of bed roughness $r/12$ above the dunes was determined. Then adding waves, for which the maximum orbital velocity at the bed U_{max} was known, a total shear stress τ_T in the direction of wave propagation was similarly computed. Bijker graphed the ratio τ_T/τ_c against a parameter which included variables of roughness coefficient for the bed, and a ratio of orbital velocity (U_{max}) at $r/12$ to free velocity (v). The evaluation of bed roughness was difficult to obtain and was reportedly subject to large errors. Results of τ_T/τ_c plotted against U_{max}/v only (see Fig. 1-16) do not contain any more scatter than they do with the more complex parameter. It is seen that where the mean current v equals the maximum orbital velocity U_{max} (i.e., $U_{max}/v = 1$), the combined stress in the direction of the current is up to three times that of the current alone. This implies that the shear stress due to the wave is double that due to the current. It is not until $U_{max} \doteq v/5$ that the shear stress due to the wave can be considered negligible.

Bijker conducted experiments with sediment in a large wave basin, in order to measure rates of transport. For this purpose shallow trays were inserted whose top edges were flush with the movable bed. These trays were either square or rectangu-

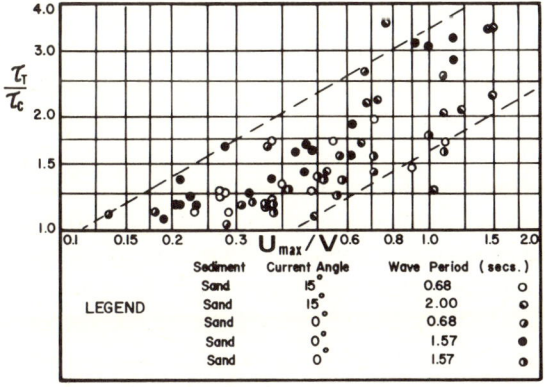

Fig. 1-16. Ratio τ_T/τ_c for ratios of U_{max}/v. (Ref.[101].)

lar in plan (the latter was termed "narrow" since the width was one tenth the length, which was transverse to the current). From rates obtained for current alone and then current plus waves, Bijker states: "Apparently, the correction of the bed shear for wave influence is not sufficient to cover all results... The physical explanation of the fact that there is nevertheless an appreciable transport, is that the bed material is stirred up by the wave motion, and that a very low velocity, or bed shear, is sufficient to move the material in the direction of the current." He then suggests two current systems, one for stirring the material, the other main current for transportation. This latter did not include the mass-transport component due to the waves, except in a qualitative sense.

Reference could also be made to the work of Goddet and Jaffry [102] on laws of similitude and of Collins [103] on the influence of currents on mass transport. It would appear that sediment transport rates and time scales for erosion and accretion in coastal models, which involve both waves and currents, will have to depend upon calibration tests for many years to come. However, such prototype comparisons are not easily made. It is not sufficient to record trends over months or years and relate these to some mean wave characteristic. Even when cognisance is taken of the different influences of storm waves and swell, not all incidences of either can produce results commensurate with their energy input. The temporal variability of beach processes has been stressed by Hom-ma et al. [104]: "It must be recognized that an absolute topographical change occurring in a short-term interval can be far greater than the resultant change over a yearly cycle."

The macroscopic approach

The loss of faith by workers in what has been the microscopic approach to sediment transport is exemplified in the statement by Kemp and Grass [105]: "The mean shear stress has long been recognized as an unreliable parameter for the determination of the initial movement of grains. Bed material can move under conditions of zero mean shear stress... It has equally been realized that turbulence characteristics, and notably the intensity of turbulence, are of great importance." They refer to Townsend's study [106], where it was concluded that for the same shear stress, velocity and pressure fluctuations vary considerably for different flows.

The significance of turbulence intensity in hydraulic structures and on downstream beds of sedimentary material was stressed at the 12th Congress of IAHR in 1967, when several papers reported exploratory studies in this direction. These have similar relevance to oscillatory flow at the sea bed.

Spectra of turbulent pulsations have been derived theoretically and verified experimentally [107]. Measurements made in models [105,108–110] and in rivers [109,110] have shown intensity to increase where the steady velocities are least, and also where depths are greater, both observations of which are particularly relevant to the deeper ocean situation. The instantaneous velocities and pressures recorded at the beds of rivers have been found to fluctuate considerably. A flood flow in the Mississippi River [111], for example, was shown to have the capacity for lifting a 6" thick slab of concrete, due to fluctuations of 0.5 lb./sq. inch over a wide zone. Similar capacities have been reported for waves by Carey [21], who states: "Many incidents may be cited of the almost incredible effects of storm action on a coastline of extreme exposure. At the harbour works at Peterhead, concrete apron blocks weighing 47 tons each, and placed in position at a depth of about 40 ft. below low water, were rooted out of their bed in the work and drawn seawards for a considerable distance."

The stochastic models being suggested for sediment transport in alluvial channels [112,113] provide a lead for the coastal engineer. It is well to recall that the great advances in ocean wave forecasting were made when it was treated as a stochastic process. So that finally, for all the complexities of the surface undulation in a fully arisen sea, it is now possible to compute a spectrum of waves from a mean wind velocity at a known height above the sea. As discussed in Ch. 2 *Coastal Engineering, I* the area under this spectral distribution is the variance of the water profile, from which a useful level of energy can be computed, such as $H_{1/3}$. A similar value might be obtainable for the turbulence spectrum, which could then be correlated with sediment suspension. Once concentrations were known at different levels, the application of water-velocity distributions would permit rates of transport to be computed. Near the bed itself, where concentration is optimal the mass-transport due to the waves would supply the maximum effect.

Description of particle motion

Although quantitative relationships are difficult to derive for sediment transport, or even sediment suspension, it is instructive to examine the modes of movement of granular bed particles. These cover the stages from incipient motion on a flat bed, through the formation of ripples and dunes, to the limit where the bed becomes smooth again and appears to oscillate back and forth. The picture below emerges from reports on experiments in flumes, oscillating sand beds, oscillating water flows in tunnels, and vortex-generating equipment. It will be seen that in all cases the role of vortex generation remains paramount.

Initiation of motion. After some minimum orbital speed has been reached by the water particles, the bed grains commence to roll. This movement is randomly distributed over the whole surface. Initial motion occurs at the peak velocity. As this is increased, grains move for a longer proportion of the wave cycle, and therefore over longer distances.

TABLE 1-I

Maximum orbital velocities (cm/sec) for various granular bed developments

Bed development		Sand (s.g. 2.47)	Glass (s.g. 2.62)	$U_{max}D_{50}/\nu$
	D (mm):	0.297	0.585	
Boundary layer transition commences		19.4	–	56.3
Incipient bed motion, bed undisturbed		24.6	29.2	71.3
Spontaneous appearance of ripples		32.4	73.2	94.0
Boundary-layer transition commences [1]			25.8	74.8
Fully turbulent boundary layer [1]			41.1	119

[1] Impervious smooth bed.

Soon after this incipient motion, particles begin to align themselves transversally to the flow, or parallel to the wave crests. Bagnold [114] records for this stage: "More grains reach these zones than leave them, so there is a progressive congregation of grains in them, and the zones soon become little wavy ridges a few grains high, whose crests sway from lee side to lee side during successive stroke reversals." The nature of these "rolling-grain ripples", as Bagnold termed them, will be described in the next section, but first it is necessary to enquire as to why they should form.

Carstens and Neilson [115] conducted tests in a water tunnel, in which a constant period of oscillation was maintained (3.54 sec), whilst the amplitude, and hence the maximum orbital velocity, could be varied. By using silicous sand or glass beads, they studied the ripple and dune development. By also using dye issuing from the bed they could predict the change from a laminar to a turbulent boundary layer. The maximum orbital water velocities required for various stages is recorded in Table 1-I, modified slightly from the original.

Sleath [137] in his velocity measurements within the boundary layer at the bed of a wave tank, found that the parameter $U_{max}D_{50}/\nu$ could define the transition in the velocity profile from laminar to one still laminar but influenced by vortex formation around the grain of sand. A value of around 50 was noted for this critical condition. Substitution of appropriate values in Table 1-I, using $\nu = 11.0 \cdot 10^{-6}$ ft.^{2}/sec, gives the result as listed. The closeness of the results under the different testing conditions is encouraging.

The point of interest in Table 1-I is the occurrence of incipient motion and of ripple formation after the boundary-layer transition commences but before the fully turbulent layer exists. The process was described [115] as follows: "The beginning of transition was judged to occur when line vortices were first noticed to form parallel to the bed and perpendicular to the direction of the water motion. Vortices formed about at the time of flow reversal. The axes of the vortices were a small distance from the bed. This distance was estimated to be about 5 mm. The flow was laminar both above and beneath this visual row of vortices. In the beginning the vortices persisted for a very short period in the cycle. With increasing amplitude, the vortices persisted for longer durations in the cycle. The boundary layer was deemed to be fully turbulent when the vortices persisted throughout the entire cycle. Even with the turbulent boundary layer, the main body of the flow remained laminar."

Sutherland [116,117] complements such observations by proposing an entrainment hypothesis, from observations of grain motion induced by turbulent flows, based on the concept of turbulent eddies disrupting the viscous sublayer and impinging directly onto the grain surface. It is suggested that entrainment results from the interaction between fluid elements within an eddy and the sediment grains [116]. He recalls that Vanoni [118] had observed particles "embedded" in a boundary layer 7–50 grain diameters thick, moving in gusts or bursts, even though the shear stress was not sufficient to cause general movement.

Sutherland was concerned with uni-directional flow and accepted the thesis of Runstadler et al. [119] that an eddy consists virtually of a sphere of rotating fluid. In oscillatory flow, on the other hand, the vortices generated would tend to be cylinders of fluid with their axes transverse to the flow (i.e., parallel to the wave crests). Whatever form these small masses of water take, the main difference between the two flow conditions is the interaction of the rotating mass with its surrounding fluid.

Rolling-grain ripple. As the velocity of flow increases so does the strength of the vortices. They reach an intensity where the majority of the grains on the bed are being rolled back and forth. This results in ripples being formed. Bagnold [120] has reported, from his studies with an oscillating sand bed in still water, that, once the ridges form, a degree of shelter is afforded particles in the troughs. For a given maximum velocity the crest heights reach an equilibrium value, dictated by the angle of repose for the sediment. This height then determines the "shadow" length and hence the pitch of the ridges. The trough profiles vary for fine and coarse sediment, the former being flat and the latter crescentic. As the ridge height grows so the particles cease to oscillate, first in the trough and finally at the crest, but at no stage do they leave the bed.

As the oscillatory speed increased, Bagnold found that the rolling-grain ridges assumed new heights and pitch lengths. This was accomplished by some ridges expanding and absorbing others, until a new equilibrium series was complete. From this it can be inferred that during the transition period more material is in motion and therefore subject to any mass-transport or other current existing at the bed. The above ripple development occurs for oscillatory speeds up to double that required for incipient particle motion [115,120].

Vortex ripples and dunes. At the limiting orbital velocity mentioned above the height to pitch length (steepness) of the rolling grain ripple reaches a maximum value of 1:5, or about 20 grain diameters height. This condition is not reached simultaneously all over the bed, but takes place randomly in patches. This critical bed roughness causes the streamlines to separate from the crest and establish a new vortex-generating regime. This causes radical changes in the bed pattern which Bagnold [120] termed "vortex ripples", because it was at this stage he noticed the influence of vortices. Since, as already noted, vortex generation plays a part even before the boundary layer is fully turbulent, it appears advisable to denote the new bed undulation as a "dune" as was done by Carstens and Neilson [115].

The development of a duned bed is triggered from the one or two spots where a ridge has reached the critical steepness. Bagnold records that the new regime "spreads rapidly over the rest of the surface like a disease". This bed "projection" generates turbulent fluctuations of velocity and pressure which eject particles away from the "shadow" zone. Soon the adjoining ridge being constructed then reaches a limiting height of around 20 grain diameters and, in turn, sponsors another trough and ridge. In this oscillatory situation, growth of the dune system takes place simultaneously on both sides of the first dune.

Carstens and Neilson have measured the speed of spreading of such dunes and derived the equation:

$$\frac{V}{U_{max}} = 1.7 \cdot 10^{-4} \left[\frac{U_{max}^2}{(S-1)gD} - 1.7 \right]^{3/2} \tag{1-7}$$

where V = velocity of dune propagation; U_{max} = maximum oscillatory water velocity; S = sediment to fluid density ratio; g = acceleration due to gravity; D = grain diameter.

This equation is dimensionally homogeneous and, in fact, contains Carsten's [121] sediment number:

$$N_s = \frac{U_{max}}{(S-1)^{1/2} g^{1/2} D^{1/2}} \tag{1-8}$$

which will be referred to later. Eq. 1-7 is depicted in Fig. 1-17 for sands and seawater $S = 2.6$.

Example. An obstacle triggers the formation of dunes on a sandy bed ($D = 0.1$ mm) when 12-sec waves 1 m high are passing over this 12.6 m deep zone. Compute how many waves are required to spread this formation over a 100-m zone either side of the obstacle.

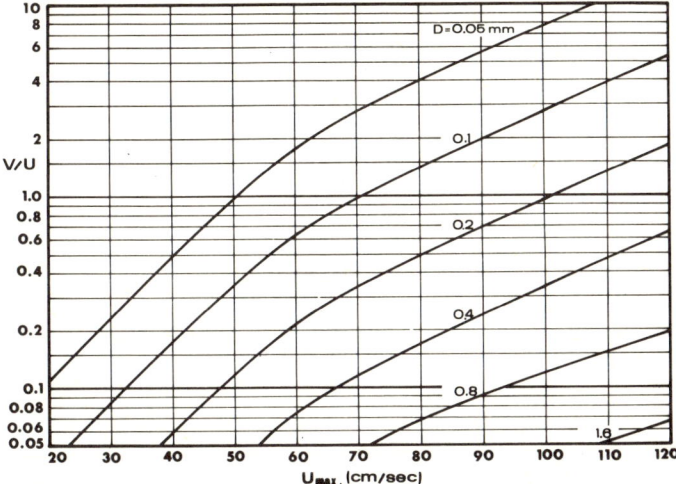

Fig. 1-17. Speed of dune development in terms of maximum orbital water velocity.

$$d/L_o = \frac{12.6}{1.56 \; 12^2} = 0.056 \; \sinh 2\pi d/L = 0.67$$

$$U_{max} = \frac{\pi H}{T \sinh 2\pi d/L} = \frac{\pi \; 100}{12 \; 0.67} = 39 \; cm/sec$$

From Fig. 1-17 $\frac{V}{U_{max}} = 0.16 \quad \therefore V = 6.2 \; cm/sec$

Time to spread 100 m $= \frac{10,000}{6.2} = 1610 \; sec \; (= 27 \; min)$

Number of waves passing $= \frac{1610}{12} = 135$ waves.

It is noteworthy that dunes can be initiated artificially by providing a bed undulation, either as a ridge or a trough which is greater than 20 particle diameters in height or depth. The ridge can be a single object or a mound of sediment. In this way dunes can be promoted by oscillatory velocities less than those required for a rolling-grain ridge formation. This has particular reference to collecting trays in model work and will be discussed more fully later.

When the critical steepness of the ripple has been reached the water velocities are such as to carry particles from the crest, whence they fall into the trough. Some may arrive at the bed with sufficient velocity to impell other grains to jump up the slope of the next ridge. The vortex formed on the leeside of the crest (each side becomes the leeside alternately) traverse the almost stagnant water in the "shadow" zone (see Fig. 1-18), and hits the trough or the slope of the adjacent ridge. The process of diameter reduction and increased velocity of the eddy upon impinge-

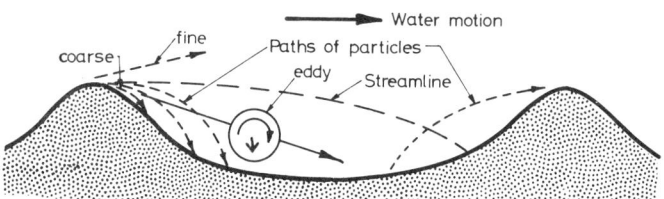

Fig. 1-18. Water vortex and sand grain motions near a sand dune.

ment, causes sand particles to be ejected from the trough zone. The significance of this turbulence was stressed by Sutherland [116], who conducted tests in a flume that had previously developed a stable-duned bed; then, from zero flow a uni-directional flow was slowly built up and the initial motion of sediment observed. He reported: "It occurred first in the troughs and at the base of the upstream slope of the dune. Motion was intermittent with five or six grains moving simultaneously from a small area. In the trough there did not appear to be a preferred direction of motion even within one burst. On the dune slope the preferred direction was downstream, but there were many instances in which grains moved laterally and upstream."

It is pertinent to note that activity was observed in the trough before particles commenced moving from the crest. Turbulent eddies therefore appear to be of prime importance in sediment motion. It was also noted that the mean flow velocity at which this trough disturbance occurred was about 70% of that required for incipient motion of a flat bed.

The stronger vortices, developed as velocities increase, may retain their identity long enough to be swept from the adjoining ridge into the main flow. This was accompanied by bed particles being drawn into a trajectory by the upward velocities, and thrown onto the dune crest. This is particularly so for the coarser material, the finer fractions being carried into suspension.

Sutherland studied the influence of eddies over various sections of the dune profile by means of dye seeping through the granular bed. Although his observations were made for channel flow, they are applicable to wave conditions with little reservation. In the trough dye would accumulate and then suddenly disperse and be carried away into the main flow. Intrusion of fluid into the trough zone was also observed. Sutherland records: "The observations showed convincingly that the lee of a dune is not a region of quiescent fluid, but one in which frequent bursts of motion occur. These bursts had a very definite effect on the sediment grains." Another pertinent observation was the patchy appearance of the dye on the surface: "A possible explanation is that the dye concentrates in places where the vorticity is high. These would be the ends of vortex lines which extend out into the flow. Another possible reason is the existence of very local pressure fluctuation which suck the dye out from between the grains and cause a filament of dye to form."

With the more cylindrical shape of vortices in oscillatory flow it would be of interest to investigate whether such filaments would assume streaky shapes parallel to the wave crests when dunes exist. Carstens and Neilson [115] noted this streakiness for the initiation of vortices on a flat bed.

On the upstream slope of the dune the dye moved upwards close to the surface, but spasmodically filaments would rise and mix with the main flow, accompanied by greater grain-motion intensity. At the stagnation region, where the streamlines from the crest re-contacted the bed, dye would move either downstream or into the trough. At the crest the dye flowed swiftly, some curling down into the trough to form high concentrations, some curling upwards to disappear into the mainstream. Occasionally the streaks jetting from the crest dipped sharply, indicating the intrusion of an eddy down the lee-side of the dune.

Carstens [121] found that the dunes enlarged as orbital velocity increased, whilst maintaining a constant steepness of 1/5.75. He related dune height h to sediment number N_s (see eq. 1-8). A comparison with channel flow conditions [122] is illustrated in Fig. 1-19. Height in both cases rose until $N_s = 8.5$, after which it fell. Hom-ma et al. [123] derived a relationship from wave flume tests which varied with about the 1/5 power of P/x where P is the pitch length of the dune and x is the horizontal displacement of the water particle from its mean position.

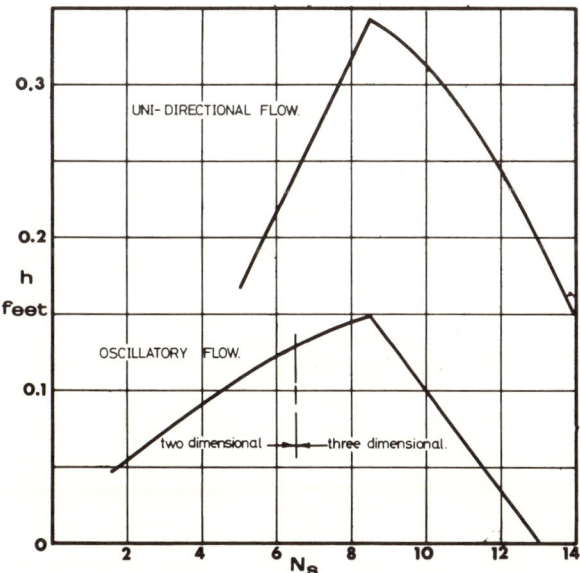

Fig. 1-19. Dune height versus N_s for uni-directional and oscillatory flow. (Ref.[121].)

Carstens [121] observed that his dunes initially contained long crests that were transverse to the flow (N_s = 6.5, see Fig. 1-19), all of similar height and pitch. He records: "In this range, the fluid motion appears to be two-dimensional with line vortices being formed in the lee of the dune crests. Two vortices are formed each cycle in the trough between a pair of adjacent crests. On reversal of motion, the previously formed vortex is moved back toward the crest on which it was formed and is ejected into the main flow above the dune system." He questioned whether similar characteristics should have been reported for uni-directional flow. However, from the previous discussion on cylindrical versus spherical masses of rotating fluid it would seem that lack of such evidence has not been an oversight.

Bagnold [120] has also reported the change from transverse ripples to a three-dimensional system, by the addition of longitudinal ripples alternating to form a "brick" pattern. This occurrence, for water amplitudes of about P/b, gave him confidence in his oscillatory bed technique, since the same bed form had been observed in flumes for similar small water oscillations. Manohar [82] also observed this ridge structure but offered no explanation for it. Under the considerations of small water amplitude, the cylindrical vortices would have little chance to deviate from their straight axes until they hit the bed. Their consequent reduction in diameter and axial elongation could well result in a systematic snaking as shown in Fig. 1-20 [143]. Accretion and erosion would occur in alternate loops, which could produce the linking ridge between transverse crests. Once such an undulation were initiated in a trough it would amplify the action, through the differential speeds of the vortex across the varying trough depths. The reason for the alternate spacing of the connecting ridges, whose heights need not be related to those of the transverse crests,

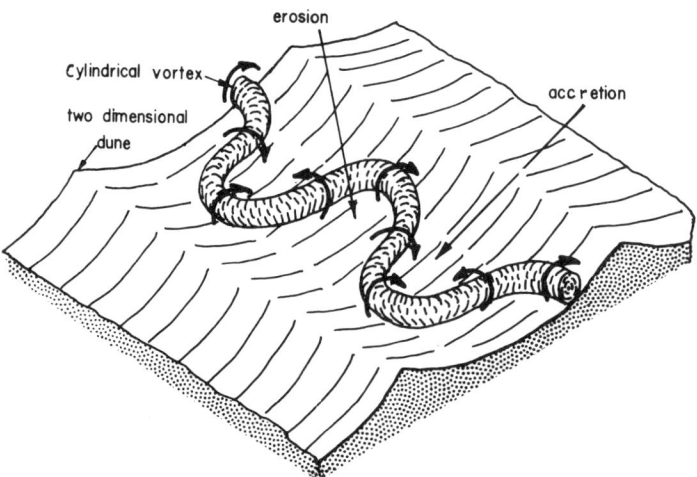

Fig. 1-20. Probable snaking of cylindrical vortices when impinging on the bed.

may possibly be determined from tests using pulsating jets from a submerged slit orifice, as distinct from the circular orifice employed by Sutherland [116].

The dunes, having thus become three-dimensional at $N_s = 6.5$, grew in height until $N_s = 8.5$, beyond which they were reduced, disappearing completely at $N_s = 13$ approximately (see Fig. 1-19). This attrition of the dune is caused by the erosive forces at the crest exceeding the accretionary influence of the vortices. As velocities increase further, the decreased dune height restricts the generation of vortices, until ultimately the whole bed is in oscillation. In this flat bed condition turbulence is at a premium, together with high shear stress at the bed. This velocity gradient tends to distort and dissipate the eddies as they approach the bed, so appearing to reduce their effect. However, on this point Sutherland comments as follows: "Comparison of observations made of critical conditions with those made at higher velocities over a flat bed, show that noticeable increases in grain agitation take place as the flow velocity is increased. Based on this, it seems that the effect of increased turbulence intensity would be more important than the increased resistance to eddy impingement by the faster flow." The previous references to large-scale pressure fluctuations [21,111] in rivers and on the coast would appear to substantiate this observation.

At this high degree of turbulence it could be expected that sediment concentration would vary from the bed upwards, but it also suffers temporal fluctuations with the passage of each wave. Hom-ma and Horikawa [123,124], using a photo-electric meter, have recorded these fluctuations synchronously with the wave profile. Typical records obtained by them are reproduced in Fig. 1-21, for measurements made above the crest (A) and above the trough (B) when $x/P = 1.1$ and $h/P = 1/7.33$. Fairchild [125] had stated that concentrations at equal distances above the crest and trough of dunes would differ, but Hom-ma and Horikawa [124] did not find this, except for small concentrations, when motion would be initiated in the trough in any case. Whilst x/P exceeded 0.7, four peak concentrations were present, indicating that the two high-velocity peaks during each wave period were complemented by two others supplied by vortex impingement on the bed. In

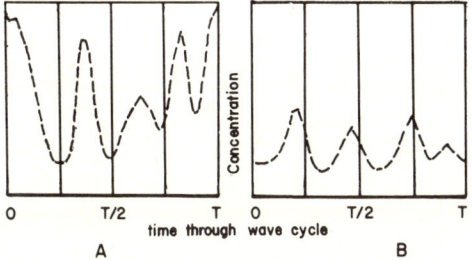

Fig. 1-21. Typical record of sediment suspension during cycle of model wave. A. for 0.1 h above trough; B. for 0.16 h above crest of dune. (Ref.[123].)

presenting design data on sediment concentration Hom-ma et al. [123] have expressed heights above the seabed in dune pitch lengths. The relationships possibly require some rationalisation, but are a major advance in the understanding of this complex phenomenon.

Effects of permeability

Permeability of the bed material could exert an influence by permitting the fluid to pass down between the particles when a downward pressure gradient exists near the bed, and upwards when the gradient is reversed. Fig. 1-22 depicts the likely sequence of events when an eddy approaches a sedimentary surface. At A the inertia of the fluid mass, plus the downward peripheral velocity, causes a flow into the bed, which passes out at B and C. Immediately afterwards a "suction" is applied at A and C as the vortex core contacts the bed and the upward components of the new distorted peripheral velocities exert their influence. The probable path of the particles, commencing in a near vertical direction is depicted in Fig. 1-22. The forward curvature is caused by the strong curvature of water particle motions in the

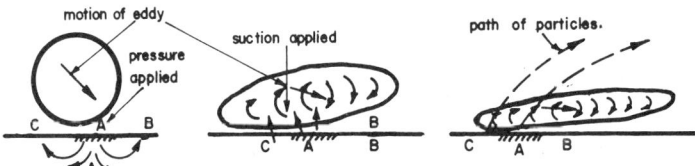

Fig. 1-22. Stages of impingement of eddy on a permeable bed.

flat eddy. The ejection of sand grain from eddies has been noted previously. This jumping action has been termed "salt-action" and has been reported by a number of workers [126–128], some of whom have put it down to grain particle impact. However, Bagnold [126] contends: "When the solids are sheared over a gravity bed the dispersion must necessarily be upward, against the normal component of the gravity force. To maintain such shear thus requires the maintained exertion of an upward supporting stress of some kind on every granular shear layer."

The case with which particles are loosened and lifted will depend upon their buoyant weight and upon the fluid velocities developed in their voids. Bagnold [120] reports: "Owing to the rapid decrease in the permeability as the grain-size of the same material is reduced, the depth of the surface movement of the material as measured by grain-diameters, is far smaller for fine sands than for coarse. The finer the sand, the slower becomes the rate of sand drift under a given wave movement. With very fine quartz sand the grain movement is confined to the topmost layer. When the grain movement of quartz and of perspex sands of the same grain-size is compared, it is found that, whilst only one or two grain layers of quartz sand are disturbed, the movement extends down to a depth of ten or more layers of perspex." Vincent [71] has reported that tests using thicker beds or sand facilitated "the onset of turbulence" due to the increased porosity. Carstens and Martin [129] have discussed this topic also.

Reach of waves

Closely allied with the problem of sediment transport is the limit to which ocean waves can disturb the bed, termed the "reach" of the waves. Much of the technical literature, particularly that in geology, concedes only an active shoreline to a depth of 30 ft. or so. Hence the magnitude and importance of wave energy in submarine geological processes has been severely discounted. The fault here lies as much with the engineer as with the scientist, as evidenced by the following quotation from Dietz [130]: "Unfortunately the literature does not seem to clearly provide us with parameters about wave-base such as its depths or range of depths, its depth relative to pertinent wave characteristics such as wave length, or the ambient water velocity at wave base. The assumed depth of wave-base especially is treated in a vague and non-commital way in more recent papers."

Working on the basis of shear stress alone, engineers have also restricted their view to depths which are a small proportion of those existing near the edge of the continental shelf (400 ft. approximately). However, as replication of the water motions near the seabed has become more realistic, so the apparent capacity of the waves to roll and suspend sediment has increased. From the following discussion it should become evident that the whole continental shelf serves as a highway for longshore sediment transport.

Both Manohar [82] and Carstens [122] have used a dimensionless parameter whose value could be identified with various stages of sediment motion and bed formation. These can be expressed as:

$$\Psi = \left[\frac{\pi H}{T \sinh 2\pi d/L}\right] \frac{1}{(S-1)^{0.4} g^{0.4} \nu^{0.2} D^{0.2}} \quad \text{(Manohar)} \quad (1\text{-}9)$$

$$N_s = \left[\frac{\pi H}{T \sinh 2\pi d/L}\right] \frac{1}{(S-1)^{0.5} g^{0.5} D^{0.5}} \quad \text{(Carstens)} \quad (1\text{-}10)$$

where S = specific gravity of the sediment, and D = median diameter of the particles.

The linear expression for U_{max} in the brackets is used in each case. The major difference in these equations is the inclusion of kinematic viscosity (ν) in one and not the other. In Table 1-II are listed the values of the dimensionless parameters ψ

TABLE 1-II

Parameters ψ and N_s for various bed events (References [82, 115, 121])

Event	ψ	%	N_s	%
No movement even with projections	–	–	1.3	10
Initial movement on smooth bed	7.45	35.6	3.04–3.76	24–29
Initial movement on duned bed	–	–	2.3	18
Spontaneous rolling-grain ripples	8.2	39	5.0	38.5
Initiation of dunes	9.23	44	6.5	50
Maximum dune steepness	13.0	62	8.5	65
Disappearance of dunes	20.9	100	13.0	100

and N_s for various stages of sediment movement and bed development. For ready comparison percentage values of these are given based upon 100% for the condition of dune disappearance. The agreement in the percentage values of the Table would indicate that a general solution is near at hand.

The empirical formulae derived for incipient motion of sand particles on a flat bed, derived by many workers, can be put into similar dimensionless form as in eq. 1-11 to 1-23.

The relationship by Abou Seida[83] cannot be written in such a form but a modification and iterative process carried out by Mogridge [131] permits it to be plotted as in Fig. 1-23. The graph of Bonnefille and Pernecker [92] consisted of two curves which have been modified into one for the presentation in the figure. The condition of the boundary layer has been indicated in the equations or been presumed.

Bagnold [120] (laminar)
$$\frac{U_{max}}{(S-1)^{2/3}g^{2/3}D^{1/3}D^{0.1}T^{1/3}} = 3.18 \quad (1\text{-}11)$$

Bonnefille and Pernecker [92] (laminar)
$$\frac{U_{max}\nu^{1/6}}{(S-1)^{5/6}g^{5/6}D^{1/2}T^{1/2}} = 0.072 \quad (1\text{-}12)$$

Bonnefille and Pernecker [92] (turbulent)
$$\frac{U_{max}\nu^{19/30}}{(S-1)^{16/15}g^{16/15}D^{6/5}T^{1/2}} = 0.01 \quad (1\text{-}13)$$

Bonnefille and Pernecker [92] (modified)
$$\frac{U_{max}\nu^{5/18}}{(S-1)^{8/9}g^{8/9}D^{2/3}T^{1/2}} = 0.069 \quad (1\text{-}14)$$

Carstens et al. [115,132, 133] (turbulent)
$$\frac{U_{max}}{(S-1)^{1/2}g^{1/2}D^{1/2}} = 3.5 \quad (1\text{-}15)$$

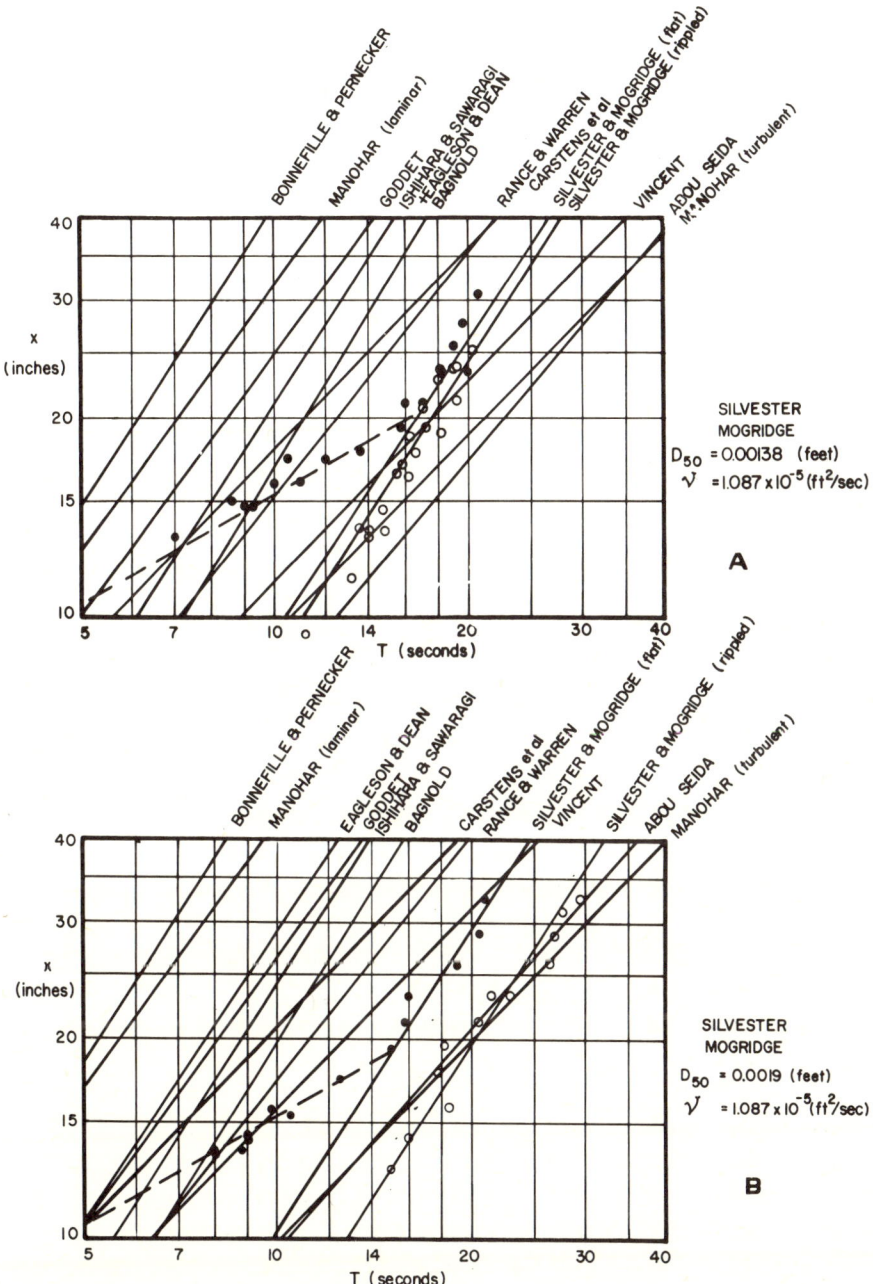

Fig. 1-23. Wave periods (*T*) and amplitudes (*x*) at the sea floor to produce incipient motion on a flat bed with sand particles of mean diameters: A. 0.00138 ft. (0.42 mm); B. 0.0019 ft. (0.58 mm).

Eagleson and Dean [88] (laminar)	$\dfrac{U_{max}\nu^{1/2}}{(S-1)gDT^{1/2}} = 0.131$	(1-16)
Goddet [90] (turbulent)	$\dfrac{U_{max}}{(S-1)^{2/3}g^{2/3}\nu^{1/24}D^{1/4}T^{3/8}} = 3.0$	(1-17)
Ishihara and Sawaragi [68] (turbulent)	$\dfrac{U_{max}}{g^{3/4}(S-1)^{3/4}D^{1/4}} = 0.093$	(1-18)
Manohar [82] (laminar)	$\dfrac{U_{max}\nu^{1/3}}{(S-1)^{2/3}g^{2/3}D^{2/3}} = 0.159$	(1-19)
Manohar [82] (turbulent)	$\dfrac{U_{max}}{(S-1)^{0.4}g^{0.4}D^{0.2}\nu^{0.2}} = 7.45$	(1-20)
Silvester and Mogridge [134] (flat bed)	$\dfrac{U_{max}\nu^{1/18}}{(S-1)^{7/9}g^{7/9}D^{1/3}T^{1/2}} = 0.034$	(1-21)
Rance and Warren [135] (sand)	$\dfrac{U_{max}}{(S-1)^{3/5}g^{3/5}D^{2/5}T^{1/5}} = 0.69$	(1-22)
Vincent [71] (turbulent)	$\dfrac{U_{max}}{D^{1/2}} = 19.1$ (in f.p.s. units)	(1-23)

The U_{max} can also be expressed in terms of amplitude of motion when the wave period is known or assumed, in which form these equations have been graphed in Fig. 1-23A,B for $D_{50} = 0.00138$ and 0.0019 ft., respectively. Also assumed in the figures is $\nu = 1.087 \cdot 10^{-5}$ (ft^2/sec), $S-1 = 1.6$ and $g = 32.2$ ft/sec^2. It is seen immediately that these relationships are not compatible, probably through being extrapolated beyond the zones of verification.

Also included in Fig. 1-23 are the results of tests conducted on an experimental rig [136] which essentially was a block of water oscillated over a sand bed with prototype amplitudes and periods. Black dots represent flat bed conditions and the open circles the rippled surface produced from prior oscillatory motion [131]. For the flat bed two relationships were found, one in which x varied approximately as $T^{1/2}$ for smaller x and T values, and the second in which x varied as $T^{3/2}$. The transition from the one to the other occurred at $x = 20.5''$ and $T = 17$ sec in Fig. 1-23A and $x = 19''$ and $T = 15$ secs in Fig. 1-23B. As seen in these figures, the presence of ripples on the ocean floor will produce incipient motion at smaller amplitudes of the water particles for a given wave period, or at longer periods for a given amplitude. However, the traces for this more realistic condition do not match

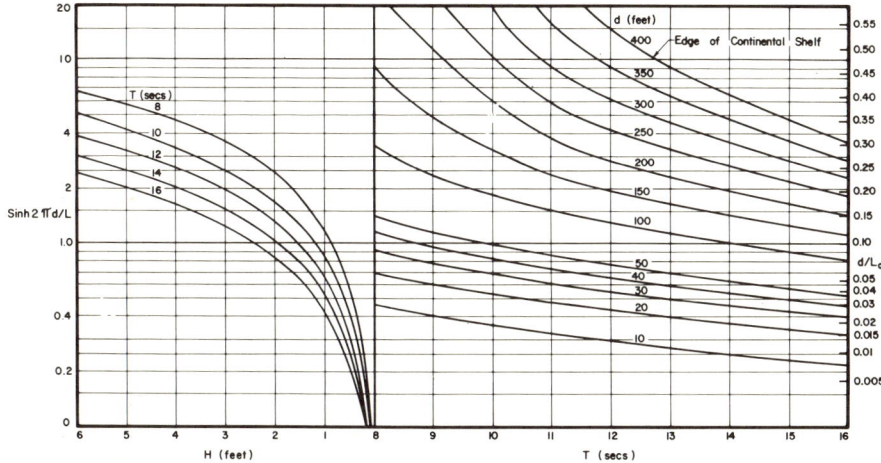

Fig. 1-24. Wave conditions for incipient motion on a flat bed for sand (D_{50} = 0.00065 ft. = 0.2 mm).

the one equation, so the conservative flat-bed relationship of eq. 1-21 has been put in terms which are graphed in Fig. 1-24. With this diagram a wave of any specific period in a certain depth of water will have to have a minimum height in order to initiate particle movement on a flat bed.

TRANSPORT RATES IN SURF AND OFFSHORE ZONES

Various means of measuring sediment transport rates along the coast have been attempted. In the surf zone and on the beach itself this has taken the form of sediment tracers [138,139], sediment sampling associated with littoral current evaluation [140,141], and assessment of silting or erosion over specific periods. From the previous discussion it would seem necessary to distinguish between early-summer and late-summer beach profiles when measuring littoral drift in the surf zone. Johnson [142] has summarized some of these computations from around the world, which vary from thousands to millions of tons per annum.

Measurements become more difficult in the offshore zone, due to the greater depths involved and the increased hazards, especially when storm conditions are trying to be recorded. Radioactive tracers have been used with moderate success, which essentially indicate only the direction of net movement, the actual evaluation of transport being open to question. Such attempts require a large expenditure of money and manpower and hence can be undertaken only by large organisations. Complete wave data must be taken over the period of sediment tracing. With radioactive tracers there is the constant risk that the wave conditions are not suitable at the time the material is available at the site. If the location is anywhere close to shore, assurances must be given that radioactive particles cannot pollute the beach itself.

Once the particles have begun to be disturbed offshore the mass-transport of the water at the bed will carry them in the direction of wave advance. When the waves are arriving obliquely to the shelf such advance will have a longshore component. Wave theory has shown that the mass-transport velocity within the boundary layer at the bed is predictable and is operative as soon as waves commence, even though net motion at higher levels takes time to develop.

Computations will be carried out below for a series of wave bands of 2 sec width which make up the fully-arisen sea of a 35-knot wind, the average optimum wave spectrum. As can be verified from Chapter 3 in *Coastal Engineering, 1,* the $H_{1/3}$ heights of trains centred on 8, 10, 12, 14 and 16 sec periods are 3.2, 4.9, 5.3, 4.2 and 2.0 ft., respectively. It will be considered that these are all angled at 50° to the outer edge contour of a uniform plane shelf with a slope of 0.002:1 (see Fig. 1-25). In Fig. 1-25 each d/L_o value represents the mean for a certain width of shelf, from which it is seen that $(\Delta d/L_o)/(b/L_o) = 0.002$, so that $b = (\Delta d/L_o)\ 5.12\ T^2/0.002$. At intermediate depths the differential $\Delta d/L_o$ is the difference between the values mid-way between adjacent points, whilst at the extremities of $d/L_o = 0.01$ and 0.5 the outside half-width is assumed the same as the inner half (e.g., $\Delta d/L_o = 0.015 - 0.005 = 0.01$).

The computations for the five wave bands are summarized in Table 1-III. The longshore component of the mass transport velocity (U_L) has been determined by substituting height, period and angle for each train at specific d/L_o values. The laminar or turbulent state of the boundary has been taken into account. Although the calculation has been carried out to $d/L_o = 0.5$ the limit of reach of the wave has been used in the sum $\Sigma\ bU_L$. The double lines in the Table indicate approximately this limiting depth ratio. Assuming the suspended and rolling sediment is confined to a half-inch layer at the bed, the total discharge of water across a plane normal to the shore has been computed (*MT*).

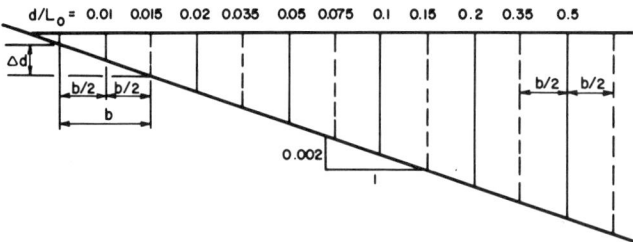

Fig. 1-25. Assumed uniform continental shelf for purposes of computing longshore drift.

TABLE 1-III

Longshore mass transport in various wave trains across portion of the continental shelf (see Fig. 1-2)

		8	10	12	14	16	T (sec)
$\alpha_0 = 50°$		3.2	4.9	5.3	4.2	2.0	H (ft.)
$\sin \alpha_0 = 0.766$		2.10	1.75	2.59	6.56	43.0	$gT^3/H_0^2 \sin \alpha_0 \, 10^3$
		2.3	2.49	3.63	7.13	24.4	$gT^{2.6}/H_0^{1.2} \sin \alpha_0 \, 10^3$
		14.35	26.9	26.3	14.1	2.78	$H_0^2 \pi / T \nu \, 160^2 \geqslant A$
d/L_0	0.01	–	–	–	–	–	$U_L \cdot 10^3$ (laminar) [1]
$\Delta d/L_0$	0.01	78.1	72.1	49.5	25.2	7.4	$U_L \cdot 10^3$ (turbulent) [1]
A	0.048	1.64	2.56	3.69	5.02	6.55	$b \cdot 10^{-3}$ (ft.)
$A_{0.6}$	0.161	128.0	183.0	183.0	127.0	48.5	$b \cdot U_L$
d/L_0	0.02	–	–	–	–	–	
$\Delta d/L_0$	0.02	41.2	38.0	26.1	13.3	3.9	
A	0.138	3.28	5.12	7.40	10.04	13.10	
$A_{0.6}$	0.305	135.0	195.0	193.5	133.8	51.0	
d/L_0	0.05	–	–	–	–	–	
$\Delta d/L_0$	0.04	18.3	16.9	11.6	5.9	1.7	
A	0.530	6.56	10.24	14.80	20.08	26.20	
$A_{0.6}$	0.685	120.0	173.0	171.7	123.0	44.5	
d/L_0	0.10	–	–	–	–	–	
$\Delta d/L_0$	0.075	9.8	9.1	6.2	3.2	0.9	
A	1.52	12.30	19.20	27.70	37.70	49.10	
$A_{0.6}$	1.28	120.5	174.8	172.0	121.0	44.2	
d/L_0	0.2	–	–	–	–	1.4	Note. For laminar
$\Delta d/L_0$	0.2	4.8	3.8	3.0	1.5	–	conditions it is as-
A	5.10	32.9	51.2	73.8	100.5	131.0	sumed that $D/T^{1/2}$
$A_{0.6}$	2.64	158.0	195.0	221.5	150.5	183.2	= 0.007 equivalent
d/L	0.5	1.1	1.3	0.9	0.3	0.05	to ripples 1/4" to
$\Delta d/L_0$	0.3	–	–	–	–	–	1/3" high on bed.
A	140	49.2	76.8	110.5	150.5	196.0	
$A_{0.6}$	19.4	54.2	99.9	99.5	45.0	9.8	
(1) U_L (ft./sec)		0.075	0.20	0.3	0.325	0.31	limiting d/L_0
(2) $MT = \Sigma b U_L/24$		1.00	0.50	0.75	0.87	0.80	prop. of last b
(ft.^{3}/sec)		120	97.5	166.0	131.0	146.8	final $b \times U_L$
(3) G littoral drift		383	823	886	636	335	$\Sigma b U_L$
(ft.^{3} sand/sec)		16	34	37	27	14	MT in 1/2" layer [2]
(1 ft.^{3} = 110 lb)		0.37	1.08	1.26	0.61	0.76	G [3]
(4) Concentration by volume		2.3	3.2	3.4	2.3	5.4	$C\%$ [4]

Each train will ultimately reach the beach where the waves will break at some slight angle to the shore and generate a littoral current. The transport of sediment in the surf zone (G) can be computed by means of eq. 1-5 in cubic measure of dry sand. For equal overall rate of longshore drift within and beyond the surf zone, a volumetric concentration can be determined from $C = G/MT$. As seen in Table 1-III, the required concentrations are extremely small and could readily be exceeded some five-fold.

The above analysis contains many assumptions which the reader may wish to modify for his own satisfaction. However, the results are considered to be conservative since the following prototype phenomena tend to give more reach to the waves, so extending the shelf width over which mass-transport could be active and increasing the suspension of sediment throughout:

(1) Velocities of water particles near the ocean bed will be higher than those derived by linear theory and used in experiments to date. In fact, it is the differential in forward and reversed flow that provides the net motion of mass-transport.

(2) The interaction of wave trains of slightly differing period can generate greater instantaneous velocities than are implied in the above analysis. Similar arguments hold for standing waves which can occur during storm sequences.

(3) Wave trains angled to each other produce vortices of large dimensions, with associated turbulence, which may disturb the bed more readily.

(4) Currents and internal waves associated with tidal action near the edge of the continental shelf could assist waves in disturbing the bed at these larger depths.

(5) Shell debris on the sea floor can cause water oscillations to disturb sand particles sooner than smooth or even rippled beds. Values of N_s in Table 1-II, which are directly proportional to maximum velocity and hence wave height, indicate the more reasonable values to employ. As Carstens and Neilson [115] state: "Since any natural bed is almost certain to be littered with flow obstructions, a disturbed flow, deformed bed, incipient motion criterion is the logical lower limit for the development of a duned bed, rather than the undisturbed flow, flat bed, incipient motion criterion."

(6) It is possible that marine fauna are concentrated heavily at the shelf bed. Any motions by them on or near the bed will continually disturb material which the mass-transport of the waves will carry into and along the shore.

Finally, another approach to the optimum reach of waves may be gleaned from the tests of Bagnold [114], who reported: "The turbulence which takes the form of large vortices, formed in the troughs and thrown off upwards into the water at the end of each stroke, is a maximum when amplitude equals natural ripple-pitch and falls off when the water amplitude is greater or smaller."

This vortex-ripple or dune pitch he found to vary with sediment diameter D, such that:

$$P = 64 D^{1/2} \tag{1-24}$$

for P and D in cm. Equating P to water amplitude results in:

$$H/\sinh 2\pi d/L = 128 D^{1/2} \tag{1-25}$$

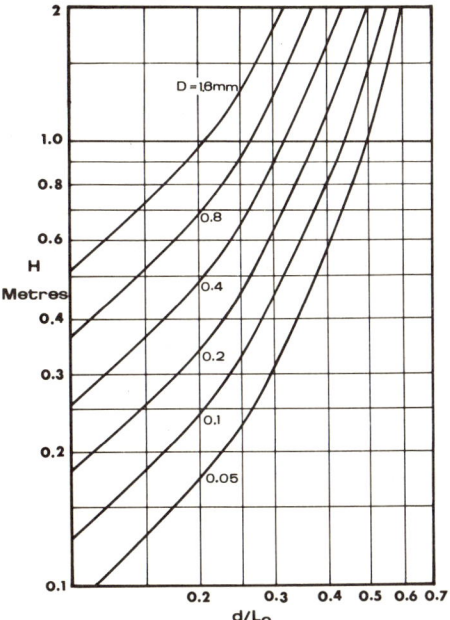

Fig. 1-26. Wave characteristics to produce maximum turbulence at the seabed.

which has been graphed in Fig. 1-26. This represents wave conditions which produce maximum turbulence at the bed and hence optimum suspension within the wave boundary layer, where the mass-transport velocity is also optimized. Thus, for a 12-sec wave to produce maximum bed disturbance in 100 m of water (d/L_o = 0.43) for a fine sand (D_{50} = 0.2 mm), its height would need to be only 1.4 m, much less than is implied in Fig. 1-24 for incipient motion on a flat bed. Other values read from Fig. 1-26 would indicate that the required wave heights are well within the realms of possibility, even for swell waves arriving from distant storms.

The discussion above may appear too inconclusive for the practicing engineer to employ for design purposes. However, it is only meant to illustrate the importance of the complete continental shelf as a stage for the sediment drama of the sea. Whilst waves of the longer-period end of the average oceanic spectrum can disturb and transport material slowly near the edge of the shelf, a very much greater level of energy is applied to zones nearer the shore, but beyond the breaker line. Coastal engineers concerned with offshore structures should therefore not discount the ability of waves, even of modest height, to erode and accrete sand around structures resting on the seabed. Now that long-life facilities are being planned for the outer reaches of the continental shelf it is imperative that verification of transport rates in these regions be carried out.

PROBLEMS

1

In what ways can two wave trains angled to each other exert an influence on sediment particles at the bed which do not exist when they are aligned? List the situations in which such short-crested systems could occur.

2

What evidence is there that swell waves are the predominant distributing agent for sediment reaching the coastline via the river systems? Cite examples for all cases.

3

Describe the sequence of events from the time storm waves first arrive at a swell-built beach to a week or so after the passage of the cyclone from the area. What conditions worsen the influence of such storm sequences?

4

If you were to institute the observation of waves breaking at the beach each day, wnat wave characteristics would you demand? State whether such observations could be useful for sediment transport studies.

5

Why do swell waves cause accretion on a beach? When do they not do so?

6

It has been observed that sediment movement offshore during the transition from one wave alignment to another is optimum. Give reasons for this.

7

Discuss the proposition that storm waves have little influence in the longshore transport of sediment. In what circumstances would this view be incorrect?

8

List reasons for the movement of material along the coast fluctuating both in the short run and the long term. What precautions should a coastal engineer observe to minimize errors of judgement associated with such variations?

9

How does the average beach profile on the margin of an enclosed sea differ from that on oceanic margin? Consider this out to the edge of the continental shelf and give reasons for your observations.

10

Discuss the difference in nature of the sediment problems usually encountered on coasts with and without large tidal ranges.

11

Why can tidal currents normally be ignored in the study of silting or erosion on the open coast? List the conditions when such currents could exert an excessive geomorphological influence.

12

Why should estuaries or similar indentations of the coast suffer continual siltation? What tests should be carried out before remedial measures are recommended?

PROBLEMS

13
It is possible to calculate the profile for swell-built beaches and those resulting from erosion by storm waves. What useful purpose can such computations serve?

14
A persistent swell arriving at a coast has an average height of 1 m offshore where the sand has a mean diameter of 0.3 mm and density of 2.65. Draw the resulting beach profile from the breaker line to a depth of 15 m. Determine the profiles separately for an increase of 50% in wave height and of sediment diameter. If the sand on the beach face in each case is double that offshore, compute the width of the surf zone for the three conditions above.

15
A beach face with sand ($S = 2.65$) whose mean diameter is 0.7 mm is attacked by storm waves, the significant height of which is 6 ft. and the significant period 10 sec. Calculate the width of the surf zone once the offshore bar has been constructed. Determine also the average depth of water from the eroded beach line to the bar. Compare this with the given height of 6 ft.

16
The following information is available respecting a beach zone:
sediment $S = 2.65$; D_{50} offshore = 0.4 mm; D_{50} beach face = 0.8 mm; waves $H_{1/3}$ swell = 2 ft.; $H_{1/3}$ storm waves = 8 ft.; $T_{1/3}$ swell = 10sec; $H_{1/3}$ storm waves = 9 sec.
Determine the recession in the water line when the storm waves arrive, by drawing profiles and equating the cut and fill sections of it.

17
Compute the volume of littoral drift per annum for the average wave characteristics of $H_o = 3$ ft.; $T = 9$ sec, $\alpha_o = 45°$, using the sediment characteristics of eq. 1-4. What error is involved in this calculation if the wave height is wrong by 20%?

18
What kind of records would you demand for a shoreline if an annual rate of littoral drift were required? List the difficulties involved in obtaining these.

19
Discuss the differences between river and marine conditions for disturbing bed material, putting it into partial or full suspension, and transporting it in a lateral direction.

20
The application of the shear stress approach to sediment movement by waves is intricate and is based upon unproven assumptions. List the actions in which errors could arise.

21
What are the essential differences between the action of waves disturbing a bed of sand and a unidirectional flow of water in a channel?

22
List papers in the literature that deal with turbulent structure of fluids in relation to their ability to suspend sediment. Discuss their general conclusions.

23
Discuss the stages through which sediment particles pass before the maximum dune height is developed under the action of waves. Note conditions which promote the initiation of this motion.

24

Waves of 5 ft. height and 13 sec period are propagating over an area of the ocean bed which is 300 ft. deep. A shell resting vertically at the bed causes the sand ($S = 2.6$) of diameter 0.2 mm to form a dune on either side. Determine how long it will take for these dunes to spread 1/4 mile. In terms of swell duration, is such dune proliferation feasible?

25

Compute the depths at which the waves listed will initiate motion of sediment of which the characteristics are noted. Find values for flat and rippled bed conditions.

H(ft.)	T(sec)	D_{50}(mm)	S
5	13	0.1	2.65
3	10	0.2	2.6
2	16	0.1	2.62
1	20	0.1	2.59

26

Assuming the wave bands of $H_{T_{max}}$ and T_{max} for fully arisen sea conditions with $U_{19.5}$ wind velocities of 25, 30 and 35 knots arrive separately at the edge of the continental shelf at an angle of 45°, compute the longshore component of the mass-transport velocity beyond the breaker line. Also compute the littoral drift using the normal assumptions in the text. From this determine the volumetric concentration of sand within 1/2″ of the bed for the longshore drift to be similar offshore and in the surf zone. Make similar assumptions to the example in Table 1-III.

27

What evidence can you supply, besides that noted in the text, for ocean waves being able to disturb sediment out to the edge of the continental shelf? Has such knowledge any engineering significance?

REFERENCES

[1] U.S. Grant, 1943. Waves as a sand transporting agent. *Am. J. Sci.*, 241: 117 – 123.
[2] M.A. Mason, 1950. Geology in shore control problems. In: *Applied Sedimentation*. Wiley, New York, N.Y., pp.276–290.
[3] J.W. Johnson, 1959. The supply and loss of sand to the coast. *Proc. ASCE*, 85 (WW3): 227–251.
[4] R.O. Eaton, 1950. Littoral processes on sandy coasts. *Proc. Conf. Coastal Eng.*, 1st, 1950: 140–154.
[5] W.N. Bascom, 1953. Characteristics of natural beaches. *Proc. Conf. Coastal Eng.*, 4th, 1953: 163–180.
[6] R. Silvester, 1959. Engineering aspects of coastal sediment movement. *Proc. ASCE*, 85 (WW3): 11–39.
[7] J.B. Schijf, 1959. Generalities on coastal processes and protection. *Proc. ASCE*, 85 (WW1): 1–12.
[8] J.W. Johnson, 1966. Nearshore sediment movement – central California coast. *Proc. Santa Barbara Conf. Coastal Eng., 1965*: 537–560.
[9] F. Snodgrass et al., 1966. Propagation of ocean swell across the Pacific. *Phil. Trans. R. Soc.*, A 254: 431–497.
[10] P.D. Trask, 1955. Movement of sand around southern California promontories. *Beach Erosion Board, Tech. Mem.*, 76.

REFERENCES

[11] F.P. Shepard, 1963. *Submarine Geology.* Harper and Row, New York, N.Y., 2nd ed., 557 pp.
[12] D.L. Inman, 1949. Sorting of sediments in the light of fluid mechanics. *J. Sed. Petrol.,* 19: 51–70.
[13] P.S. Eagleson and R.G. Dean, 1959. Wave-induced motion by bottom sediment particles. *Proc. ASCE* 85 (HY10): 53–79.
[14] R.L. Miller and J.M. Zeigler, 1964. A study of sediment distribution in the zone of shoaling waves over complicated bottom topography. In: R.L. Miller (Editor), *Marine Geology.* McMillan, London, pp. 133–153.
[15] P.S. Eagleson, G. Glenne and J.A. Dracup, 1963. Equilibrium characteristics of sand beaches. *Proc. ASCE,* 89 (HY1): 35–58.
[16] U.S. Grant, 1948. Influence of the water table on beach aggradation and degradation. *J. Mar. Res.,* 7: 655–660.
[17] K.O. Emergy and J.F. Foster, 1948. Water tables in marine beaches. *J. Mar. Res.,* 7: 644–654.
[18] W.N. Bascom, 1951. The relationship between sand and size and beach face slope. *Trans. Am. Geophys. Union,* 32: 866–874.
[19] L. Greslou and Y. Mahé, 1954. Etude du coefficient de réflection d'une houle sur un obstacle constitué par un plan incliné. *Proc. 5th Conf. Coastal Eng.,*: 68–84.
[20] C. Kidson and A.P. Carr, 1959. The movement of shingle over the sea bed close inshore. *Geogr. Gaz.,* 145: 380–389.
[21] A.E. Carey, 1907. The protection of sea shores from erosion. *J. Soc. Arts,* 1907; 650–663.
[22] J.B. Herbich and S.C. Ko, 1969. Scour of sand beaches in front of seawalls. *Proc. 11th Conf. Coastal Eng.,* 1: 622–643.
[23] P. Bruun, 1955. Migrating sand waves or sand humps, with special reference to investigations carried out on the Danish North Sea coast. *Proc. 5th Conf. Coastal Eng.,* 269–295. 269–295.
[24] R. Silvester, 1956. The use of cyclonicity charts in the study of littoral drift. *Trans. Am. Geophys. Union,* 37: 694–696.
[25] *Atlas of Climatic Charts of the Oceans* – U.S. Dept. Agr., Weather Bur., Rep. 1247, 1938.
[26] G.M. Watts, 1954. Laboratory study of the effect of tidal action on wave-formed beach profiles. *Beach Erosion Board, Tech. Mem.,* 52.
[27] P.J.H. Unna, 1942. Waves and tidal streams. *Nature (Lond.),* 149: 219–220.
[28] Y. Yu, 1952. Breaking of waves by an opposing current. *Trans. Am. Geophys. Union,* 33: 39–41.
[29] J.W. Johnson, 1947. The refraction of surface waves by currents. *Trans. Am. Geophys. Union,* 28: 867–874.
[30] J.N. Hunt, 1955. Gravity waves in flowing water. *Proc. R. Soc.,* A 231: 496–504.
[31] R.A. Bagnold, 1963. Beach and nearshore processes, 1. Mechanics of marine sedimentation. In: M.N. Hill (Editor), *The Sea, 2.* Wiley, New York, N.Y., pp.507–528.
[32] J.A. Sitarz, 1963. Contribution à l'étude de l'évolution des plages à partir de la consistance des profile d'équilibre. *Trav. Centre Etud. Rech. Oceanogr.,* : 10–20.
[33] W.N. Bascom, 1951. The relationship between sand size and beach face slope. *Trans. Am. Geophys. Union,* 32: 866–874.
[34] P. Bruun, 1963. Longshore currents in one and multi-bar profiles, relating to littoral drift. *Proc. 8th Conf. Coastal Eng.,* 211–247.
[35] T.Y. Chiu and P. Bruun, 1964. Computations of longshore currents by breaking waves. *Univ. Fla. Eng. and Ind. Exptl. Sta., Tech. Pap.,* 279.
[36] J. Larras, 1959. Les profiles d'équilibre des fonds de sable sous le mer. *Ann. Ponts Chaussées,* 129: 391–404.
[37] D.R. Wells, 1967. Beach equilibrium and second order wave theory. *J. Geophys. Res.,* 72: 477–504.

[38] C.J. Sonu, J.M. McCloy and D.S. Arthur, 1967. Longshore currents and nearshore topography. *Proc. 10th Conf. Coastal Eng.*, 1: 525–549.
[39] J.A. Putman, W.H. Munk and M.R. Taylor, 1949. The prediction of longshore currents. *Trans. Am. Geophys. Union*, 30: 337–345.
[40] D.L. Inman and W.H. Quinn, 1951. Currents in the surf zone. *Proc. 2nd Conf. Coastal Eng.*, 24–36.
[41] S. Nagai, 1954. On coastal groins. *Proc. Conf. Coastal Eng. Japan, 1st, 1954.*
[42] A. Brebner and J.W. Kamphuis, 1963. Model tests on the relationship between deep-water wave characteristics and longshore currents. *Queens Univ. Can., CE Rep.*, 13.
[43] C.J. Galvin Jr. and P.S. Eagleson, 1965. Experimental study on longshore currents on a plane beach. *Coastal Eng., Res. Center, Tech. Mem.*, 10.
[44] D.L. Inman and R.A. Bagnold, 1963. Beach and nearshore processes: littoral processes. In: M.N. Hill (Editor), *The Sea, 2.* Wiley, New York, N.Y., pp.529–553.
[45] I.F. Shadrin, 1961. Longshore currents and compensating currents on the shallow cumulative beach. *Tr. Oceanogr. Comm. Acad. Sci. U.S.S.R.*, 1961.
[46] P. Bruun, 1963. Longshore currents and longshore troughs. *J. Geophys. Res.*, 68: 1065–1078.
[47] *Monthly Met. Charts of Atlantic Ocean*, H.M.S.O. London (reprinted 1959).
[48] B.A. Pyshkin, V.L. Maksimtchouk and E.S. Zaitz, 1965. The investigation of littoral sand transport in seas and reservoirs. *Proc. Conf. IAHR, 11th, Leningrad*, 5: 151–152.
[49] A.V. Karaushev, A.S. Sudolsky and A.J. Schwarzman, 1965. Alongshore sediment transportation. *Proc. Conf. IAHR, 11th, Leningrad*; 5: 134–136.
[50] P. Bruun, 1965. Quantitative research on littoral drift in field and laboratory. *Proc. Conf. IAHR 11th, Leningrad*, 5: 211–215.
[51] J. Larras and R. Bonnefille, 1965. Quantités de sable charriées par la houle parallèlement à la côte. *Proc. Conf. IAHR, 11th, Leningrad*, 5: 233–236.
[52] K. Wyrtki, 1953. The balance of littoral transport in the surf zone. *Dtsch. Hydrogr. Z.*, 6: 65–75.
[53] Y. Mashima, 1958. Study of littoral drift and longshore current. *Coastal Eng. Japan*, 1: 85–96.
[54] P.S. Eagleson, 1965. Theoretical study of longshore currents on a plane beach. *Mass. Inst. Tech. Dept., Civil Eng., Rep.*, 82.
[55] J. Castanho, 1966. Breaking waves and littoral drift. *Lab. Nacl. Eng. Civil (Lisbon), Mem.*, 275 (in Portugese).
[56] Y. Iwagaki and T. Sawaragi, 1962. A new method for estimation of the rate of littoral sand drift. *Coastal Eng. Japan*, 5: 67–79.
[57] T. Edelman, 1963. Littoral transport in the breaker zone caused by oblique waves. *Proc. Conf. IAHR, 10th, London*, 1: 61–68.
[58] P. Groen and M.P.H. Weenink, 1950. Two diagrams for finding breaker characteristics along a straight coast. *Trans. Am. Geophys. Union*, 31: 398–400.
[59] B. Le Méhauté and R.C.Y. Koh, 1967. On the breaking of waves arriving at an angle to the shore. *J. Hydraul. Res.*, 5: 67–88.
[60] D.L. Inman and J.D. Frautschy, 1966. Littoral processes and the development of shorelines. *Proc. Santa Barbara Conf. Coastal Eng., 1965:* 511–536.
[61] W.C. Krumbein, 1944. Shore currents and sand movement on a model beach. *Beach Erosion Board, Tech. Mem.*, 7.
[62] T. Saville Jr., 1950. Model study of sand transport along an infinitely long straight beach. *Trans. Am. Geophys. Union*, 31: 555–565.
[63] E.A. Shay and J.W. Johnson, 1951. Model studies on the movement of sand transported by wave action along a straight beach. *Univ. Calif., Berkeley, IER*, 14(7).
[64] M.G. Sauvage and M.G. Vincent, 1955. Transport littoral; formation de fleches et de tombolos. *Proc. 5th Conf. Coastal Eng.*, 296–328.
[65] G.M. Watts, 1953. Study of sand movement at south Lake Worth inlet, Florida. *Beach Erosion Board, Tech. Mem.*, 42.

REFERENCES

[66] J.M. Caldwell, 1956. Wave action and sand movement near Anaheim bay, California. *Beach Erosion Board Tech. Mem.*, 68.

[67] G.W. Moore and J.Y. Cole, 1960. Coastal processes in the vicinity of Cape Thompson, Alaska. *U.S. Geol. Surv. Trace Elem. Invest., Rep.* 753: 41–54.

[68] T. Ishihara and T. Sawaragi, 1962. Fundamental studies of sand drifts. *Coastal Eng. Japan*, 5: 59–65.

[69] J.W. Kamphuis, 1966. Mathematical simulation of bottom sediment motion by waves. *Proc. 10th Conf. Coastal Eng.*, 766–792.

[70] J.W. Kamphuis, 1967. A mathematical model to advance the understanding of factors involved in the movement of bottom sediment by wave action. *Queens's Univ. C.E. Rep., Rep.*, 53.

[71] Vincent, G.E., 1958. Contribution to the study of sediment transport on a horizontal bed due to wave action. *Proc. 6th Conf. Coastal Eng.*, 326–355.

[72] W.S. Chepil, 1959. Equilibrium of soil grains at the threshold of movement by wind. *Proc. Soil Sci. Soc. Am.*, 23: 422–428.

[73] W.H. Graf, 1967. A modified venturimeter for measuring two-phase flows or particle dynamics and the venturimeter. *J. Hydraul. Res.*, 5: 161–188.

[74] A.D.K. Laird, C.A. Johnson and R.W. Walker, 1959. Water forces on accelerated cylinders. *Proc. ASCE*, 85 (WW1): 99–120.

[75] S.R. Keim, 1956. Fluid resistance to cylinders in accelerated motion. *Proc. ASCE*, 82 (HY6): 1113.

[76] G. Bugliarello, 1956. La resistenza al moto accelerator di sfere in acqua. *Ricerca Sci.*, 26(2).

[77] L.B. Torobin and W.H. Gauvin, 1959. Fundamental aspects of solids gas flow, 3. *Can. J. Chem. Eng.*, 37: 224–236.

[78] R.G. Lunnon, 1926. Fluid resistance to moving spheres. *Proc. R. Soc.*, A110: 302–326.

[79] H.W. Iverson and R.A. Balent, 1951. A correlating modulus for fluid resistance in accelerated motion. *J. Appl. Phys.*, 22: 324–328.

[80] W.S. Chepil, 1958. Use of evenly spaced hemispheres to evaluate aerodynamic forces on a soil surface. *Trans. Am. Geophys. Union*, 39: 397–404.

[81] H.A. Einstein, 1965. Hydrodynamics of sediment movement in waves. Univ. Calif. Berkeley (unpublished). Summarized in: *Proc. Conf. IAHR, 11th, Leningrad*, 5: 185–186.

[82] M. Manohar, 1955. Mechanics of bottom sediment movement due to wave action. *Beach Erosion Board, Tech. Mem.*, 75.

[83] M.M. Abou-Seida, 1965. Bed load function due to wave action. *Univ. Calif., I.E.R. Tech. Rep.*, HEL-2-11.

[84] G. Kalkanis, 1966. Transportation of bed material due to wave action. *Coastal Eng. Res. Centre, Tech. Mem.*, 2.

[85] H.A. Einstein, 1950. The bed-load function for sediment transport in open channel flows. *U.S. Dept. Agr., Tech. Bull.*, 1026.

[86] A.T. Ippen and P.S. Eagleson, 1955. A study of sediment sorting by waves shoaling on a plane beach. *Beach Erosion Board, Tech. Mem.*, 63.

[87] P.S. Eagleson, R.G. Dean and L.A. Peralta, 1958. The mechanics of the motion of discrete spherical and bottom sediment particles due to shoaling waves. *Beach Erosion Board, Tech. Mem.*, 104.

[88] P.S. Eagleson and R.G. Dean, 1961. Wave-induced motion of bottom sediment particles. *Trans. ASCE*, 126: 1162–1189.

[89] S. Sato and T. Kishi, 1952. Studies on the sand drifts, 7. Shearing stress at bottom and sand movement by wave action. *J. Res. Public Work Inst.*, 85(6).

[90] J. Goddet, 1960. Etude du début d'entrainment des matériaux mobiles sous l'action de la houle. *Houille Blanche*, 15: 122–135.

[91] A. Rubatta and F. Zoccoli, 1965. Le début d'entrainment sous l'action de la houle. *Proc. Conf. IAHR, 11th, Leningrad*, 5: 218–222.
[92] R. Bonnefille and L. Pernecker, 1965. Début d'entrainment des sédiments par la houle. *Proc. Conf. IAHR, 11th, Leningrad*, 5: 207–208.
[93] P. l'Hermitte, 1961. Movement des matériaux de fond sous l'action de la houle. *Proc. 7th Conf. Coastal Eng.*, 211–261.
[94] R.A. Bagnold, 1969. Threshold movement of shingle subject to wave action. *Wallingford Hydraul. Res. Sta., Note* 15.
[95] A. Martinot-Lagarde and A. Fauquet, 1955. Essais d'entrainment de matériaux solides sous l'effet de la houle et du clapotis. *Proc. 5th Conf. Coastal Eng.*, 383–394.
[96] S.R. Singamsetti, 1966. Diffusion of sediment in a submerged jet. *Proc. ASCE*, 92(HY2): 153–168.
[97] A. Gyr, 1967. The behaviour of the turbulent flow in a 2-dimensional open channel in presence of suspended particles. *Proc. Conf. IAHR, 12th*, 2: 9–16.
[98] A. Muller, 1967. Measurements of the influence of suspended particles on the size the vortices. *Proc. Conf. IAHR, 12th*, 4: 107–114.
[99] H. Rouse, 1963. On the role of eddies in fluid motion. *Am. Scientist*, 51: 285–314.
[100] M. Hom-ma and K. Horikawa, 1962. Experimental studies on the cooling water intake system for the atomic power plant at Tokai, Japan. *Coastal Eng. Japan*, 5: 111–133.
[101] E.W. Bijker, 1967. Some considerations about scales for coastal models with movable bed. *Delft Hydrol. Lab., Publ.*, 50.
[102] J. Goddet and P. Jaffry, 1950. Similitude laws for sediment transport under the simultaneous action of waves and currents. *Houille Blanche*, 15: 136–147.
[103] J.I. Collins, 1964. The effect of currents on the mass-transport of progressive water waves. *J. Geophys. Res.*, 69: 1051–1056.
[104] M. Hom-ma, K. Horikawa and C. Sonu, 1962. Field investigation at Tokai, Japan, conducted by combined procedure of macroscopic and microscopic approaches. *Coastal Eng. Japan*, 5: 93–110.
[105] P.H. Kemp and A.J. Grass, 1967. The measurement of turbulent velocity fluctuations close to a boundary in open channel flow. *Proc. Conf. IAHR, 12th*, 2: 201–209.
[106] A.A. Townsend, 1961. Equilibrium layers and wall turbulence. *J. Fluid Mech.*, 11: 97–120.
[107] V.M. Ljatkher, 1967. Calculation of spectra of turbulent pulsations in uniform flows. *Proc. Conf. IAHR, 12th*, 2: 17–25.
[108] M. Hino, H. Shikara and M. Nakai, 1967. Large eddies in stratified flow. *Proc. Conf. IAHR, 12th*, 2: 61–73.
[109] V. Jezdinsky, J. Cakrt, M. Rudis and R. Smutek, 1967. Macroturbulence measurements in natural streams. *Proc. Conf. IAHR, 12th*, 2: 323–332.
[110] Y. Ishihara and S. Yokosi, 1967. The spectra of turbulence in a river flow. *Proc. Conf. IAHR, 12th*, 2: 290–297.
[111] J.B. Tiffany, 1967. Turbulence in the Mississippi River. *Proc. Conf. IAHR, 12th*, 2: 349–355.
[112] W.W. Sayre and W.J. Conover, 1967. General two-dimensional stochastic model for the transport and dispersion of bed-material sediment particles. *Proc. Conf. IAHR, 12th*, 2: 88–95.
[113] C.F. Nordin and E.V. Richardson, 1967. The use of stochastic models in studies of alluvial channel processes. *Proc. Conf. IAHR, 12th*, 2: 96–102.
[114] R.A. Bagnold, 1947. Sand movement by waves: some small-scale experiments with sand of very low density. *Proc. Inst. Civil Engr.*, 27: 447–469.
[115] M.R. Carstens and F.M. Neilson, 1967. Evolution of a duned bed under oscillatory flow. *J. Geophys. Res.*, 72: 3053–3059.
[116] A.J. Sutherland, 1966. Entrainment of fine sediments by turbulent flows. *Calif. Inst. Tech. Rep.*, KH-R 13.

REFERENCES

[117] A.J. Sutherland, 1967. Proposed mechanism for sediment entrainment by turbulent flows. *J. Geophys. Res.*, 72: 6183–6194.

[118] V.A. Vanoni, 1964. Measurements of critical shear stress for entraining fine sediments in a boundary layer. *Calif. Inst. Tech. Rep.*, KH-R-7.

[119] P.W. Runstadler, S.T. Kline and W.C. Reynolds, 1963. An experimental investigation of the flow structure of the turbulent boundary layer. *Stanford Univ., Rep.*, MD-8.

[120] R.A. Bagnold, 1946. Motion of waves in shallow water – interaction between waves and sand bottoms. *Proc. R. Soc.*, A 187: 1–18.

[121] M.R. Carstens, 1966. Similarity laws for localised scour. *Proc. ASCE*, 92 (HY3): 13–36.

[122] R.A. Stein, 1965. Laboratory studies of total load and apparent bed loads. *J. Geophys. Res.*, 70: 1831–1842.

[123] M. Hom-ma, K. Horikawa and R. Kajima, 1965. Study of suspended sediment due to wave action. *Coastal Eng. Japan*, 8: 85–103.

[124] M. Hom-ma and K. Horikawa, 1962. Suspended sediment due to wave action. *Proc. 8th Conf. Coastal Eng.*, 168–193.

[125] J.C. Fairchild, 1959. Suspended sediment sampling in laboratory wave action. *Beach Erosion Board, Tech. Mem.*, 115.

[126] R.A. Bagnold, 1966. An approach to the sediment transport problem from general physics. *U.S. Dept. Int. Geol. Surv., Prof. Pap.*, 422-I.

[127] P. Danel, R. Durand and E. Condolios, 1953. Introduction à l'étude de la saltation. *Houille Blanche*, 8: 217–233.

[128] L. Escande, 1954. Some results to support the hypothesis of Danel, Durand and Condolios on the phenomena of saltation. *Houille Blance* 9(A): 275–277.

[129] M.R. Carstens and C.S. Martin, 1962. The influence of a permeable bed on sediment transport with special reference to gravity waves. *Georgia Inst. Tech., Tech. Rep.*, 1 (Proj. A-628).

[130] R.S. Dietz, 1963. Wave-base, marine profile of equilibrium, and wave-built terraces: a critical appraisal. *Geol. Soc. Am. Bull.*, 74: 971–990.

[131] G.R. Mogridge, 1965. *Sediment Transport due to Wave Action*. Thesis, Univ. Western Australia.

[132] M.R. Carstens, F.M. Neilson and H.D. Altinbilek, 1967. An analytical and experimental study of bed forms under water waves. *Georgia Inst. Tech., Final Rep.*, Proj. A-798.

[133] M.R. Carstens, F.M. Neilson and H.D. Altinbilek, 1969. Bed forms generated in the laboratory under an oscillatory flow: analytical and experimental study. *Coastal Eng. Res. Centre, Tech. Mem.*, 28.

[134] R. Silvester and G.R. Mogridge, 1970. Reach of waves to the bed of the continental shelf. *Proc. 12th Conf. Coastal Eng.*, 2: 651–668.

[135] P.J. Rance and N.F. Warren, 1968. The threshold of movement of coarse material in oscillatory flow. *Proc. 11th. Conf. Coastal Eng.*, 1: 487–491.

[136] G.R. Mogridge, 1970. Testing sediment movement due to wave action. *Proc. ASCE*, 96 (HY7): 1587–1604.

[137] J.F.A. Sleath, 1970. Velocity measurements close to the bed in a wave tank. *J. Fluid Mech.*, 42: 111–123.

[138] P.J. Rance, 1963. The determination of quantities of sediment transport in oscillatory motion by consideration of the dispersion of tracer sediment. *Proc. Conf. IAHR*, 1: 181–188.

[139] S. Sato, 1962. Sand movement at Fukue coast in Atsumi Bay, Japan, and its observation by radioactive glass sand. *Proc. Coastal Eng. Japan*, 5: 81–92.

[140] G.M. Watts, 1953. Field investigation of suspended sediment in the surf zone. *Proc. 4th Conf. Coastal Eng.*, 181–199.

[141] H. Fukushima and Y. Mizoguchi, 1958. Field investigation of suspended littoral drift. *Proc. Coastal Eng. Japan*, 1: 131–134.

[142] J.W. Johnson, 1959. Littoral drift problems at shoreline harbors. *Trans. ASCE*, 124: 525–546.

[143] R. Silvester, 1970. Sediment movement beyond the breaker zone. *Trans. Inst. Eng. Aust.*, CE12: 63–71.

Chapter 2

LONG-TERM LONGSHORE DRIFT

Having discussed the transient exchange of material to offshore and back again to the beach and the slower more persistent movement of material along the coast, it is now necessary to consider some longer-term effects. Especially does this take into account the slow changes in sediment supply as climates of landmasses change and even the landmasses themselves. Over a longer geological period the wave climate of the oceans may have changed in respect to a particular continent, especially if the latter has drifted or the poles have shifted, as contended by many geologists.

However, for the present discussion, which relates to movement of the more "recent" sediment, it can be assumed that the storm areas of oceans have remained essentially the same over thousands, if not millions of years. In this respect the following conclusion in a comprehensive article by Meyerhoff [1] appears relevant: "Axisymmetry of cool and evaporite belts about the present rotation axis, the coincidence through time of the earth's two horse latitude belts, and the occurrence of 95 percent of the world's evaporites — by volume and by area — in areas now receiving less than 100 cm of rainfall show that the planetary wind and ocean current pattern has been essentially the same for 800–1000 million years."

The implications of such repeated generation of storm waves, and their propagation across the oceans as swell, has been examined by Davies [2]. Such swell arrives at distant coasts from essentially one direction, certainly within the quadrant one side of the normal to the coast. The repetitive nature of this action has already been noted, from which the title of "persistent swell" appears most apt.

It is the sweeping action of swell, through its mass-transport broom, that dictates the resultant direction along the coast that sediment is transported, both in the surf zone and offshore area. While the supply and removal of material is balanced, the section of coast in the fluvial sense is "in regime". Large volumes of material are then by-passing it, without any undue changes being effected in the beach profile offshore. Closer to the beach the annual storm sequences produce their transient offshore bar, which in a stable section of coast is of little concern.

But now consider a sandy shoreline suffering oblique swell where replenishment does not compensate for material being removed downcoast. The immediate effect will be for the offshore area to be swept cleaner by the swell and so result in its deepening and steepening. Each year as this occurs more material is demanded from the beach by the storm waves to construct the offshore bar. The beach line thus recedes, in keeping with the falling profile of the beach zone as a whole. As noted previously, successive hydrographic surveys could indicate such shoreward movement of bed contours, and so forewarn possible beach degradation.

Whilst the instability of a coastal zone is exhibited during storm sequences, the occurrence of these high-energy periods is not essential for the long-term result. A continental margin which experiences swell waves only will react to changes in sediment supply in a similar way, by recession of the beach line. Similarly, a coast receiving storm waves alone, will have its profile shifted landwards as material runs out.

EQUILIBRIUM SHAPES OF BAY

Where there is a surfeit of movable material being transported along a coast, it may "swamp" any rock outcrops, reefs or man-made structures. But where the supply is rather limited for the longshore energy available to move it, these features act as fixed points along the shoreline around which the sediment must pass. Such by-passing is accomplished offshore in the deeper water as well as in the surf zone, as stressed repeatedly before.

In order to study the ultimate effect of non-replenishment in sediment supply on a section of beach between fixed headlands, Silvester [3] conducted some model tests. Waves were generated at an angle of 45° to an initially straight sandy shoreline, which incorporated three concrete blocks representing rocky headlands. No replenishment was permitted from up-coast of these two beach units, which were allowed to erode away to the limit of the tests. Material collecting beyond the downcoast block was continually removed, so as to isolate the sculpturing process within the three headlands.

Two tests were conducted in which conditions were identical. The water depth was 7 inches, the flat beach berm was 3 inches above the SWL, the wave height was 1 inch, the wave period 1.3 sec, and the headland spacing 20 ft. The waves were generated for intervals of several hours, after which a survey was made of water-line and underwater depths. This was continued until a quasi-permanent shape resulted in the upcoast bay so formed. At this stage any incoming wave broke simultaneously around the periphery of the bay.

The results of the two tests are depicted in Fig. 2-1A,B where duration in hours and depth contours in inches are marked in the respective diagrams. The model was not meant to represent any particular prototype condition, so that no scale could be placed on the configuration. However, the following conclusions could be drawn:

(1) The equilibrium shape for a sedimentary coastline with oblique swell is that of a half-heart, with the curved portion at the upcoast end and the tangent section

EQUILIBRIUM SHAPES OF BAY PHYSIOGRAPHIC UNITS 73

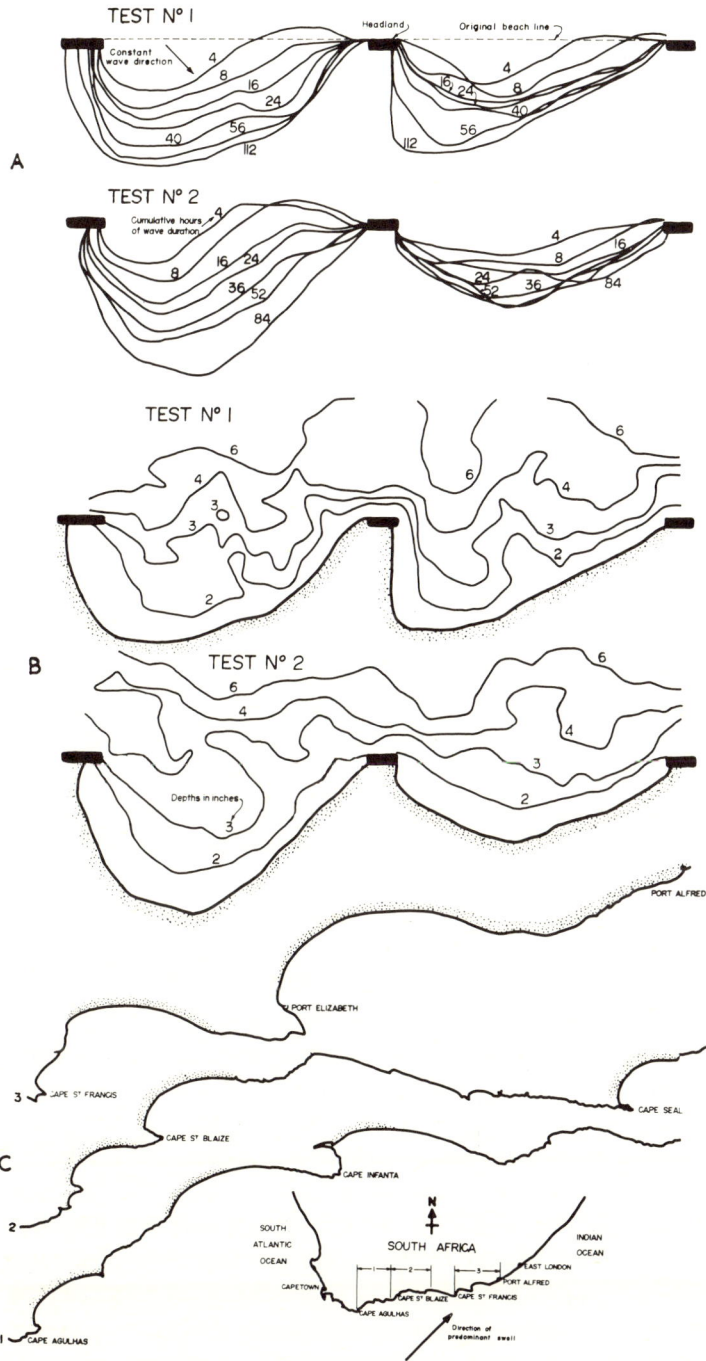

Fig. 2-1. A. Progressive development of crenulate-shaped bays in a model study. (After Ref.[2].) B. Underwater contours at the cessation of tests in the model study of A above. (After Ref.[2].) C. Bay informations along the South African coast.

at the downcoast end. These have been termed crenulate-shaped bays by geographers. Although the final configuration, as displayed in Fig. 2-1A,B approximate to this, it is believed that more smoothing of the beach line would have taken place on a larger scale. The tangent section, particularly of the upcoast bay, at the completion of the test was essentially parallel to the crests of the waves as they approached across the flat floor of the model basin. The curved coastline followed the crests of waves diffracted around the upcoast headland and subsequently refracted towards the beach line.

(2) The shape of a downcoast bay is dictated by the sediment passing to it from an upcoast adjoining bay. In Fig. 2-1A it is seen that the downcoast bay is more eroded in test no.1 than in test no.2 because less material has been removed from the upcoast bay. It can be seen how the supply of sediment to the downcoast bay is determined by the stage of stability in the upcoast bay.

(3) The underwater contours of Fig. 2-1B indicate a possible circular current that has developed from the littoral current generated in the curved bay section. Once this circulation were established it would tend to maintain itself, and so break contact with the beach at the commencement of the tangent. In Nature this is not likely to occur due to variations in wave approach and periods of calm.

The principal deduction from the tests was that an oblique persistent swell will move sediment downcoast and, in the process, sculpture the coast into crenulate-shaped bays that have a unique orientation in respect to the incoming waves or the direction of the net sediment movement. Such shapes are considered stable since the swell is so twisted as to arrive normally to all sections of beach. Under these conditions no further longshore drift was possible. Perhaps, continued duration of the waves in the model would have produced further deepening of the offshore zone, with consequent slight moulding of the water line.

Field observations

The concept of the crenulate-shaped bay as a stable physiographic feature was approached by Jennings [4], without full knowledge of wave phenomena. Later, Davies [5] realized the importance of refraction. The role of waves in this moulding mechanism has reached the stage of being in text books [6], although the association of the actual bay orientation with the direction of net sediment movement had not been appreciated.

It is of interest to note that should the wave pattern be determined by wave diffraction alone, about the upcoast headland, a bay would be formed as in Fig. 2-2. The shoaling depths cause the waves to be refracted also, modifying this near circular outline to the changing curvature as illustrated. In fact, very close to the headland in the shadow zone, diffraction dictates the shape (an arc of a circle) but this gives way very quickly to a curve which Yasso [7] has found to be a logarithmic spiral.

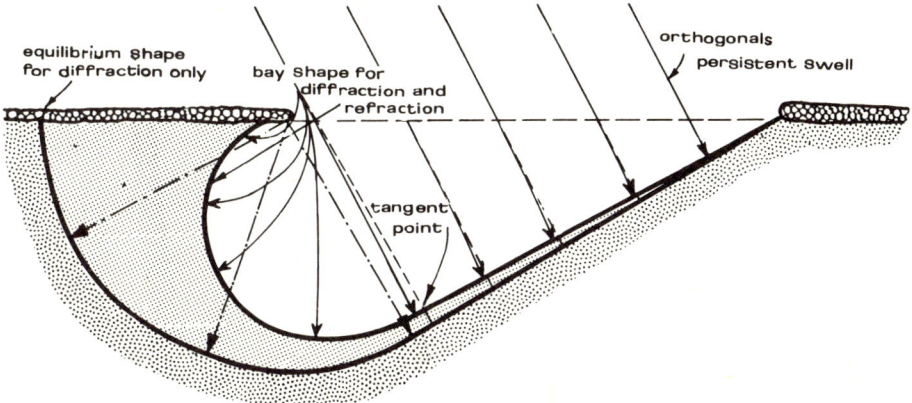

Fig. 2-2. Schematic plan of bay showing the influence of wave refraction on shape.

To verify the orientation of the bays in Nature, a continental margin must be chosen in which the swell waves have a known predominant direction. Such a coast is that of South Africa [3], which is depicted in Fig. 2-1C. Here the persistent southwest swell has formed a number of bays where the coastline is sedimentary. Whilst differing slightly in their plan and shape, they are of similar form and are oriented in a manner predicted from the model tests. A strong west to east movement of sediment is indicated. Other areas which display a similar correlation of bay orientation and persistent swell are: (a) the southwest coast of Africa where there is a northward drift; (b) the Pacific coast of North America where there is a southward drift; (c) the east coast of the Malaysian peninsula, where a southward drift is predominant.

Whilst sediment is still being supplied to, and therefore passing through, a crenulate-shaped bay, it is not in equilibrium. That is, a reduction in sediment supply will produce further changes in shape. Waves will have to break at some angle to the shore in order to propel material through the bay. Thus, it should be possible to distinguish between bays in full equilibrium, when longshore drift has ceased (except for minor annual cycles of reversed movement) and those that have not yet reached such stability.

Model studies

Silvester [8] has reported some model tests [9] which showed that the constant in the logarithmic spiral equation approached an optimum value for the equilibrium-shaped bay. A wave basin as illustrated in Fig. 2-3 was employed [9] to study the chronological development of bays. Waves were generated between a wave guide and one wall of the basin. A sandy beach was constructed which ran from the tip of

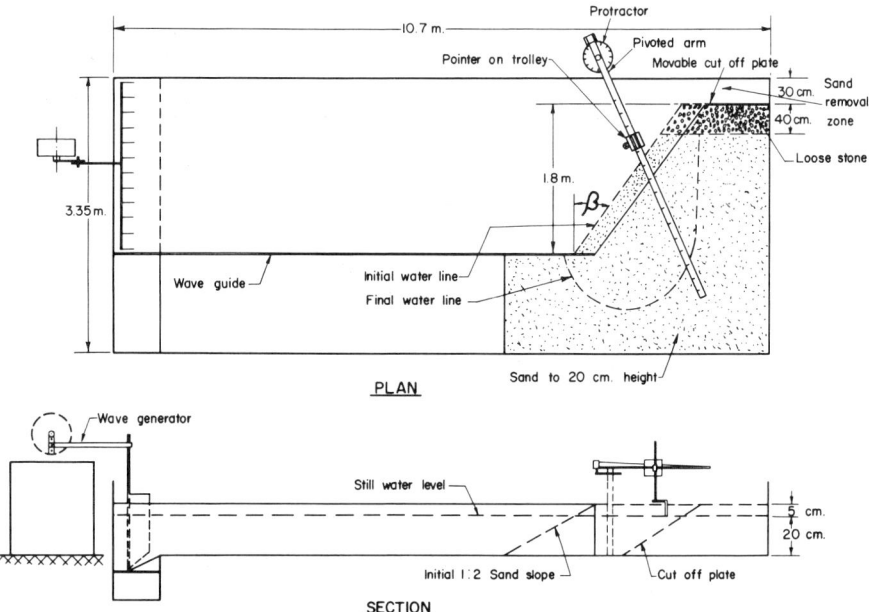

Fig. 2-3. Details of model study to determine equilibrium shapes of bay.

the wave guide to a similar but movable cut-off plate near the wall. By shifting the location of this plate, three successive straight beaches were formed at angles of 30°, 45° and 60° to the crests of the incoming waves. The water was 20 cm deep and the beach, consisting of fine sand, was sloped at 1:2 up to a berm level of 5 cm above SWL. Three wave periods were used of 0.6, 1.0 and 1.4 sec, whilst the wave height was kept sensibly constant.

As sand was eroded and collected beyond the downcoast headland (between the cut-off plate and the wall), it was removed at half-hourly intervals. The use of the wave guide as the upcoast headland, in place of the concrete blocks used previously [3], had the advantage of excluding reflections from this feature, and permitting the study of the bay shape very close to the headland. It was found necessary with the downcoast headland to provide stone for dissipating the wave energy; otherwise the cut-off board acted somewhat as a groin, which triggered a rip from the littoral current, resulting in excessive local erosion.

After 30 min duration the wave generator was stopped and the water line plus underwater contours recorded. This cessation also obviated any circular current that may have tended to form, as previously discussed. As noted in Fig. 2-3, a pointer gauge attached to a trolley running along a pivoted arm was used to record the shape in plan of the water line and bed contours. The approach to full equilibrium was gauged by the volume of sediment deposited in the downcoast trough, the

EQUILIBRIUM SHAPES OF BAY PHYSIOGRAPHIC UNITS

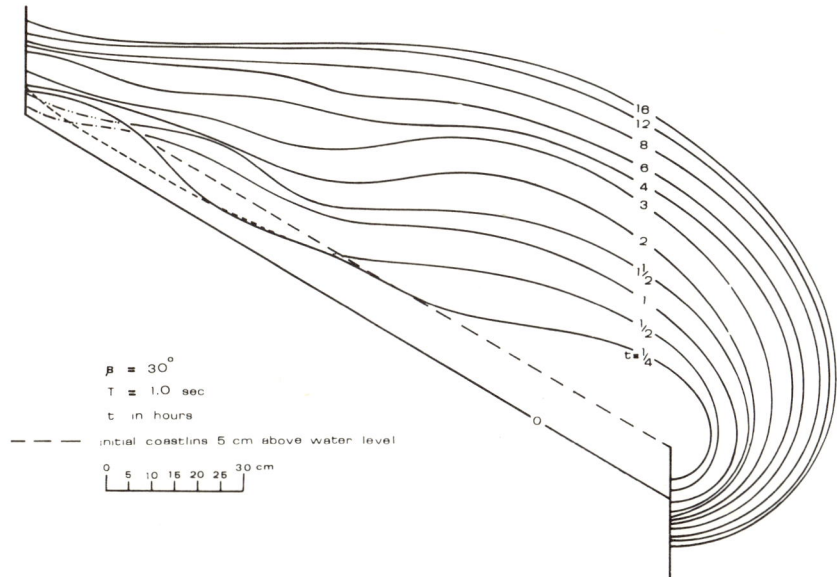

Fig. 2-4. Successive beach outlines for a given approach angle and wave period.

simultaneous breaking of waves around the bay periphery, the movement of dye dropped into the water at different points, and the decreased recession of the beach line.

One of the many beach traces recorded [9] is depicted in Fig. 2-4. The coastline was observed to have three distinct curvature zones, namely, a circular section in the lee of the upcoast headland, the logarithmic spiral zone adjoining the above, and a tangential segment running through the downcoast headland. The growth towards equilibrium was achieved by the tangential alignment changing from the original headland chord to a direction parallel to the crests of the incoming waves. This implies an angular change equal to the original obliquity of the wave (β) to the original straight shoreline. This tangent commences at the downcoast headland and grows steadily in length as equilibrium is approached. The speed with which stability was obtained varied with the wave length. The longer waves, containing the most energy per wave-length, effected the sculpturing more quickly than the shorter waves.

Equation of beach plan

The equation for the logarithmic spiral is:

$$R_2/R_1 = \exp(\theta \cot \alpha) \tag{2-1}$$

where, as seen in Fig. 2-5, R_1 and R_2 are radii from an origin that are at an angle of

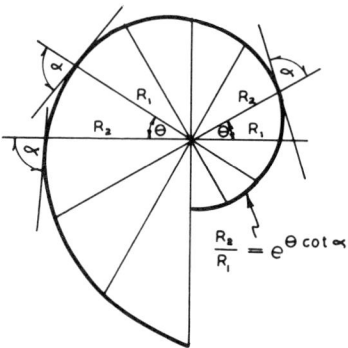

Fig. 2-5. Definition sketch of logarithmic spiral.

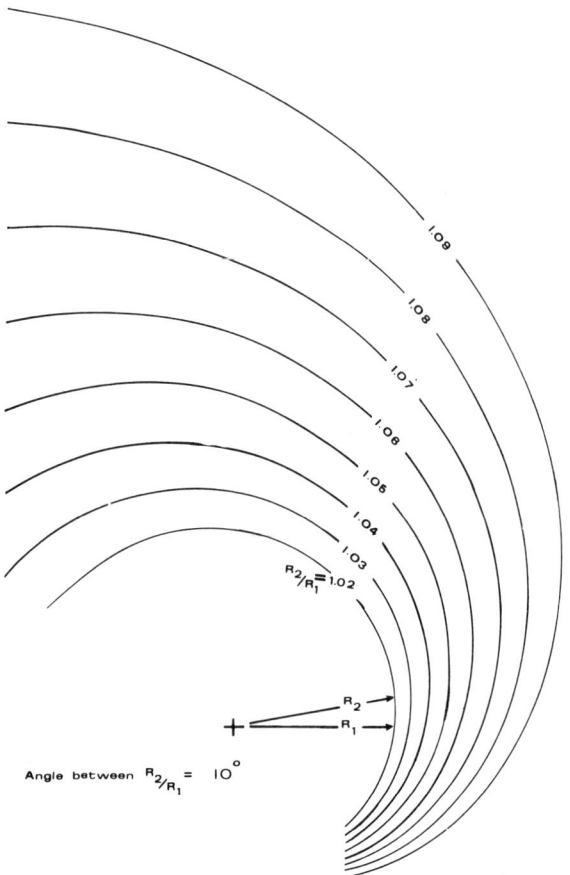

Fig. 2-6. Logarithmic spirals for $R_2/R_1 = 1.02-1.09$.

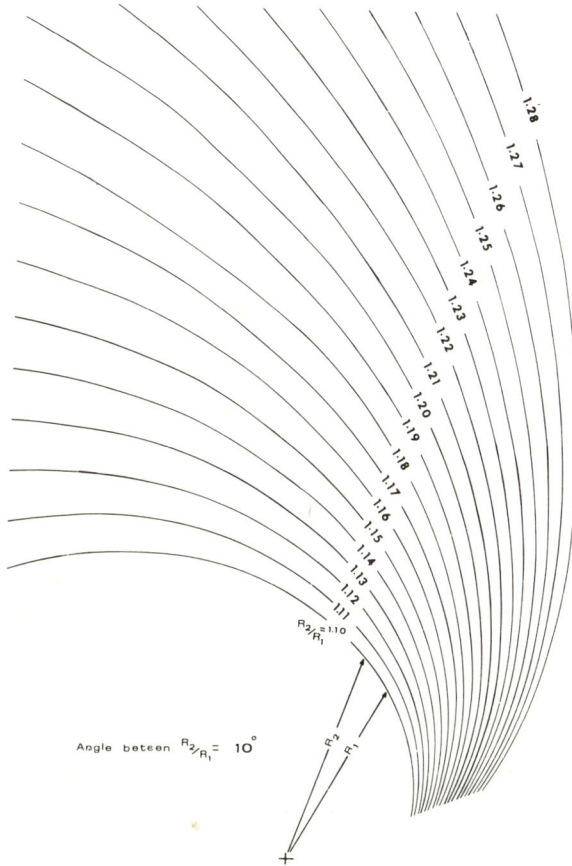

Fig. 2-7. Logarithmic spirals for $R_2/R_1 = 1.10-1.28$.

θ (radians) to each other, and angle α is the constant angle of the radii to tangents of the curve. For a given curve α is stipulated, which provides a ratio R_2/R_1 for some accepted value of θ. Curves were thus drawn, as in Fig. 2-6 and 2-7, for a range of R_2/R_1 and $\theta = 10° = 0.175$ radians. The equivalent values of α can be read from Fig. 2-8.

The above curves were plotted on tracing paper and placed over the appropriate section of the experimental beach lines, to determine its value of R_2/R_1 or α. It should be noted that the reproduction scale of the bay does not influence the constant so obtained, since a larger-scale plan will fit the logarithmic spiral further from the origin.

With the above procedure, a series of spiral constants was obtained as the bay was eroded towards an equilibrium shape. Values of R_2/R_1 have been graphed in Fig. 2-9 against duration in hours. The abscissa scale applies to the $\beta = 60°$ curve,

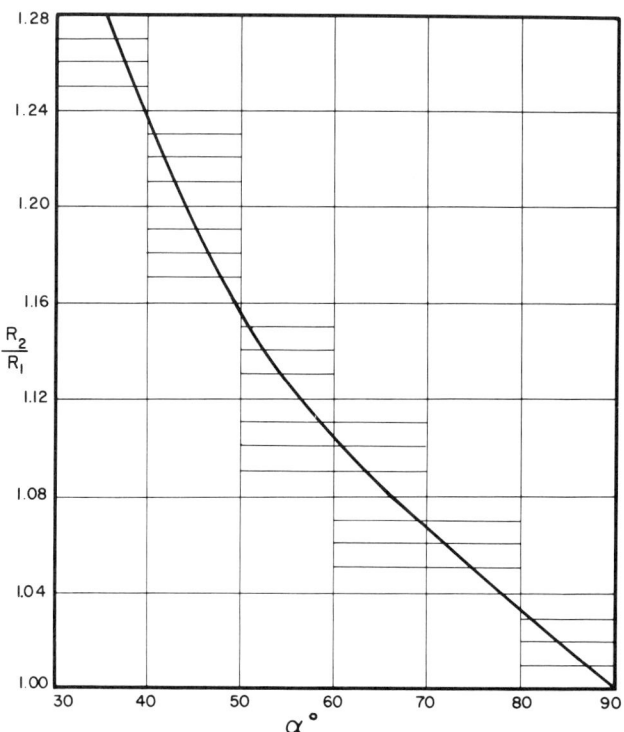

Fig. 2-8. Relationship between R_2/R_1 and tangent angle α.

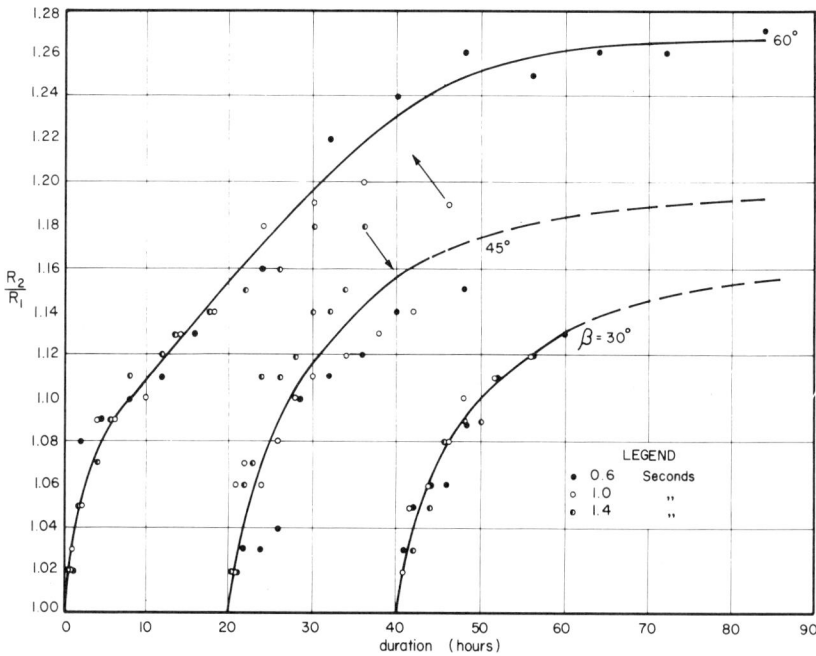

Fig. 2-9. Spiral constants R_2/R_1 versus duration of waves. (Note curves 45° and 30° have been offset from the origin.)

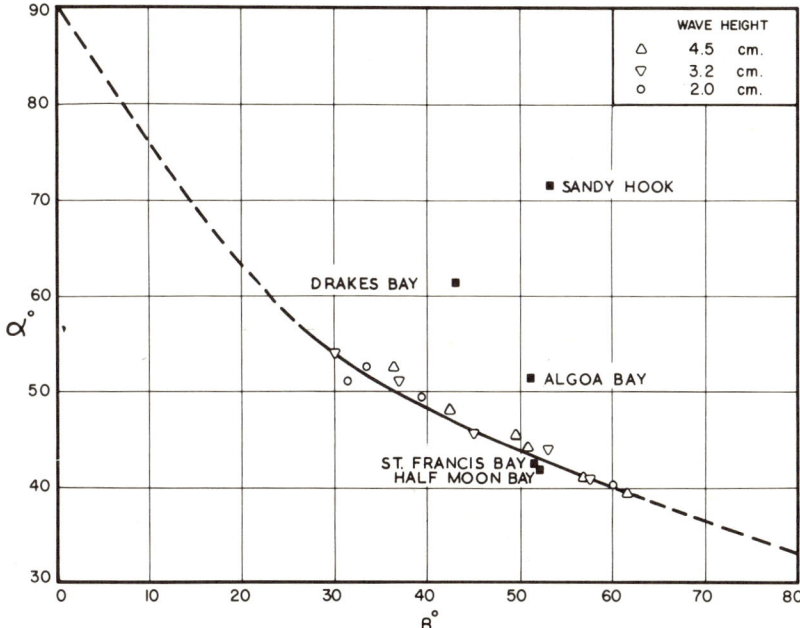

Fig. 2-10. Spiral constant α versus approach angle β.

those for 45° and 30° also commence at zero. Besides the three approach angles β to the original straight coast joining the pseudo-headlands, results for the three wave periods are incorporated into Fig. 2-9. Inspite of some scatter in the results, there is a strong indication that period or wave length does not enter the problem. Hence the model could represent any sized bay, or for any given bay it could indicate erosion by any wave period.

Reference to Fig. 2-9 also shows that the spiral curve changes swiftly at first, but approaches a maximum value of R_2/R_1 assymptotically. The lines have been extrapolated for the 30° and 45°-curves in keeping with the trend exhibited by the 60°-trace. Further tests are warranted to determine beyond dispute the R_2/R_1 or α values for the full equilibrium state, since the existence of a relationship between α and β is of prime engineering significance. This relationship is depicted in Fig. 2-10, where the resulting straight line needs further confirmation from model tests with other angles of β. It is obvious from the changes recorded in Fig. 2-9 that, prior to equilibrium, values of α measured for a bay will fall above the line in Fig. 2-10.

Test of stability

A tool is thus made available by which to judge the stability of any bay. By this is meant that, should the sediment supply from upcoast be reduced to zero, the water line of the bay will recede to some limiting curvature. All that is required is a knowledge of the approach angle of the waves to a line joining the headlands at the extremities of the bay. This angle is equivalent to that between the downcoast tangent and this headland alignment, especially for a bay approaching equilibrium. Prior to this the β so measured may be smaller than actual, due to changing orientation of this section with time. However, as noted in the model tests, the tangent in close proximity to the downcoast headland could reach this equilibrium alignment long before the whole equilibrium tangent section is fully formed. Thus the length of the straight downcoast beach should give a close indication of the closeness to full stability.

Further confirmation of the relationship depicted in Fig. 2-10 could be obtained on model tests similar to those reported, without the need for varying the wave period. It could also be carried out by measuring spiral constants per Fig. 2-6 and 2-7 on prototype bays, that were proven to be in equilibrium. For this purpose it should be established that no sediment is being fed into the bay, either from the adjacent upcoast bay or from rivers within the bay under test.

In using Fig. 2-10 to test stability the evaluation of β may prove more difficult than that of α. But assistance in this respect can be had by noting the tangential alignments of several adjacent bays. If the continental shelf is of constant width and of uniform slope and the headlands of the bays are more or less aligned, then all tangents should be parallel. However, if the shelf is curved and the headland chords are not aligned, as in Fig. 2-11, then refraction of the persistent swell to these downcoast zones must be considered. This infers that the persistent swell arriving at the various bays emanates from a common direction in deep water. In such an examination the period of this predominant sculpturing swell must be assumed. The previous discussion on the abundance of the 12- to 14-sec components of the average storm spectra is very relevant for an oceanic margin.

An example of the distinctly different orientation of two adjacent bays is illustrated in Fig. 2-11, where two are centred on Pt. Reyes, California. The persistent swell on this coast is from the NW in deep-water which would arrive at the beach north of Pt. Reyes around WNW as indicated. By the length of this tangential section it could be inferred that this bay is close to equilibrium. The second orthogonal to the south would be oriented in a similar direction whilst in similar offshore depths, but would be refracted as shown (qualitatively only) to the downcoast limit of Drakes Bay. It is possible, therefore, to derive values of β from bays in close proximity even though their orientations may differ drastically.

Yasso [7] has recorded values of α for three bays, namely, Sandy Hook, Drakes Bay and Half Moon Bay. The values obtained by use of the overlay described above differ somewhat from those given by Yasso. This is due in part by a stricter limit being set on the bay perimeter at the downcoast limit. It is believed that testing with the overlay is easier and possibly more exact than regression analysis as employed by Yasso.

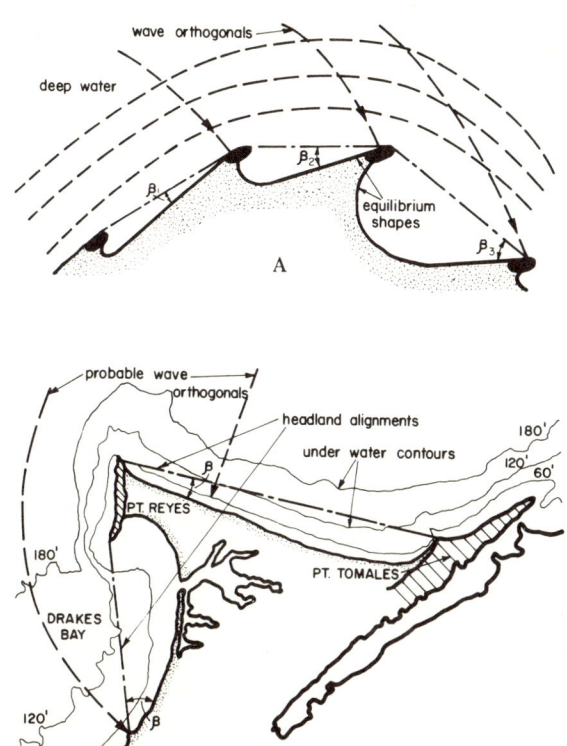

Fig. 2-11. A. Possible variation of α and β in consecutive bays; B. Bay disposition at Pt. Reyes, California.

The results so obtained for the three named bays are plotted in Fig. 2-10. It is seen that Sandy Hook is well above the equilibrium curve, indicating an abundant supply of material, as its name implies. Drakes Bay, as noted in Fig. 2-11 has an estuary feeding sand to it and hence is not likely to be in equilibrium. Half Moon Bay, on the other hand, has no major source of sediment of its own and has a minimal supply from upcoast. The point in Fig. 2-10 actually falls a little below the equilibrium line, but certainly confirms the results of the model tests.

Two further bays have been tested [8], namely, St. Francis and Algoa bays on the South African coast (34°S and 25°–26°E). St. Francis Bay is seen in Fig. 2-10 to be in equilibrium. This is supported by the fact that the upcoast headland has extremely deep water close to shore, thus precluding sediment supply from this source, and the absence of any large rivers feeding material into the bay itself. Algoa Bay just to the east of St. Francis Bay, is not in equilibrium according to Fig. 2-10, which is complemented by the presence of three rivers debouching into it and a modicum of material arriving from St. Francis Bay. Should civil works on these rivers and those in the upcoast bay prevent subsequent supply of sediment, Algoa Bay will become more indented until a fresh rock outcrop of large magnitude is exposed to the sea, or it assumes a logarithmic spiral shape for which $\alpha = 43°$. The application of this yardstick of stability is discussed in the chapter on coastal defense.

PHYSIOGRAPHIC UNITS

Where crenulate-shaped bays occur on a coastline they can be used to determine the direction of net sediment movement. The term "net" is stressed here, since fluctuations in direction could occur throughout the year, but a resultant vector is bound to exist which changes little in direction over decades or centuries. This is one major advantage of the bay-direction indicator, that it is also an integrator of the wave energy over the "recent" geologic period. Should a tracer test, or other spot measurement in time, show a longshore drift which is contrary to that concluded from the bay shape, a check should be made of the uniqueness or otherwise of the weather conditions at the time of the experiment.

Where bays are plainly displayed in maps or hydrographic charts, due to a reasonable spacing of headlands, the direction of net littoral drift can be determined over a long length of coastal margin. In the event of the coast consisting of cliffs, with no sedimentary sections for considerable lengths, other means must be devised for gauging net sediment motion. Here again the bay shape may be used as a standard, since certain underwater features along a rugged coastline can be correlated with directions as derived from adjacent sandy sections. For example, if a section of rugged coastline is bounded at both ends by sandy shorelines, both of which indicate net transport is in the same longshore direction, then it can be assumed that the same sediment motion is occurring within this zone. If this were not so there would be accumulation of material within the rocky section, which would have caused a beach or large shoal to form seawards of the indented waterline.

As swell spreads out from a storm zone, in the manner described in the section on swell propagation, its angle of approach to a long length of coast will change very little. The net drift will therefore be in the one direction over considerable lengths of continental margin, even though the rate of transport may vary in the long and short term. This is also true for swell arriving from several annual storm zones, since a vector resultant must exist for a substantial length of coastline. Even in enclosed or partially enclosed seas the movement of sediment should be consistently in one direction or be zero for considerable portions of the sea-margin.

The length of coast in which net sediment transport is in one direction, or is zero, will be termed a *physiographic unit* [10]. This is not to infer similarities of climate, geology or geography throughout the unit, nor even rate of transport or type of sediment. The limits of a physiographic unit are thus identified as places where the net movement changes direction, or becomes zero. These will generally occur at major changes in direction of the continental margin.

As noted already, certain topographic features of the seabed can be correlated with net sediment motion, through association with crenulate-shaped bays in the vicinity. Hydrographic charts, which are useful in identifying these bays, are also helpful in this latter study, since they contain depth readings and contour lines. They are available in a vast variety of scales, so that fine detail of coastal features can be studied. The following submarine features on these charts may assist in the assessment of direction of net sediment movement: (a) contour spreading; (b) shoals; (c) end conditions.

Contour spreading

If successive surveys of shorelines made at long intervals of time indicate a long-term stability of same, it can be concluded that sediment is passing along it at a steady rate. Thus material passing through or across one bay is doing likewise in the adjacent downcoast bay. It will be passing at the same rate unless accretion or denudation is occurring somewhere along the unit.

This implies that the *overall rate* across the active profile is the same, not that the rate per unit width is constant, because the active zones differ in profile, width and supply of wave energy. As noted previously, wave energy around an embayment varies due to the diffraction and refraction taking place. Hence, for a given deep-water energy input, the active shoreline in the sheltered part of a bay must be wider than that at the exposed tangential downcoast end. This infers that all the material is passing within the confines of this active zone; it has been noted already that material can traverse the bed from the upcoast headland to the tangential section almost directly. However, the variation in distance from the beach to the 20-fathom contour in Fig. 2-12 would indicate the changing needs of the waves in these zones to transport sediment at a constant overall rate.

Also illustrated in this figure is the substantial reduction in the width to the 20-fathom contour in the vicinity of cliffs. In these zones no beach exists at all, so that all transport is effected at the bed and none in a surf zone. The double application of wave energy due to reflection, with its concomitant production of turbulent eddies, expedites the passage of material in these areas. The choice of the 20-fathom width in Fig. 2-12 should not be taken to mean that all wave activity is confined to this depth limit. Transport can occur out to the limit of wave reach, but the sediment contained in this band of the bed could well remain within it for reasonable lengths of shoreline.

As sediment is fed to the coast at intervals along the length of a coastline, in

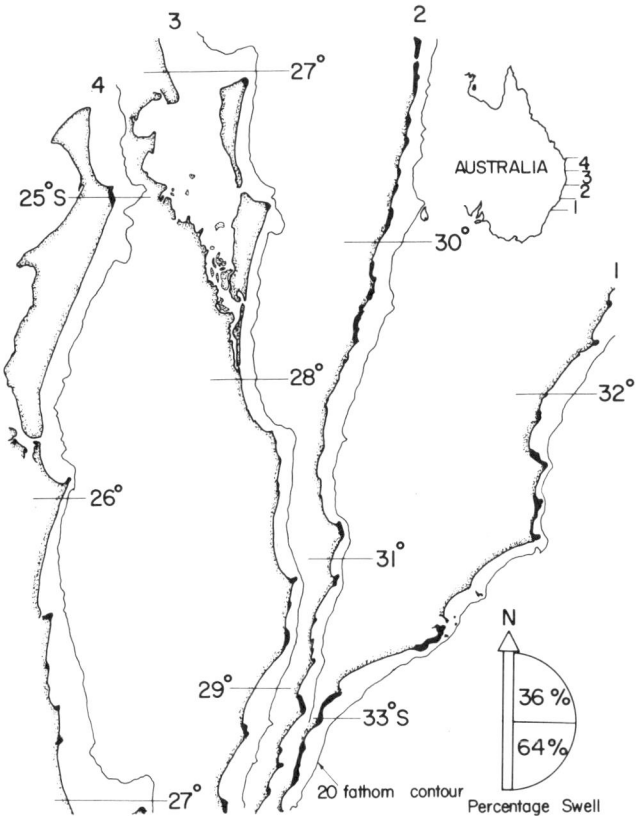

Fig. 2-12. Crenulate-shaped bays along the eastern coast of Australia.

which the direction of net movement is the same, more material must be transported by the given wave energy if no accretion is to occur. Thus, for a uniform distribution of longshore wave energy along the coast, the active zone must widen to cope with the enlarging load. This general increase in width of the continental shelf is not to be confused with the more local changes described above, which depend more on variations in wave energy closer inshore.

Over long lengths of transport, sand particles are abrased to smaller general dimensions. This increased fineness results in a flatter profile of the shelf, causing the contours to be spread at the downcoast end of a physiographic unit. For example, most west coasts of land masses receive wave energy from the predominant storm zones in 40°–60° latitudes. Physiographic units could thus extend from these latitudes to the equator, since the waves are from a more-or-less common zone of generation. However, as the equator is approached so the wave energy is decreased. Thus, for equal transport rates, the shelf width must increase to enable the transfer of wave energy via a sufficient expanse of sea floor.

Shoals

The case has been discussed earlier of the shoaling that occurs just downcoast of a headland. This is occasioned by the sudden change of energy available to transport material. Along the cliff face the reflected waves expedite the delivery of material past it. Beyond the headland, the only energy available to the sediment is that from the incoming progressive waves. They cannot handle it until the bed is built up sufficiently for them to more effectively oscillate the water particles and so effect a net sediment movement. Under these conditions the finer particles get separated from the coarser ones, which then constitute the majority of the shoal. Such batches of coarser sediment have been taken to "prove" that bed material does not become progressively finer away from the coast.

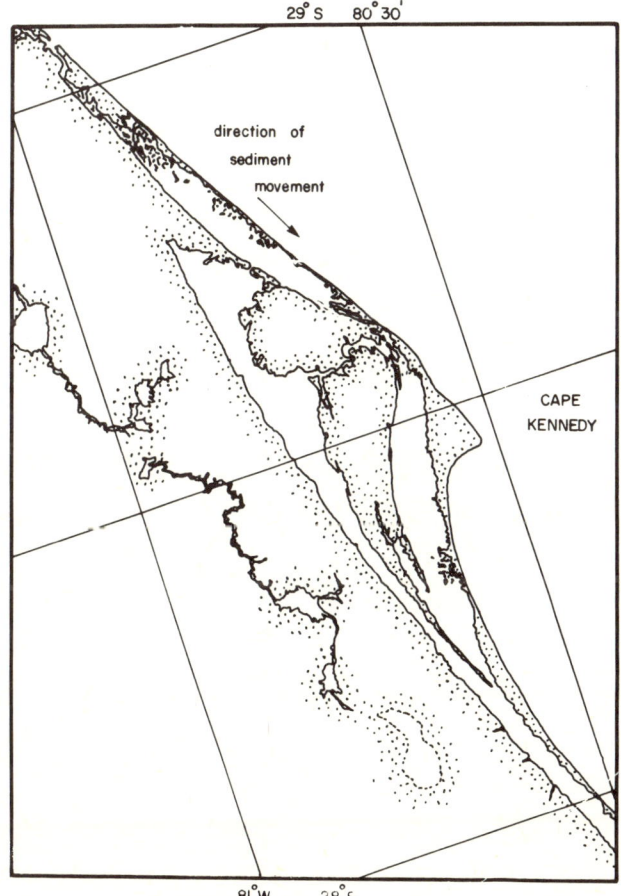

Fig. 2-13. Lagoons formed by oblique waves — example from Florida.

A rugged coastline is likely to have similarly rugged islands offshore from it. If the persistent swell is oblique to the coast, and therefore producing a net longshore drift, this will be exhibited by the orientation of the leeward shoal behind the island. If it is directly behind, it means that little or no sediment is available in the area, because otherwise continuous accumulation with no net drift would result in the formation of a plain around and ultimately seawards of the rock outcrops.

Where there is a surfeit of sediment available to a coast it will have formed a coastal plain beyond the rock outcrops which were originally at the sea margin. In fact, it may be difficult over long lengths of shoreline to identify crenulate-shapes of bay. In such cases, offshore shoals may be identified whose tongue shape will clearly indicate the direction of net drift. River mouths will have been diverted downcoast, sometimes considerable distances. These coasts also generally contain

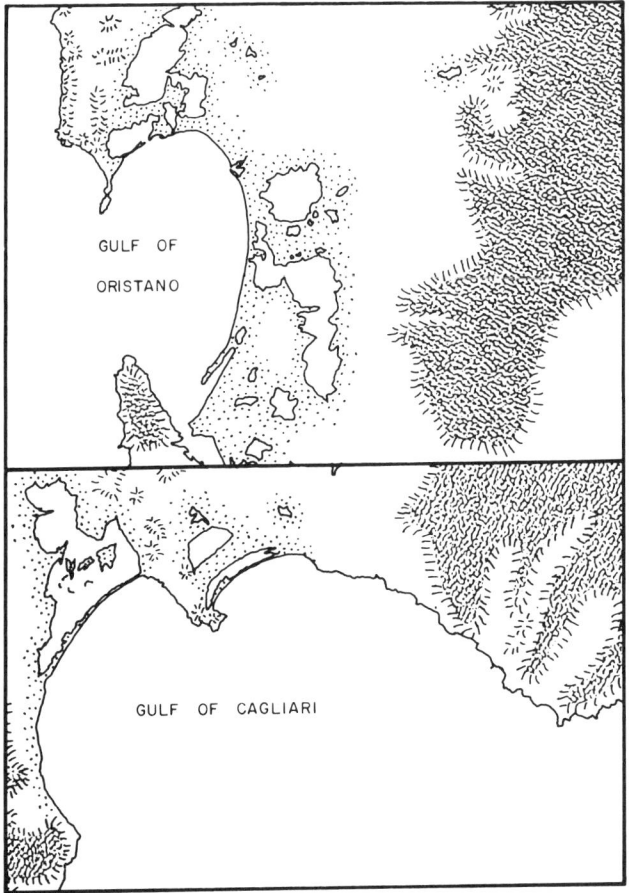

Fig. 2-14. Lagoons formed by normal wave approach — two examples from Sardinia.

elongated lagoons, sometimes two or three wide, which run for scores of miles along the coast. Fig. 2-13 gives an example of this. Such lagoons, constructed in a previous geologic period, may now be under attack due to changed climatic conditions, which have interrupted the sand supply to the coast.

Another condition conducive to spit formation is when the persistent waves arrive normal to the coast. Lagoons are again partially or completely enclosed by spits which build up from either side. In this case the lagoons are not necessarily elongated in the direction of the shoreline, but may have meandering shapes as part of a river mouth so silted up. This is illustrated in Fig. 2-14. Where any such shoals or spits occur, with beach lines evenly curved between the limiting headlands, it can be assumed that the net sediment movement is zero. This does not preclude annual oscillations of material up and down the coast.

End conditions

Another general indicator of net sediment movement along a physiographic unit is that the downcoast region must contain an abundance of sediment, whilst the upcoast region suffers a dearth of it. This will be reflected in the width of the continental shelf and width of coastal plain. Accumulation of longshore drift over thousands or even millions of years, can result in vast volumes of accretion, either as landmass or as submarine shelf. For example, if the conservative estimate of

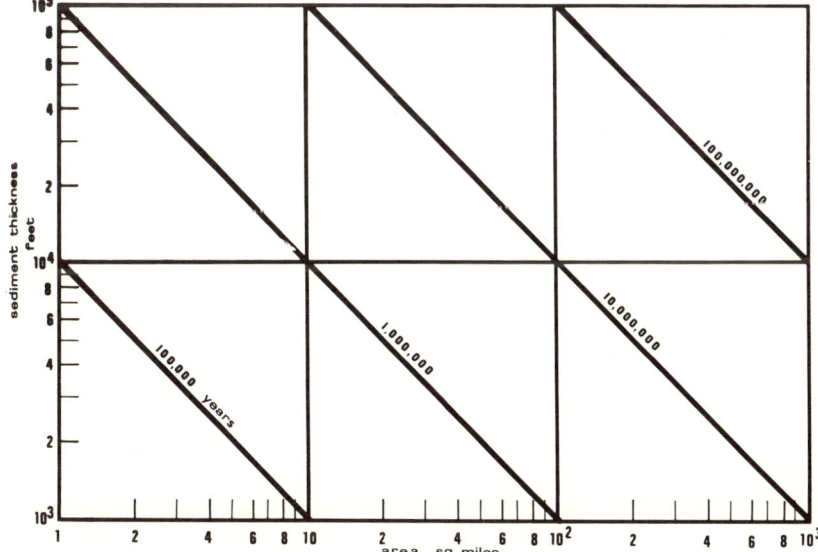

Fig. 2-15. Areas and thickness of accretion resulting from an accumulation of 100,000 cubic yards per annum.

100,000 cubic yards per annum of littoral drift were accumulated, the thicknesses and areas of deposition are as specified in Fig. 2-15. A duration of 1,000,000 years, which is short, geologically speaking, could construct a 1000 ft. thick plain or shoal covering 100 sq. miles.

At the upcoast ends of physiographic units rocky headlands will predominate. Any bays will be in full equilibrium and beaches will have steep offshore profiles.

CONTINENTAL PATTERNS

By use of the crenulate-shaped bays and other tools discussed above, Silvester [10, 11] examined the coastlines of all major continents, in order to determine the direction of net sediment movement. Correlation was continuously carried out with the ocean swell data available from a variety of sources [12–14]. Although hydrographic charts distinguish between sedimentary and rocky sections of shoreline, or these can be inferred from the smoothness of the outline, a check of coastal land forms was possible by the map produced by McGill [15]. This provided general

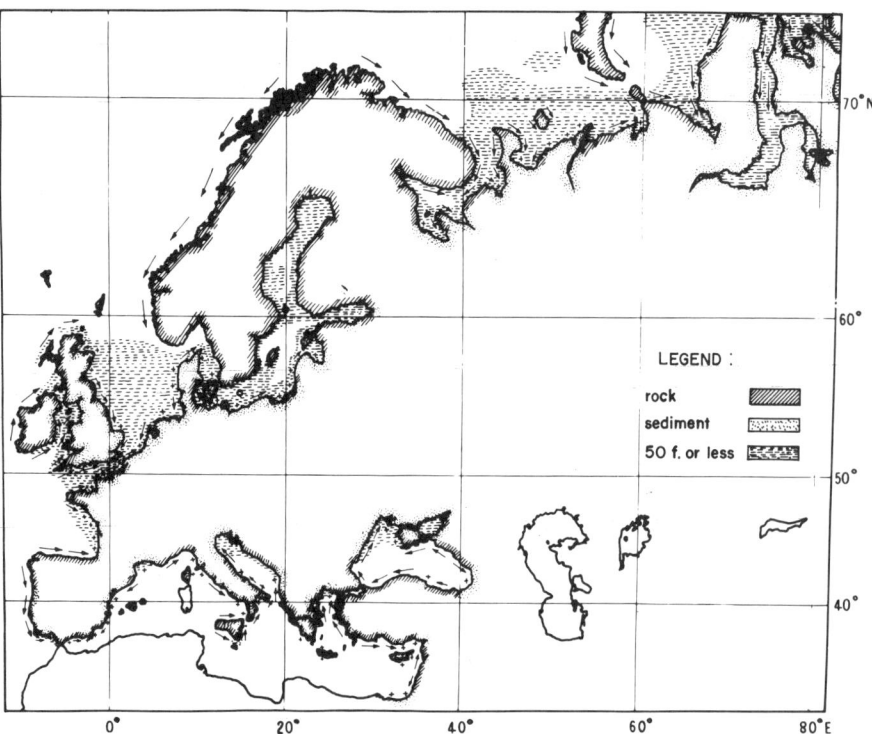

Fig. 2-16. Net sediment movement around European and Arctic coasts.

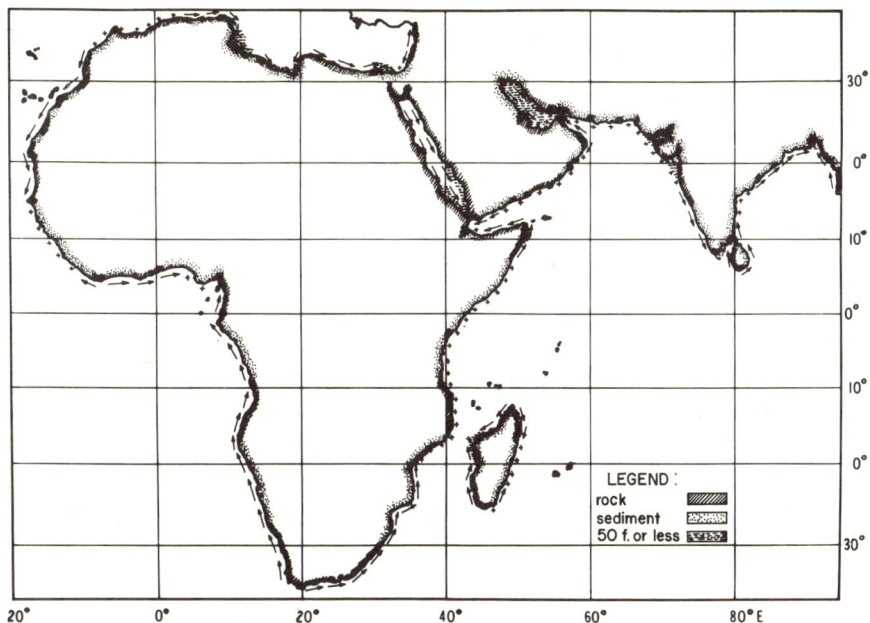

Fig. 2-17. Net sediment movement around the African coast.

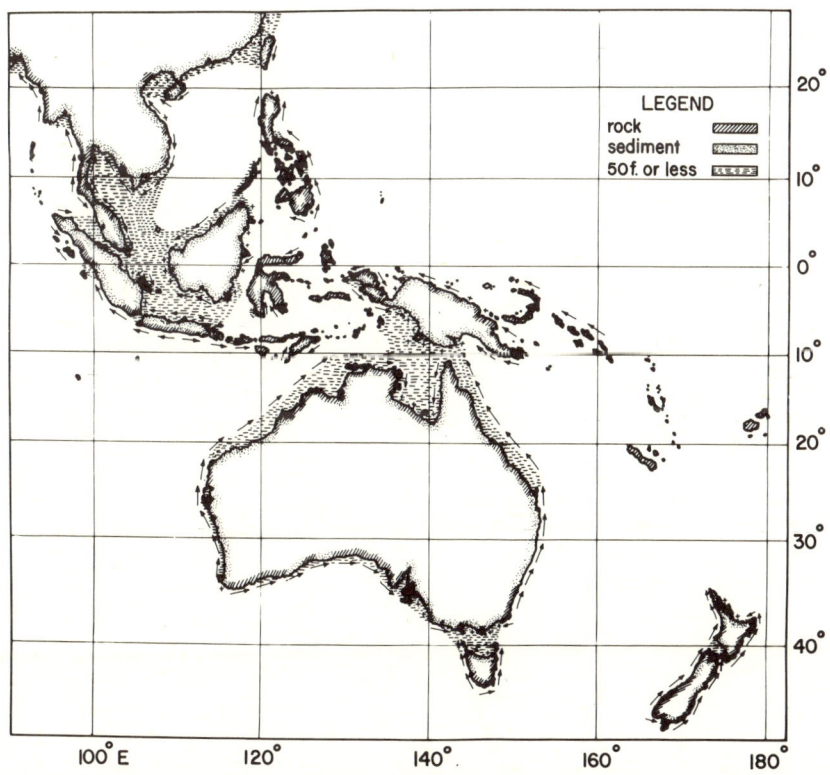

Fig. 2-18. Net sediment movement around south-east Asian coasts.

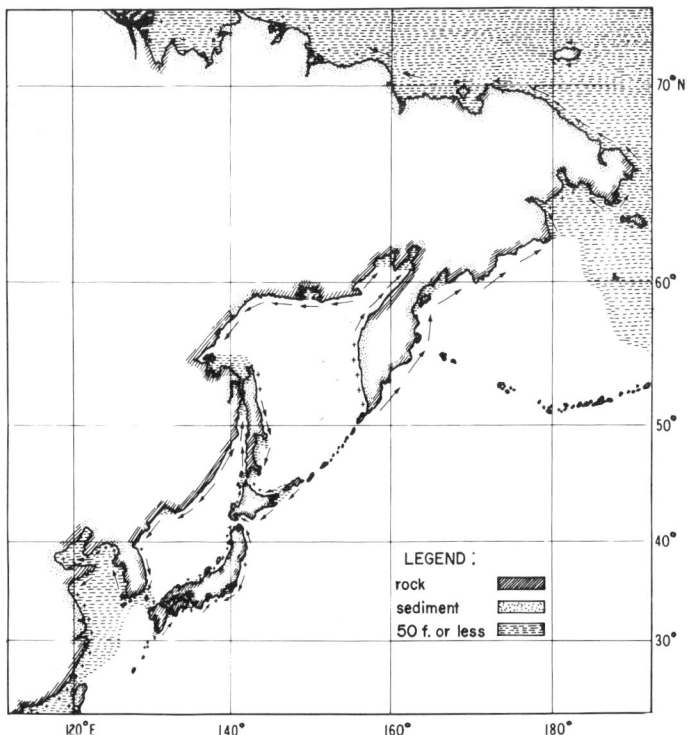

Fig. 2-19. Net sediment movement around northwest Pacific coasts.

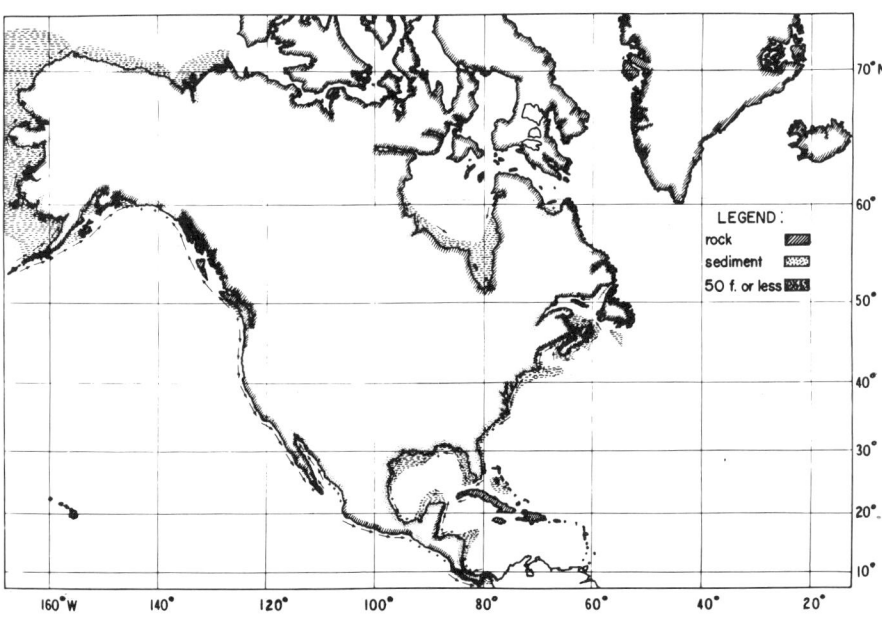

Fig. 2-20. Net sediment movement around North American coasts.

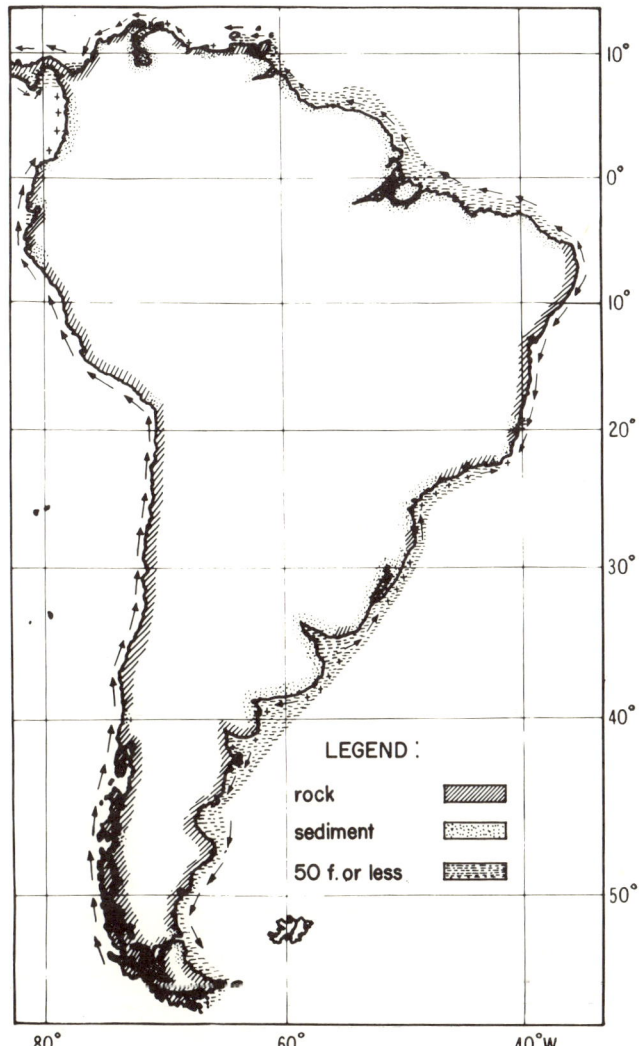

Fig. 2-21. Net sediment movement around South American coasts.

geographical information about the coastal plain as a whole, whereas the present differences relate to the occurrence of rock outcrops at the water line, or movable material that can be transported along the shore by waves.

Information is accumulating on the nature of each section of continental shelf. However, a ready assessment of the accumulation of sediment to form or fill out a shelf to 50 or 60 fathoms is provided by the smoothness of the bed contours. If these lines are indented then an undulatory surface is indicated which is unlikely to be produced by sand or silt dispersed by wave action. The criterion will more specifically apply in the deeper zones of the shelf than near-shore, where influences of rivers and islands are greater in the disposition of material.

The maps of the continents in Fig. 2-16 to 2-21 contain arrows where a net drift has been presumed from the various indicators listed above. Where no clear indication is available and therefore no net motion apparent a plus (+) sign has been used. These figures also have marked the areas of continental shelf out to a depth of 50 fathoms. It will be recalled that the shelf edge is around 65 fathoms for most oceanic margins, so that the majority of the wave-active zone is indicated.

In a general picture so painted it is necessary to simplify drastically the data displayed. Certain restrictions must be observed, therefore, when utilizing these figures as follows:

(a) The rate of sediment movement is in no way specified. The arrows indicate only that if material is available it will move in the direction shown.

(b) Net-longshore transport is inferred. Minor reversals of drift, due to storms or secondary swells, may occur spasmodically or periodically.

(c) Where drift is substantially retarded, or equal in both directions, plus (+) signs have been used. This does not necessarily presume a progradation of the beaches in the area, since little sediment may be available for this.

(d) The general coastline only is considered. Specific patterns of sediment movement occurring within or behind minor physiographic features, such as islands or promontories, are excluded. However, on all coasts facing the major ocean the drift is as shown.

(e) The stippled zones on the maps do not differentiate between shingle, sand or silt, for which the reader is referred to McGill's chart [15].

(f) The reference to rocky shoreline applies strictly to the water line and not to the coastal plane in general, to which reference to McGill's publication is again suggested.

WORLD-WIDE PATTERN OF NET SEDIMENT MOVEMENT

The net drift around the various continents has been combined by Silvester [16, 17] into a global pattern, as depicted in Fig. 2-22. Similar criteria for utilizing this map are applicable to those in Fig. 2-16 to 2-21. The shelf out to 100 fathom is displayed in this case. Incorporated into the figure are the general wind systems for the months of January and July. Whilst this does not provide an all-year picture of storm and wind conditions (to which the reader is referred to Ch. 2 *Coastal Engineering, 1*), it shows a strong correlation of wind and hence wave directions with the longshore drift derived essentially by other means.

From Fig. 2-22 it can be seen that on continental margins where the wind blows predominantly in the same direction throughout the year, a net drift extends for

NET SEDIMENT MOVEMENT

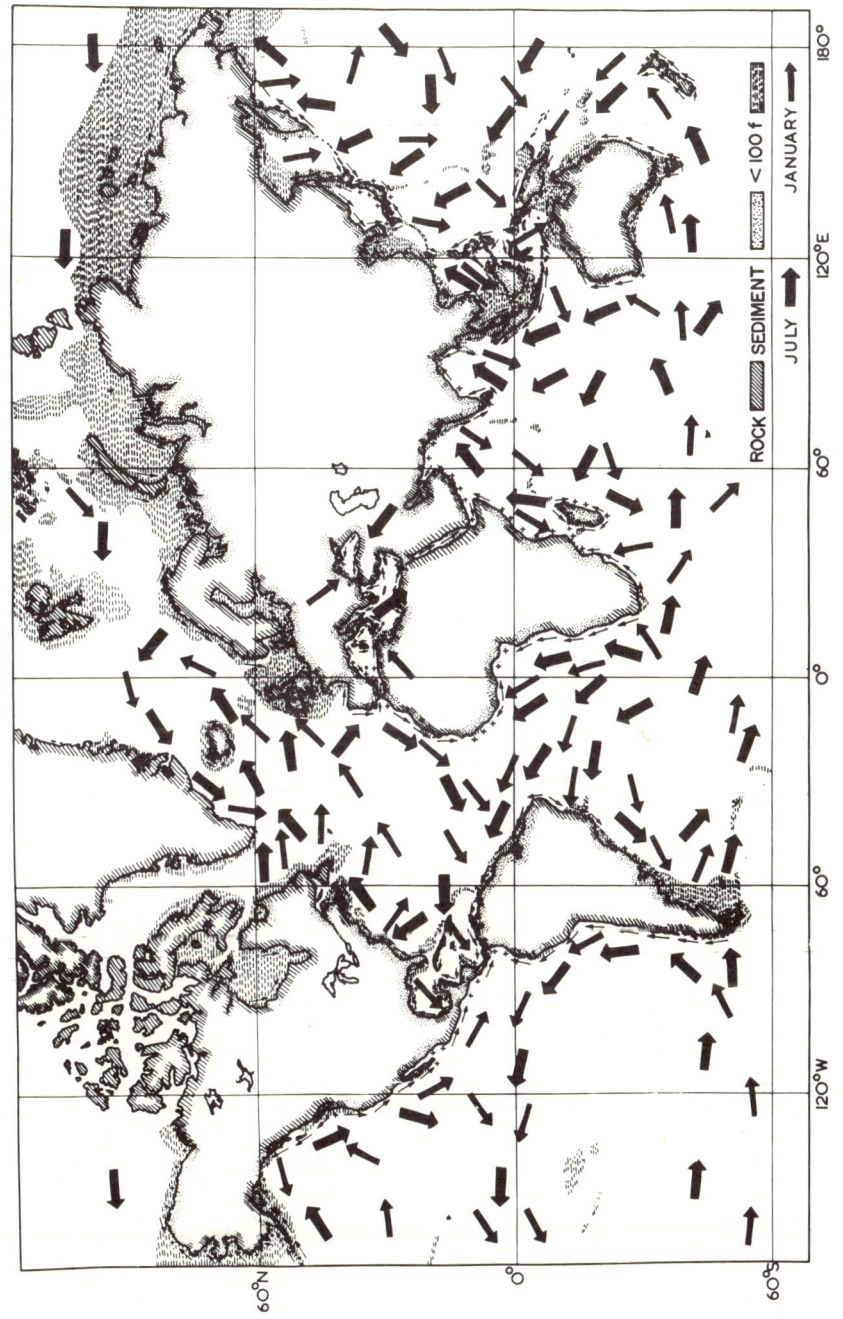

Fig. 2-22. Net sediment movement around the coastlines of the world and the global wind pattern which is its major energy source.

complete lengths of the landmass. The following examples are worthy of note: on the western margin of South America drift is towards the equator; on the southwestern coast of the African continent there is again a strong equator-ward movement; on the northwestern coast of the same land mass is a similar trend; on the western margin of Australia is a northerly net drift.

The emphasis on western margins of land masses and a distinct movement towards the equator confirms Davies' [2] thesis of the role of the strong wind systems in the $40°-50°$ latitudes. On the eastern coasts of continents the longshore drift is dictated more by the local storm genesis, such as tropical cyclones travelling from the equator polewards or extra-tropical cyclones passing across the sea margin on their way eastwards. Also, major indentations of the coastline influence the local net movement of material.

It is not proposed in this treatise to discuss the geologic implications of net longshore drift, but it should be remembered that wave energy has been available from the inception of the oceans, which Stoddart [18] quotes as possibly $2 \cdot 10^9$ years. As soon as sedimentary material became available to the waves, through weathering of base rock and its subsequent carriage to the sea, it has been retained against the continental margins by the waves and transported towards some trap, there to be deposited, so to accumulate a large land mass.

The wind systems of the world would have remained sensibly the same as they are today for as long as the present global temperature distributions existed. No matter where the ancient equator of poles have shifted, there have always been zones of high storm intensity that would have provided the swell waves to all continental margins, there to effect the distribution of sediment. It is not surprising, therefore, that around 75% of the land masses are sedimentary and that around 75% of this (as judged only by surface samples) has been deposited under marine conditions.

It is submitted, therefore, that the present land masses in the main have been accreted in a horizontal direction and could have done so with very little change of sea level. The height distribution of these continents, as illustrated in Fig. 2-23, certainly indicates that the majority of this deposition has occurred close to the present sea level. The cycle from weathered rock, through river discharge, to final sedimentation under marine conditions is a process which is not at variance with any known laws of nature. The forces required for its completion have been available throughout geologic time and hence eustatism does not have to be invoked as a process for land building either above or below the surface of the sea. Such a land accretion process has not been accepted generally by geologists, and the absence of terrigenous material, expected to be spread across the oceans generally, has been termed "the great paradox of marine sedimentation" [21].

The same process that provided the original sea floor would have also provided the zones of the crust that remained above the sea. The former have been protected

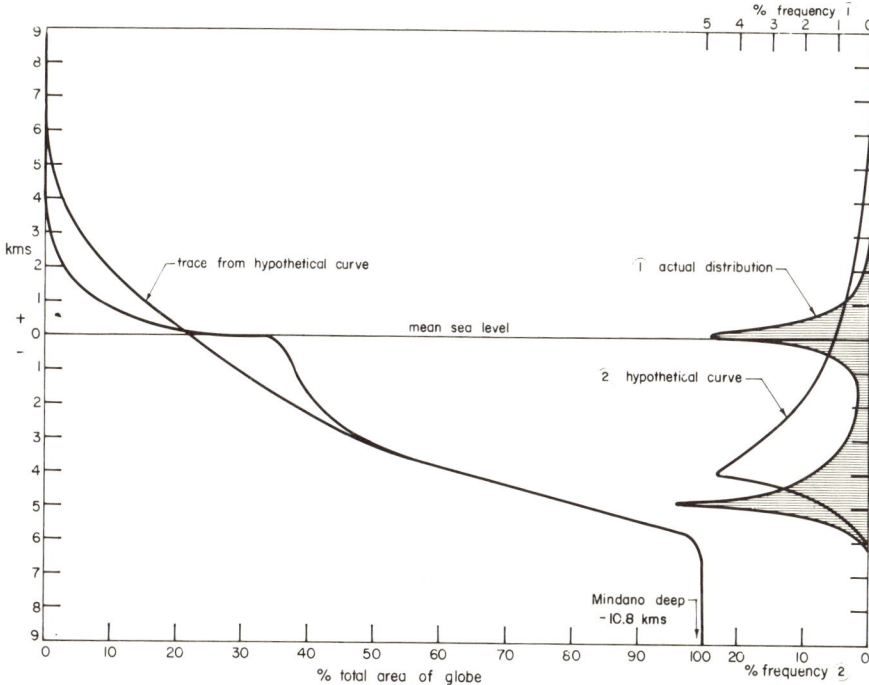

Fig. 2-23. Height distribution of surface of globe.

from erosive forces by the oceans themselves, whilst the latter have suffered all the weathering effects possible. The precipitous slopes still exhibited on some coastal margins, generally termed submarine canyons, indicate the probable ruggedness of the initial land areas. Many of the steeply sloped margins have been buried in the sediment spread along them by the waves. Where filling is not complete the precipitous valleys still exist, although their beds contain vast accumulations of material deposited in them from the upcoast direction.

This implies that the majority of the continental shelves are sedimentary. This is confirmed by figures provided by Hayes [22] which for the known shelf areas are as follows:

sand	43.5	rocky	13.2
mud	28.0	coral	6.4
shell	4.8		
gravel	4.1		
	80.4%		19.6%

Hayes subdivided shelves into various climatic categories, but in all these the percentage of sedimentary material was never below 74%.

GEOMORPHOLOGICAL STUDY OF A COASTLINE

The process of sedimentation can be traced from its genesis as weathered material from rock, transport to the ocean by rivers, and its distribution there by waves. The sediment supplied by a river is dependent on its catchment characteristics, of topography, precipitation, vegetation, geology, and temperature. The grading, as well as other qualities resulting of this sedimentary material will change over geologic time.

As has been discussed already, the transport and final deposition by waves is of paramount importance in this overall movement, to a destination of either coastal-plain or offshore shoal. Accepting that the wave climate of an oceanic region or enclosed sea has not altered much in the more recent geologic periods, it is possible to picture gradual construction of coastal plains out to the present continental outline. Such a geomorphological study can help in the prediction of future coastal developments, such as zones of siltation or erosion, as well as changes that could be brought about by river and maritime structures.

Sediment supply

Weathering of rock is effected through chemical breakdown, temperature effects and the erosive forces of wind, rain and runoff. Abrasion by ice or waves would appear to provide an insignificant percentage of material from which river and coastal plains are constructed. Kuenen [23] has derived erosion figures which indicate that shoreline degradation is approximately 1% of the volume from fluvial sources.

The volume of sediment being fed annually to the ocean at any point will be determined by the area and type of river catchment, the annual precipitation, the geologic structure, and the degree of protection afforded by vegetation. The catchment of a river can change over geologic time due to the construction of its sedimentary plain, affecting as it does both the area and the topography. The rainfall and natural vegetation are dictated by geographical location as well as topography. The weathering and erosion of land structure is a function of source material and the intensity of elements producing this breakdown. In the case of sedimentary strata being elevated above sea level for redistribution some famous formations result, such as the Grand Canyon of Colorado.

The material being transmitted by a river will change as it grows in length and flattens its hydraulic gradient. Whilst it traverses rugged rock structure it will flow swiftly and so contain sufficient energy to transport coarse sediment or even large boulders. In fact, most of the material will be of this character due to the extremes of weathering resulting from lack of protective cover by soil or vegetation. As the sedimentary plain is constructed, the river suffers reduced velocities and its capacity

GEOMORPHOLOGY OF A COASTLINE   99

for carrying is restricted to the finer material. Although a catchment may be enlarged by the plain, the river flow becomes more uniform due to water infiltration into the porous basin. The higher velocities are reserved for the few occasions when energy is distributed over wide flood plains. Thus, where short steep rivers debouche on a coast sand should predominate. Longer rivers, on the other hand, should supply the finer elements of the sedimentary spectrum.

Wave climate

The wave climate of a coastal region will encompass the energies and duration of storm waves and swell. Where the continental shelf is wide the presence of large tidal ranges should also be considered for their influence on coastal processes. The spasmodic inundations of coastal zones by storm surges are an additional factor. It

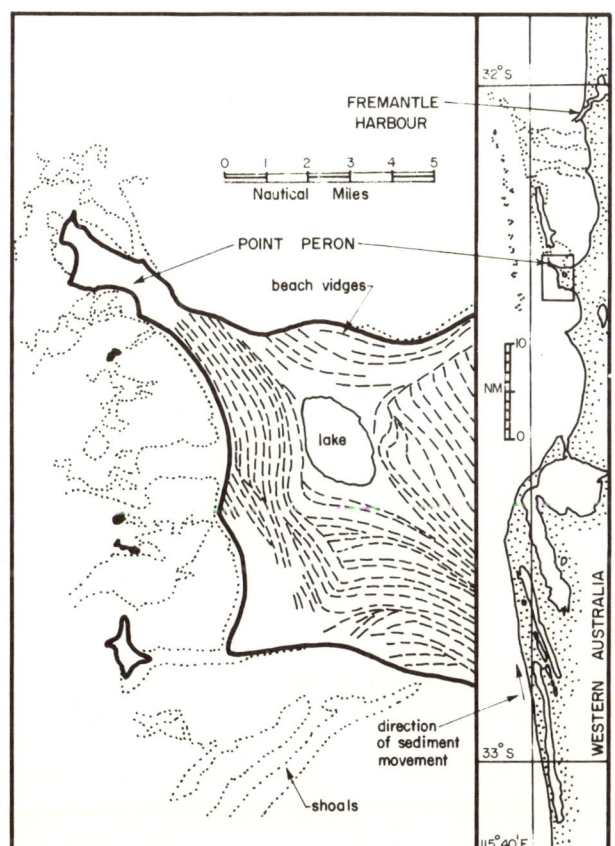

Fig. 2-24. Chronological development of coastal plain as exhibited by ancient beach ridges. (Modified from Ref.[36].)

is reasonable, therefore, to envisage the present-day wave forces being applied to some original rocky land mass, modified as required by past refraction or diffraction patterns, whilst sediment accumulated around its boundaries. The development of river and coastal plains can be gauged from the probable past curvature of the wave crests as they sculptured the beaches. Some of these ancient shorelines can be traced from sand-dune structures, which have been fixed in position by vegetation. Such coastal undulations run for hundreds of miles, with crests parallel to the shoreline which existed at the time of their genesis. In an accreting situation, the continual addition of sand to a beach results in sand dunes, forming mounds that are essentially parallel to the shoreline. Knowledge of the crenulate shapes that ancient bays could have assumed will aid in the tracing of these previous marine margins. A small example is given in Fig. 2-24 of such beach ridges.

This enquiry into the geologic past of coastal areas, and areas that are not now so coastal, even though they were at time of deposition, might be termed an "astronautical" look at the landmass. It should provide the whole picture of the past over a continent, or large segment of it, at the one time. This contrasts with the survey of a fairly local region, or the more-or-less microscopic inspection of materials along the beach in order to derive its origin. What is involved is an analysis of climatic forces that play their part in the transfer of sedimentary material from the interior of a landmass to its margins, resulting in aggradation of the latter.

Silvester [24] has made a geomorphological study of the coastal margins of Japan. The four major islands involved are relatively young geologically, so that all processes can be termed "recent". Japan also has a complete island coastline, which is subjected to waves from all directions, so providing a variety of wave climates to be analyzed. For these reasons it is considered a good example for displaying the type of study which is recommended for comprehensive zones under the jurisdiction of a coastal engineer. The major sections of this analysis are physiography and climate, wave climate, and sediment movement.

PHYSIOGRAPHY AND CLIMATE OF JAPAN

Japan consists, in the main, of four islands, which, by any comparison, are extremely mountainous. The proportion of reasonably level plain being in the order of a quarter [25]. The coastal outline features prominent peninsulars and headlands, with many minor promontories making up an altogether rugged shoreline. However, along some stretches of coastline, plains have been constructed. These vary in width from a kilometre or two to a hundred or so kilometres. These have provided valuable sites for road and rail communication as an alternative to the tortuous hinterland.

Due to the topography, the watersheds of Japan are, in general, numerous and diminutive, but some of the embayments into which they have fed, and in some cases still feeding, are of a size commensurate with landmasses of a much larger size. In forming coastal plain at the head of some of these basins, much higher tidal ranges could have been experienced in previous geologic periods than at present, due to the co-oscillation set up in these bodies of water. It is reasonable to expect, therefore, that these plains will vary in height above sea level as the reach of the waves has altered. This is plainly exhibited in Japan by the terraces sculptured in the volcanic tephra and ash at levels which are higher at the previous heads of bays [26]. These higher-level terraces do not appear to have formed on the open coast where the tidal range has remained sensibly constant.

Maps generally show land elevations in steps of 100 m, for the lowest two increments at least. Since sandy coasts at the present time can be built up into dunes two or three hundred metres high it should be expected that these lowland areas (at least to 100 m) are sedimentary plains accreted by river and wave action. Within this height limitation will be zones provided by other geologic phenomena, but these will be minimal.

The steepness of the mainland structure of Japan generally continues below sea level, resulting in extreme depths close inshore at many points. However, the zone known generally as the continental shelf deviates greatly in width and in many places rises above the water level to form islands. This type of structure has been termed "continental borderland" [19], and implies submarine topography of equal ruggedness to that of the adjacent land, within the limit of 100 fathoms. Where sediment has been deposited on this underwater surface to smooth its profile, the area may be termed a shelf. But only in one location (to be cited) is there evidence of accretion forcing the 100-fathom line out from the coast and therefore forming a complete shelf in this respect.

The rainfall of Japan varies from 1000 mm (40 inches) to 3000 mm (120 inches); it could therefore be considered a wet climate. To this run-off could be added the snow melt in the northern regions of Honshu and in Hokkaido, as well as on the higher mountain structures of all islands. A plentiful supply of energy exists, therefore, for the erosion and transport of sediment to the coastal margins. The lack of larger plains may be explained in part by the relative geologic youth of the country and the harnessing of rivers by man.

Wave climate

The coastlines of Japan may be divided into three zones in respect to wave climate. The major one is that section bordering the Pacific Ocean, which comprises the southern and eastern shorelines of all four islands. The second in order of magnitude is that bounding the Seas of Japan and Okhotsk, facing generally north or west. The third comprises the coastal margins of the Inland Sea, being the northern coastlines of Kyushu and Shikoku and approximately half the southern shoreline of Honshu.

Pacific. The waves reaching the Pacific shores of Japan emanate from six distinct wind systems. These are: (*a*) the southeast monsoons; (*b*) the northeast trades; (*c*) the southeast trades; (*d*) the tropical cyclones (typhoons); (*e*) the north Pacific low-pressure system; (*f*) the Polar cyclonic centres.

In the following discussion reference should be made to Fig. 2-25.

The southeast monsoons occur during the Japanese summer months and consist of southerly winds extending from near the equator to north of Japan. They generate waves which arrive at the coast from a southerly to southeasterly direction.

The northeast and southeast trade winds blow throughout the year and commence from the west coasts of North and South America, respectively, extending across most of the Pacific just north and south of the equator. The waves generated in the strong wind zones of this system spread westward across the Pacific and arrive almost continuously at the Japanese coast as swell from a southeast direction.

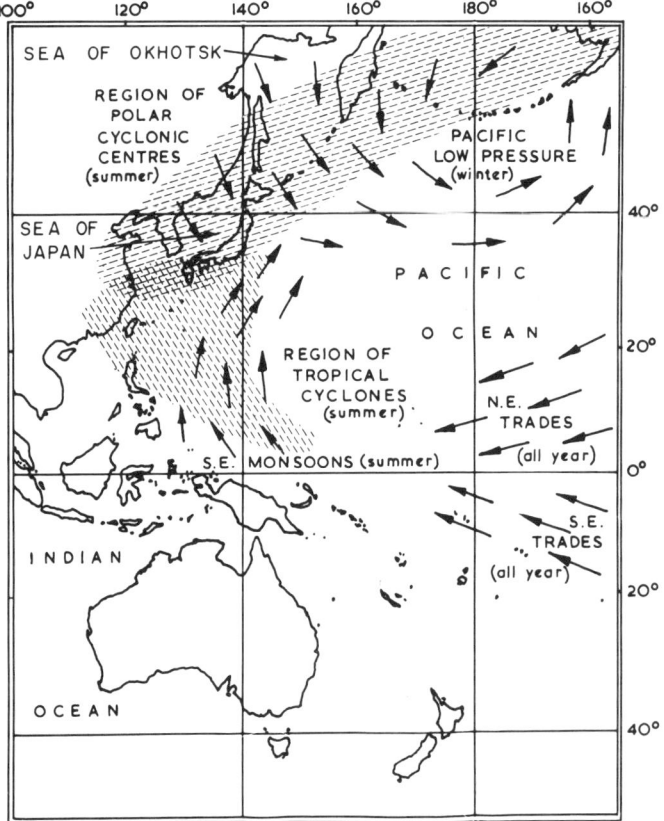

Fig. 2-25. Wind systems which supply ocean waves to the coasts of Japan.

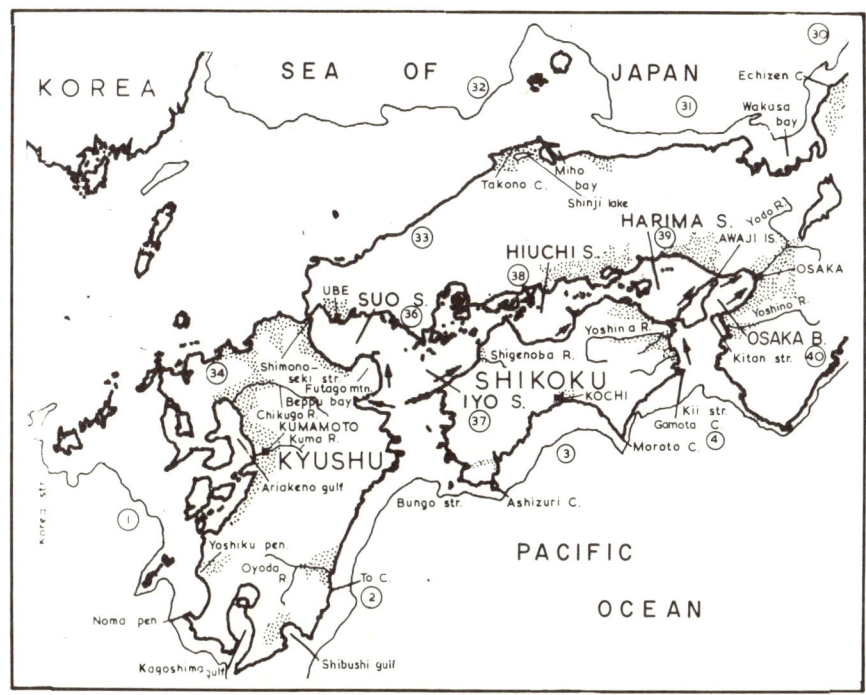

Fig. 2-26. Details of coastal features in southwestern Japan.

The typhoons form in the western Pacific at about 5°N and traverse paths to the 35° latitude, stretching from the Chinese mainland to east of Japan. They are a summer phenomenon. Because of the anticlockwise circulation of air in these low-pressure centres and their northward movement, the strongest waves are generated in a northerly direction. As typhoons reach the coast, however, the storm waves being generated at the time can hit the coast from the southeast to east and even northeast.

The low-pressure system of the north Pacific occurs during the winter months of Japan. Its anticlockwise circulation of air produces cold winds from the north across the Bering Sea and the northern Pacific. Their wave-generating capacity may be limited by the formation of ice during these colder months.

The Polar system of cyclonic centres, occurring in summer, consists of low-pressure centres travelling from the north Chinese and Manchurian landmasses, in a northeasterly direction, across the Seas of Japan and of Okhotsk towards the Bering Sea and Alaska. When they are so located as to affect the Pacific coast of Japan, the resultant waves will arrive from a northerly direction.

Considering Fig. 2-25, 2-26 and 2-27, where the above information is summarized, it would appear that the southern shores of Kyusho, Shikoku and Honshu

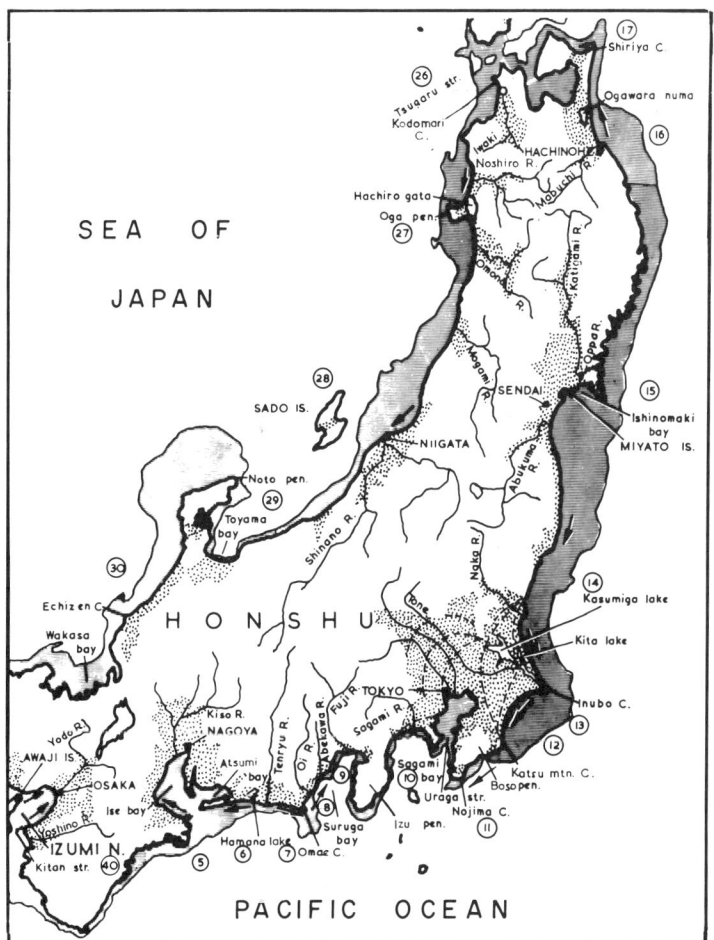

Fig. 2-27. Details of coastal features in northeastern Japan.

will receive persistent swells whose resultant will be from a southeasterly direction. The eastern shores of Honshu and Hokkaido will be mainly influenced by waves from the north. Where parts of a landmass intercept these waves and cause diffraction, the approach direction could be changed significantly.

Seas of Japan and Okhotsk. Referring again to Fig. 2-25 to 2-28, it is seen that the Sea of Japan experiences northerly winds during winter and southerlies during summer. The former appear to be of much greater significance to the Japanese coast [27]. The waves generated towards the north could only be influential on the west coast of Hokkaido, since the majority of the western shoreline of Honshu has greater fetch lengths for the southward directed waves. The Polar cyclonic centres

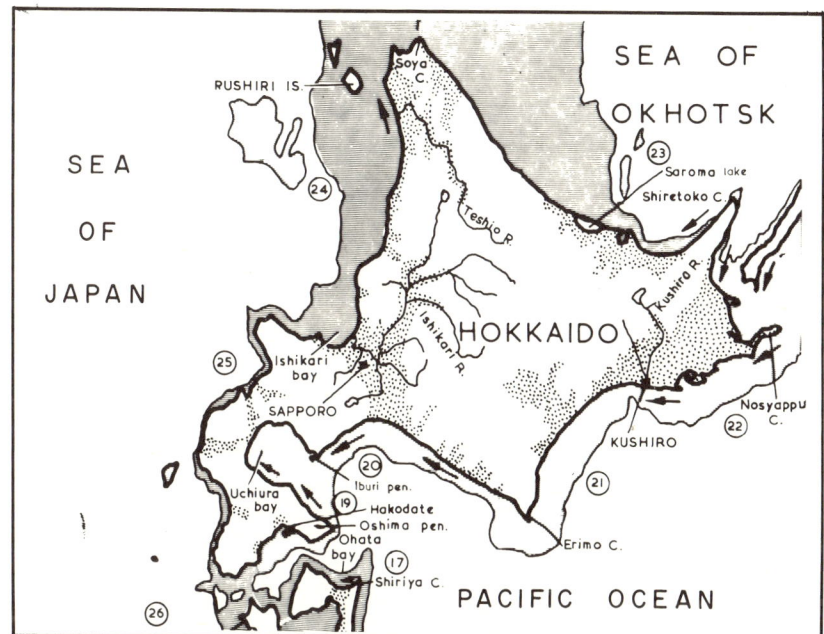

Fig. 2-28. Details of coastal features in Hokkaido, Japan.

traversing this body of water in a northeasterly direction would generate waves toward the Japanese coast from the north to west quadrant. The resultant from this system is therefore sensibly normal to the coast except for the northern tip of Honshu.

Hokkaido has a stretch of coast bounding the Sea of Okhotsk, which experiences similar winds to the Sea of Japan. The generation of waves by the winter northerlies, however, may be impeded by the ice cover over large tracts of this sea for approximately three months. Waves of any consequence reaching this northern coast must necessarily arrive normal to the coast. In the western region, which runs NW—SE, the waves again arrive normally due to their diffraction around the southern tip of Sakhalin Island.

The waves on both these seas are of storm type, lasting only as long as the winds generating them. In spite of the confused nature of these waves, the longshore component would be fixed by the mean wind direction, or its "diffracted" value.

Inland Sea. The Inland Sea, between Honshu and the northern coasts of Kyushu and Shikoku (see Fig. 2-26), is divided by island groups and promontories into five basins, whose major axes run in a NE—SW direction, the winds blowing across them being partly influenced by the surrounding mountain structure. The basins most likely to suffer the strongest winds are the two westerly ones, namely, the Suo

Nada and the Iyo Nada, the latter even permitting the penetration of swell from the Pacific through Bungo Strait. The waves generated over these bodies of water are of a storm category. Their persistence cannot be compared with the shorelines discussed in the previous two sections, but could be sufficient to concentrate littoral drift in certain zones.

Sediment movement

For the case of easier presentation, the coastlines of Japan have been divided into a number of sections, each of which has the same wave climate. This is not to imply that near-shore conditions are similar along the whole stretch of coast, but that the waves arriving in the open stretches of ocean, away from the coast, are sensibly the same. The distribution of sediment will be greatly influenced locally by promontories, offshore islands and submarine features.

The sections will be discussed in order, running around the coastline in an anticlockwise direction, starting at the west coast of the southernmost island of Kyushu. Reference should be made to Fig. 2-26, 2-27 and 2-28, where circled numbers represent the region being discussed in the paragraph of the same number.

Southern coastlines. These consist of the Pacific margins of Kyushu, Shikoku and the section of Honshu from Kii Strait to Uraga Strait, which forms the entrance to Tokyo Bay.

(1) The west coast of Kyushu could be considered open to the Pacific swell, but many sections of coast cannot receive it because of island protection. Sediment deposited by rivers in these deep indentations has formed tidal shoals, as exemplified in Ariakeno Gulf, which serves the city of Kumamoto. Where waves are able to penetrate, small pocket beaches exist, as between the peninsulas of Noma and Yoshiku. In the case of a dearth of sediment the indentation preserves its original rugged profile, as instanced by the extreme depths of Kagoshima Gulf.

(2) At the southern tip of Kyushu the continental borderland rises to form several islands. No sediment is likely to be present in this offshore area since there is no notable supply point, and any available material would be swept northwards up the coast or into an embayment. The ruggedness of this southern region is interrupted only by a pocket beach within Shibushi Gulf. Further north, beyond To Cape the Oyoda River provides sufficient material for a modest coastal plain to have smoothed off the shoreline. Towards Bungo Strait the coast again becomes rugged, with little or no sediment in evidence.

(3) The southern coast of Shikoku consists mainly of two mountainous peninsulas, ending in the capes of Ashizuri and Muroto. Between these a limited amount of sediment has been accreted to form a coastal plain, on which Kochi is located. East of Muroto Cape the submarine slopes are extremely steep, but inside Kii Strait

(past Gamota Cape) an extensive water shed has brought material to the coast. Two deltas exist here, the major one being that of the Yoshima River. Sediment transport is necessarily into the Strait because of the penetration of southerly swell and as evidenced by the bay shapes and sand spits present.

(4) Honshu provides the eastern boundary of Kii Strait, at the northern end of which the Kino River has filled a V-shaped indentation. Further accretion is probable, with the shoreline being fashioned by the swell diffracting around headlands to the south. Otherwise the peninsular east of Kii Strait is rugged and depths of hundreds of fathoms are close inshore.

(5) Further to the east this mountain structure almost seals off Ise Bay from the Pacific. Under the normally calm conditions ensuing a multiple river system has constructed a deltaic plain on which Nagoya is sited. A more modest delta exists in Mikawa Bay, which connects with the larger unit.

(6) East of these bays the southern coast has been abundantly supplied with sediment by the Tenryu River, which also exhibits a deltaic mouth. The volume of sediment has exceeded the power of the waves to distribute it west and east, resulting in a slight protuberance of the coast. Mildly curved coasts extend either side, that to the west having formed an inlet known as Hamana lake. The coastline further west, which was originally very rugged, has been silted to a smooth outline.

(7) Sand is transported east from the Tenryu delta, but much is lost on the steeply sloping submarine shelf. The projection of the continental borderland southward causes refraction of the ocean swell so that Cape Omae has been constructed. Sediment is transmitted around this prominence with related shoaling. On the eastern side a curved beach has smoothed off a previously rugged shoreline.

(8) Suruga Bay is extremely deep and but for its width might be termed a submarine canyon. But in spite of its precipitous underwater slopes the west coast of this bay has a coastal plain. The Oi and Abe rivers have been the major sources of sediment, which has been distributed by the southern swell. It is along one of these beach lines facing south that the accretion from this process has probably pushed the 100-fathom line seawards.

(9) At the head of Suruga Bay a modest plain has been shaped by the southerly swell from material supplied by the Fuji River at the western extremity. This sediment has been deposited into a deep ravine and the offshore shelf is extremely narrow. Since it follows the shoreline in a smooth curve, deposition to the 100-fathom limit is indicated. To the east, Suruga Bay is bounded by the Izu peninsula, which is extremely mountainous. Because it has little watershed to its ocean margins the submarine slopes have retained their original precipitous character.

(10) To the east of the Izu peninsula is the entrance to Tokyo Bay, through Sagami Sea and Uraga Strait. This sea and Sagami Bay is extremely deep and the strait itself contains the notable Tokyo submarine canyon. At the head of Sagami

Bay an extensive coastal plain exists, which joins further north with the Tokyo plain. Most sediment in this region is presently supplied by the Sagami River. Unlike Suruga Bay, the 100-fathom line here deviates in such a manner as to preclude the possibility of accretion out to this limit. The eastern boundary of Uraga Strait is provided by the Boso peninsula. Its mountainous structure and lack of watershed is reflected in its precipitous submarine slopes where several canyons have been recognized and mapped.

Eastern coastlines. These contain the Pacific coast of Honshu from Nojima Cape (enclosing Uraga Strait) to its northern tip of Shiriya Cape.

(11) On the eastern side of Boso Peninsular the previously rugged coast has given way to narrow beaches, the material for which has been supplied from the Tokyo-plain sources further north. This material has traversed the cliffed coastline of Katsu Mountain and formed crenulate-shaped bays around the southern tip of the peninsula. The direction of transport is westward due to the waves from the south to east quadrant. Even swell from the north will be diffracted and refracted around capes Inubo and Katsu to force material westwards around Cape Nojima.

(12) In Fig. 2-27 is depicted a suggested development of the Tokyo plain over geologic time. Initially two embayments were filled until one large bay was formed. Further accretion caused this bay to protrude behind the Boso peninsula, which at this time would have been an island. A relatively sudden addition of sediment caused the transfer of the coastline to the peninsula, with material feeding to Katsu Cape. The presence of an island (Inubo), with its related shoaling of the continental borderland caused the shoreline to attach itself to this fixed point and so form the present cape. The undulatory nature of the borderland and its weathered volcanoes is indicated by the lake system incorporated into the plain. Kasumiga Lake is extremely deep and yet Kita Lake system is shallow. The elongated form of the latter suggests its origin from a southward growing sandspit, material being mainly supplied from the Naka River. The progressive filling of the Tokyo Plain suggested above is exhibited by the concave nature of the sedimentary strata apparent in the basin [26].

(13) Considering the present shoreline, sediment south of Inubo Cape moves rapidly southwest, as evidenced by the deflection of the river mouths and the sand-spit formations, The deviations of the 100-fathom contour would indicate that saturation by sediment has not been reached.

(14) North of Inubo Cape the coast has a straighter outline, indicating a slower movement of material in a southward direction. North of the Naka River is an extensive length of rugged country. Short, steep rivers have supplied sufficient sand to form a narrow coastal plain, mainly in the form of pocket beaches, whose slight curvature indicates the near balance of wave energy from the north and the south.

(15) Further north the coastline takes on a sudden new north—south alignment

due to a mountain range projecting into the sea. This peninsular would formerly have been a series of islands and might still be considered separated from the mainland due to the double outlet of the Oppa River, into which the Kitagami flows. The elongated embayment formerly existing between this mountain structure and the mainland has been accreted by the Katigami River to the present shoreline limit of Ishinomaki Bay, which curves in equilibrium with the waves refracted from the south and those diffracted from the north. The Miyato Island group, which limits this beach, has on its southern side another coastal plain provided by the Abukuma River, which also is in equilibrium with the waves of this region. Sendai is situated on this plain and canal systems have been constructed in many places.

(16) To the north the aforementioned mountain structure extends to the coast, resulting in rugged shoreline topography with deep water close inshore. There is practically no watershed to this coast until the Mabuchi River is reached, with its sundry smaller streams. These have constructed a coastal plain, together with a large lagoon known as Ogawara Numa. The beach line is in near equilibrium with the waves, there being only a slight trend northwards. It is seen from Fig. 2-27 that Hokkaido causes waves from the north to be diffracted and to arrive at northern Honshu from the east. Southerly swell, on the other hand, will be slightly oblique at the shore, in spite of its refraction across the shelf.

(17) The continental borderland in this region varies greatly in width and extends northward beyond Shiriya Cape. This shoal refracts the waves arriving from the east and south, so that in Ohata Bay, to the west of the cape, the available sediment has accumulated in a curved coastline and formed the isthmus linking this narrow stretch of land to the mountainous block to the west.

Hokkaido. This section encompasses all the coastlines of this uniquely shaped island, which bounds the Pacific Ocean and both the Sea of Japan and of Okhotsk.

(18) There are strong indications from the topography that Hokkaido originally consisted of a number of islands which, over geologic time, have been welded together by sedimentation, particularly in the western regions. Accretion is still taking place between the many prominent peninsulas, limited only by the modest nature of the watersheds and the prevalence of ice and snow.

(19) The Oshima peninsula, north of Tsugura Strait, is rugged and the limited volume of sediment available is swept westwards by the swell. On the southern side this has formed the isthmus or tombolo on which Hakodate is sited. On the northern side the sediment has formed a coastal plain around the head of Uchiura Bay, the shape of which has been sculptured by the swell.

(20) Iburi peninsula, at the entrance to Uchiura Bay, supports a coastline to which the swell approaches almost normally. Sediment brought on to this ocean margin, by local rivers or from supply sources to the east, is thus accreted in a

sligthly curved beach line, with a moderate tendency for drift in a southwest direction.

(21) The mountainous block culminating in Cape Erimo also provides an obstruction against which sediment is retained in equilibrium, by waves from east or south. A number of indentations of this originally rugged coastline have been enclosed by sand spits. To the east a submarine canyon is indicated by the deviation in the 100-fathom contour. This would have penetrated the present mainland and provided the embayment into which the Kushiro River flowed. This valley now exhibits extensive swamp land. The sediment has filled out to the 30-fathom line where the canyon head now commences.

(22) From this river outlet, to the eastern limit of Hokkaido (Nosyappu Cape), the coastline is rugged, with the small amount of sediment available being swept westwards into pockets formed by promontories. The continental borderland extends beyond Hokkaido to support island chains running northeast. A number of rivers shed water to this region, which experiences waves from the one possible direction of northeast. This provides specific littoral drift forces, which have resulted in lagoon formation and predominant sand spits.

(23) The northern coast of Hokkaido faces the Sea of Okhotsk. The continental borderland varies from practically nothing at Cape Shiretoko in the east to a substantial width at the western extremity of Cape Soya. This is due to this submarine feature providing support for the southern regions of Sakhalin Island north of Hokkaido. It should be noted that the 50-fathom contour follows the coast of Hokkaido fairly uniformly. Waves generated across the Sea of Okhotsk would arrive at the shoreline normally, being diffracted around the southern tip of Sakhalin Island in order to approach at right angles in the region of Cape Soya, where a NW–SE alignment exists. The whole sweep of shoreline is smoothed, with equilibrium being exhibited by such features as the double spit almost enclosing the lagoon of Saroma Lake.

(24) The west coast of Hokkaido bounds the Sea of Japan and because of its position experiences strong wave action from both the northerly and southerly quadrants. The latter appears more influencial as shoreline features indicate a slight northerly drift on the main expanse of coast. At the northern tip a coastal plain has been constructed from sediment supplied mainly by the River Teshio. The shoreline protrudes in the lee of Rishiri Island.

(25) To the south the coast becomes very rugged, with many promontories and indentations edged by steep submarine slopes. One notable feature is Ishikari Bay where a smooth beach line exists. The associated lowland appears to extend through the valley containing Sapporo to the Pacific Ocean shoreline. This together with an extensive plain has been built by the Ishikari River system. This initial sea passage appears, by the river pattern, to have been blocked near the Pacific margin and progressively accreted to the existing Ishikari Bay.

Northwestern coastlines. These contain the entire length of Honshu facing the Sea of Japan, plus the small section of Kyushu also adjoining this body of water.

(*26*) The northern tip of Honshu is extremely rugged with no sediment in evidence. South of Cape Kodomari, however, a substantial coastal plain exists, the beach shape indicating near equilibrium. The main supply of sediment has been the Iwaki River, which is at present feeding a delta within an estuary that is almost closed from the sea by sand spits.

(*27*) Beyond a stretch of rugged coastline to the south, another zone of coastal plain exists. This is the northern boundary of a large tombolo that has formed in the lee of the former island of Oga. In so forming, this tombolo has entrapped a large body of water known as Hachiro Gata, which discharges to the sea south of the Oga peninsula. Material for this massive sand-spit has emanated mainly from the Noshiro River. To the south the coastal plain has been furnished mainly by the Omono River and is limited in the south by another mountainous prominence. To the south of this the Mogami River has built a low-level plain, the beach line of which protrudes slightly at the mouth. This indicates a surplus of sediment for the wave energy available to distribute it.

(*28*) Further south, sheltered by Sado Island, the Shinano River has built a plain which is protruding. Silting has been in evidence at the river mouth, upon which the port of Niigata has been developed. Coastal problems with this port have been discussed elsewhere [27]. The Shinano is among the longest rivers in Japan and is in a zone of high rainfall. Material brought down to the sea can only be spread southward as waves from the southerly quarter are blocked by Sado Island and Noto peninsula to the south. A number of small rivers has provided coastal plain north of the present mouth of the Shinano.

(*29*) From this river to the Noto peninsula the mountain structure adjoins the sea and contains practically no submarine shelf. Small rivers in this high-rainfall zone bring sediment into pockets, which is retained there by the normally approaching waves. In the hook formed by the Noto peninsula sedimentation has taken place. The narrow continental shelf in Toyama Bay indicates previous penetration of this marine depression into the present plain area.

(*30*) On the southwestern side of the Noto peninsula, the watershed is sufficient to supply a coastal plain. Near-equilibrium with the incoming waves is evidenced by the slightly curved outline and a number of lagoons. The slight southerly drift is contained by the mountainous block containing Cape Echizen.

(*31*) At Wakasa Bay the general coastline turns westward. Its mountainous structure provides practically no beach material and hence the shoreline is rugged for some 100 kilometres. The continental borderland varies greatly in width. Towards the western limit two small plains exist with their east—west alignments indicating stability with respect to the waves approaching from the north.

(*32*) A headland then occurs on the coast known as Takono Cape. This appears to have been an island originally, located fairly close to the mainland. It might still be considered an island, since it is separated by lakes, rivers, lagoons or deltas from the mainland. At the eastern end is a circular sand spit separating Milo Bay from a lagoon. The central accretion contains a lake called Shinji, into which a river is now discharging through a deltaic mouth.

(33) From Takono Cape the coastline runs southwest with a rugged outline. Only minor beaches occur in embayments of this region. The northerly waves would be reflected from the cliffs and so create standing waves which would readily dispose of any sediment down the steeply graded submarine slopes. The western tip of Honshu is similar to the above.

(34) The northern coast of Kyushu, across Shimonoseki Strait, is significantly different in that several rivers have formed pocket beaches, which, nevertheless, are still contained between the headlands. The northwestern boundary of Kyushu consists of a complex of hundreds of islands. These rise steeply out of Korea Strait, with depths of 30 to 50 fathoms close inshore. Their deep indentations could be considered as miniature submarine canyons.

Inland Sea coastlines. These consist of the southern shore of Honshu from the Shimonoseki Strait to Kii Strait and the northern boundaries of Kyushu and Shikoku.

(35) Five basins in essence constitute the Inland Sea, the major axes of which all run northeast. The strength and direction of waves in them is dictated mainly by the lengths of fetch and the control of wind direction by the surrounding mountain structure. There are indications that the more effective wave generators are the winds from the southwest.

(36) The most westerly basin is the Suo Sea which has a wide connection to the adjacent Iyo Sea. Its southern coast consists of a wide bay stretching from Shimonoseki Strait to the Futago mountain structure to the east. Shoals are caused by a number of short rivers, mainly around their mouths. No particular direction of drift is apparent. The northern shore is deeply indented, except in the vicinity of the strait where a protruding plain has been provided on which Ube is now located.

(37) The Iyo Sea has contact with the Pacific Ocean by means of Bungo Strait, through which swell penetrates to beaches in the vicinity. These mainly surround Beppu Bay, on the southern boundary of which a deltaic plain has formed. This material is swept westwards by the diffracted waves. The northern edge of Beppy Bay contains a modest coastal plain aligned with the crests of the incoming swell. A northward longshore drift is indicated around Futago Mountain, but the rate would be meagre due to the lack of watershed. The southeastern shore of the Iyo Sea has multiple indentations with steep underwater slopes. Material for transport is available from about one river and is swept eastwards towards the mouth of the Shigenoba River, which has constructed a plain from a formerly long embayment.

(*38*) The Hiuchi Sea is a smaller basin than the two previously mentioned and has islands scattered around it, particularly on its northern boundaries. Shoaling can readily be correlated with the larger rivers bringing material down to the waterline. The only littoral drift evident from shoreline features is on the southeastern region where a net movement to the northeast is indicated.

(*39*) The Harima Sea has a fairly uniform depth of about 18 fathoms, with few islands to break up the fetches for wave generation. Only in the western region do islands provide calm conditions for in-situ sedimentation. The southeast boundary of the basin is Awaji Island, which displays pronounced longshore drift to the northeast. Because of steep underwater slopes and lack of material the associated coastal plain is very narrow.

(*40*) Osaka Bay is the most eastern of the Inland Seas. It is the smallest in area of all five. It varies in depth from about 30 fathoms in the southwest to around 5 fathoms in the northeast. This excessive shoaling is probably due to the silting of the Yodo River complex, which has also constructed the plain on which Osaka is sited. Osaka Bay has contact with the Pacific Ocean through the Kii and Kitan straits. However, little swell enters the bay because the entrance is blocked by two islands. If this were not the case much of the material from the Kino River would have been carried through and along the southeastern shoreline of the bay. Both this boundary and that to the north display littoral drift towards Osaka. The construction of break-waters along these well-developed coastal plains have been designed to cope with this net movement.

DELTAS

Whilst the transport of sediment along the shoreline is accomplished by waves alone, the zones where the bulk of it is put under their influence is equally influenced by the supplier, namely the rivers. Accretion at river mouths is controlled as much by the flow in and out as by the waves arriving at this part of the coast. It would seem appropriate therefore to treat this problem separately from the normal shoreline. Emphasis will be placed on the determination of the major forces that produce the distinctive character of these zones.

Even though they may extend inland for some distance they still provide the coastal engineers with problems, since the waterways are subject to tidal action and navigability is of primary importance. Should it be possible to solve the major factors in deposition for any one or a series of deltas, it may be possible to complement natural processes so that problems of flooding and of navigation can be reduced. Even if this should not prove economical, means might present themselves for predicting future tendencies respecting silting or erosion.

The magnitude of the siltation problem can be gauged from figures quoted by

Stoddart [18] that the twenty-one largest drainage basins of the world discharge 60% of the solid load of all rivers of the world, even though they constitute only 30% of the total land area. These sediment loads range from 10 to 1800 million tons per annum and not necessarily in proportion to their water discharges.

The importance of deltas in the life of mankind needs no stressing. Civilisations were founded on them and populations have been concentrated on them ever since. Man has required the vegetation that grows in soil for his basic needs. But besides supplying food, deltaic plains provide sources of water and economical means of communication. However, these zones of great human activity suffer some of the greatest natural disasters, in the form of flooding from the river system or of inundation from the sea when storm surges occur. The economic importance of deltas varies from continent to continent, but as the population of the world increases at the present rapid rate, greater demands will be placed on these food-producing areas. The highest concentrations of population exist in the Ganges delta and The Netherlands, where large sums are being spent in reclaiming or protecting extensive areas from the sea.

The reader is referred to geological and geographical treatises for classifications of deltas, and discussion of specific features [28–34]. Here the pragmatic view will be expressed for the engineer who is concerned with maintaining navigable channels, inhibiting salt-water intrusion, building land areas above flood levels and studying strata for their load-bearing capacities. However, firm answers cannot be given to such complex questions, there is much work required before this can be done. But at least an indication can be given of fruitful lines of research or useful data collection [35].

The comprehensive system, from which a delta emerges, consists of:

(*1*) the catchment; (*2*) the river system; (*3*) the delta region; (*4*) the receiving body of water (generally the sea).

These can be considered as independent variables, together with the climatic variables of precipitation, temperature and wind, which affect all the above areas differently. The dependent variables in the regression analysis to follow are those associated with the delta itself, although rainfall, temperature and vegetation in the delta is part of the independent input.

Pattern-forming forces

Following are the variables thought to be of concern with comments on how they were included, or an explanation of their omission:

Catchment

Mean-annual rainfall. Values were taken from atlases and are therefore general in character.

DELTAS   115

Annual average temperature. Again atlases were used.

Area of mountains and plains. The catchment boundary was formed from atlases or hydrographic charts. The heads of tributaries helped in this direction. Estimates of river plains were gleaned from topographical colouring on maps. The first height division above sea level, generally 200 m, was considered to be river plain unless the next division extended over wide areas.

Topography. This expressed the mountains or plains as a percentage of the total catchment.

Perimeter. The length of the catchment boundary was measured on atlases by an opisometer.

Relief. This was the total range of elevation from sea level to the maximum elevation reached by the highest head of any tributary. This could be measured from colour divisions on an atlas (see Fig. 2-29).

Maximum length. This was the distance from the delta shoreline, in the general direction of the main river system, to the catchment boundary where the head of the main tributary was located (see Fig. 2-29).

Cyclonicity. The path of cyclones can provide information on the prevalence and distribution pattern of high intensity rainfall. Information on this topic was so scarce that its inclusion in the analysis was found impossible.

River system

Mean annual discharge. Since many of the deltaic rivers did not have discharge figures available, a correlation was carried out between catchment area, mean annual rainfall and discharge for a number of river systems. An equation was thus derived from which discharge could be computed.

Mean annual sediment yield. A similar correlation to above was carried out for this variable. The resulting equation included topography, perimeter, and relief of the catchment area as defined previously.

Length. The river length on plains and mountainous areas was measured with the opisometer. Since this information was conveniently available in atlases, an error arose due to their small scales. This was compensated for by comparing the measured value, with known lengths published in the literature for some rivers. It was

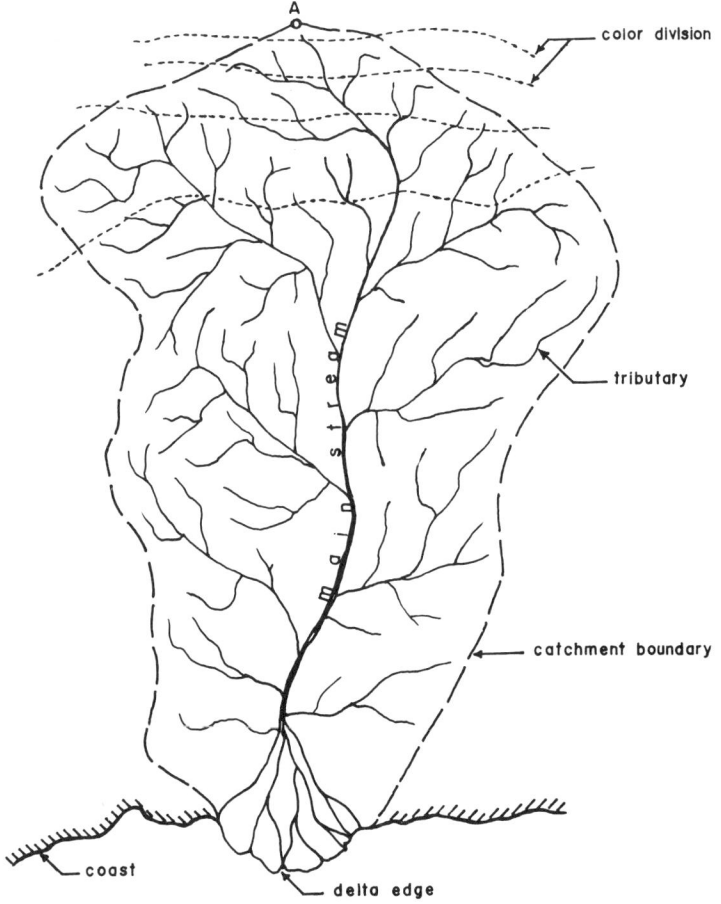

Fig. 2-29. Features of drainage basin for a large river system.

found that a 14%-addition was required to give a more exact value of length, which factor was used for all rivers.

Slope. By using the assumed limit of plain elevation (200 m) and the river length in the plain, a river slope could be determined. Values are expressed in percent to obviate decimals.

Delta

Length. The apex to sea length was measured, assuming the apex to be the point in the river where distributaries commenced and the shoreline limit where maximum protrusion occurred. (see Fig. 2-30). A length of projection was also recorded by

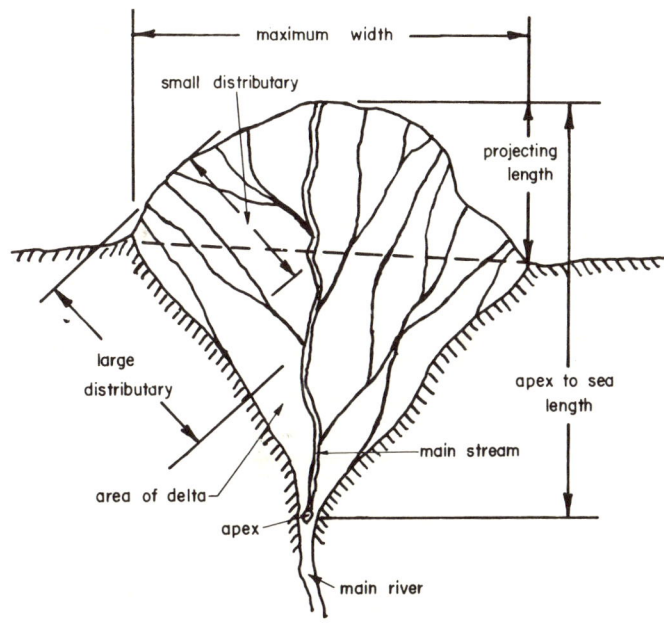

Fig. 2-30. Definition sketch of some delta variables.

drawing a line through the straightened coastline and measuring the normal distance as indicated in Fig. 2-30.

Maximum width. The section of delta having the widest distance between bounding distributaries provided the maximum width (see Fig. 2-30), which generally occurred near or on the coast alignment.

Area. The area bounded by the limiting distributaries and the shoreline was planimetered to give the delta area. The accuracy of this measurement was limited by the information provided in the charts or atlases.

Side topography. The angle between the bounding distributaries was measured. In some cases the apex section better represented a U instead of a V; these were omitted in the regression analysis.

Topography. The slope of the land from apex to sea was available for only 15 deltas [33]. It was assumed that these were measured land values and not river slopes.

Distributary system. Distributaries extending from the apex to the sea were counted, plus those termed small which were less than half the delta length (see

Fig. 2-30). The accuracy of these figures depends greatly upon the detail available in the atlases and charts. Not all distributaries reach the coast, so the number of mouths are counted. In the regression analysis the number of small and large distributaries did not appear significant as a single variable. Even the total number became significant only when used as a ratio with another variable.

Mean annual rainfall. This was taken from atlases. It could differ from the value recorded for the catchment and was likely to be more accurate due to the limited area.

Annual average temperature. As above, this was observed in atlases.

Vegetation. This type of information is also available in atlases, but a convenient way of quantifying it for the regression analysis was not found.

Presence of lagoons, lakes and marshes. A very low correlation coefficient was obtained for this variable as only the presence or absence of these features was noted. To obtain a meaningful answer the number or area of such water bodies should be used. Since these features are likely to depend strongly upon the type of sediment in the delta area, this information should be sought concurrently.

Sea

Tides. Tide tables were used, from which the mean spring tidal range was assessed.

Waves. Since wave action was considered to play a predominant role in the shaping of deltas, information on this aspect is as detailed as possible. A distinction was made between storm waves and swell for obvious reasons.

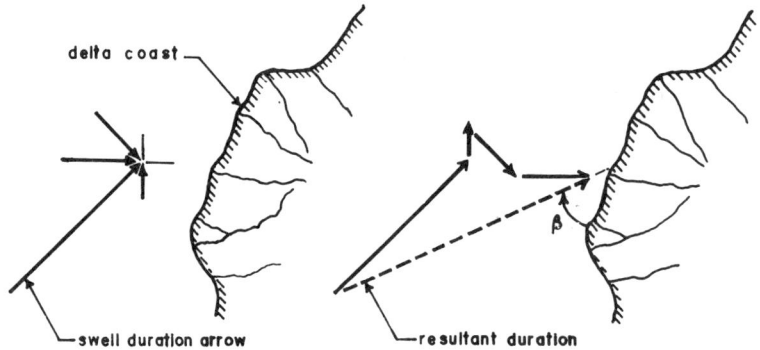

Fig. 2-31. Illustration of the swell duration polygon.

Statistical information on sea and swell, in the form of frequency roses covering sea areas of 5° latitude and longitude, was collected and integrated to an annual figure [12—14]. The percent durations of storm waves from all directions seaward of a delta were added, because the multi-directional nature of storm waves inhibits any strong longshore drift, as previously discussed. Swell, on the other hand, has a distinct direction of approach, the annual resultant of which was determined from a duration polygon as illustrated in Fig. 2-31. The approach angle β was assigned numerical values for inclusion in the regression analysis as follows: $0°-36°$: 3; $37°-72°$: 2; $73°-108°$: 1.

Continental shelf. The relatively shallow platform boundering the land masses, known as the continental shelf, has an outer limit where the slope increases significantly. The depth where this occurs has been accepted as the 100-fathom (600 ft.) contour, which limit has been used in this investigation. As noted, many times previously, more detailed hydrographic surveys have indicated that for the majority of shelves 65 fathom (400 ft.) is a more exact figure. In assessing the slope of this platform the depth of 600 ft. at the edge was divided by the width. Greater accuracy could be obtained by subtracting 30 ft. from this vertical measurement, since such a depth is quickly reached in the beach profile, which is parabolic in this zone.

Salinity. Marine atlases provided data on salinity in deltaic regions. This was available as a proportion of salts per 1000 parts of sea water. The values ranged from 15 to 39 and were used directly in the regression analysis. The majority of deltas had sea water salinities of 30 and more.

Ice. The incidence of ice cover in the sea can be a significant factor in preventing wave action and through increasing outlet velocities near the bed. The ice can also move material in various ways. However, the presence of ice was not noted in this study since it was likely to influence very few of the deltas under examination.

Regression analysis

For details of the regression analysis the reader is referred to the original work [35], only the salient points, results and conclusions will be presented here.

The independent variables were tested singly and in combination against the dependent variables, also singly and in combination. The independent variables giving the highest correlation for any given delta feature were then re-analysed to obtain a specific relationship. The features and ratios examined (see Fig. 2-30 and 2-31) were: (*a*) apex to sea length (L); (*b*) delta area (A); (*c*) maximum width of delta (W_{max}); (*d*) maximum width to number of mouths (W_{max}/M_n); (*e*) maximum width to number of distributaries W_{max}/T_D; (f) delta topography (D_T); (g) side topography (S_T).

The equations listed below for these dependent variables will not contain the constants derived, only the exponents of the independent variables. The percentage to the right of each equation is that of the total variance of the dependent variable which has been explained by the independent variables, it is the square of the correlation coefficient expressed as a percentage.

Apex to sea length (L):

$$L <= \frac{Q^{0.3}}{(R_s C_s)^{0.25}} \qquad 77\% \qquad (2\text{-}2)$$

$$L <= \frac{Q^{0.35}}{(R_s C_s T_i)^{0.22}} \qquad 77\% \qquad (2\text{-}3)$$

where Q is the average annual river discharge; R_s is the river slope on the plain to sea; C_s is the slope of the continental shelf; T_i is the mean spring tidal range.

These equations indicate that the apex to sea distance varies directly with the water discharge as might be expected. The inverse relationship with river slope on the plain and the continental shelf slope also appears natural, since the flatter these are the further material will spread from the bifurcation point. Also, the smaller R_s the greater proportion of the load is fine material which discharges further out to sea to finally elongate the delta. The smaller the tidal range the longer the apex length since accretion can occur only to a smaller height. This means that for a given volume of sediment a greater horizontal area must be utilized for delta formation. A large tidal range will result in greater mixing with the fresh-water outflow of the river and hence help disperse the sediment as offshore shoals rather than as banks upstream from the distributary mouths. Neither C_s nor T_i were significant as variables singly.

Delta area (A):

$$A <= \frac{Q^{0.8} T_e^{2.3}}{C_s^{0.4}} \qquad 79\% \qquad (2\text{-}4)$$

$$A <= \frac{Q^{0.7}}{(R_s C_s)^{0.4}} \qquad 78\% \qquad (2\text{-}5)$$

where T_e is the average annual temperature and other variables as before.

Similar arguments hold for Q, R_s and C_s as for apex length above, to which the factor of temperature is added, with an extremely high exponent. However, it was not significant as an individual variable. The influence of temperature on a deltaic area is manifold. It is an indicator of latitude. High values can be associated with dense vegetational growth in a well watered location and as a delta. This growth impedes the transport of sediment out to sea and hence assists in the growth of land, hence the delta. High sea temperatures are also associated with greater salinity, which in turn produces accretion well inland from the delta mouths. This helps in horizontal spreading of the deltaic zone. It is seen in the first equation above that increased temperature and flatter continental slope, which both tend to produce high water temperatures, tend to increase the delta area. Perhaps one of the greatest influences of higher temperatures is the precipitation of calcium carbonate from the seawater. This can provide prodigious volumes of fine sediment either directly or through debris from marine fauna that produce shell so readily in these conditions. This material is transported up river by the inward bed current associated with the saline wedge as discussed in the section on estuarine siltation.

Maximum width of delta (W_{max}):

$$W_{max} <= \frac{T_e^{2.4}}{(S_e/S_w)^{0.4}(R_s)^{0.6}} \qquad 78\% \qquad (2\text{-}6)$$

$$W_{max} <= \frac{T_e^{1.85}(S_w S_{wd})^{0.3}}{R_s^{0.6}} \qquad 78\% \qquad (2\text{-}7)$$

where S_e is the duration of storm waves; S_w is the duration of swell; S_{wd} is the swell direction (increases with obliquity to coast) and other variables as previously.

Factors tending to increase the width are temperature plus swell duration and direction. The temperature influence has been discussed above and need not be repeated. The greater the incidence of swell the more energy is available for spreading the deltaic material along the coast. This spreading is enhanced by the obliquity of the waves, which it will be recalled received higher numerical value the greater the angle from the normal to the coast.

Factors detracting from the width of the delta are river slope and ratio of sea to swell. Greater river slope implies higher river velocities and hence an elongation of the delta rather than a widening. Although R_s is in the denominator of equations relating it to L and again in the above equations relating it to W_{max}, the exponent in the latter is larger. This indicates a stronger influence on width than length. River slope, as an independent variable, had a high level of significance in L, A and W_{max}. A large proportion of storm waves to swell will tend to remove material offshore which is not returned by the swell. Also the locally generated storm waves will be accompanied by heavy rainfall and runoff. The resulting strong currents, in consort with the rough seas, will spread sediment out as submarine shoals rather than beach area.

Maximum width to number of mouths (W_{max}/M_n):

$$\frac{W_{max}}{M_n} <= Q - S_e + S_w \qquad 63\% \qquad (2\text{-}8)$$

An increase in this ratio is produced by an enlarged river discharge, greater swell duration but decreased storm wave incidence. The width per mouth should grow with discharge as this will tend to keep a few major distributaries open to the sea. Predominance of locally generated storm waves maintains more river mouths open. Swell, on the other hand tends to form bars across the mouths and spread the delta for any given numbers of deltaic openings.

Maximum width to number of distributaries (W_{max}/T_D)

$$\frac{W_{max}}{T_D} <= Q \qquad 81\% \qquad (2\text{-}9)$$

$$\frac{W_{max}}{T_D} <= \frac{T_e^{2.07}}{R_s^{0.48} S_a^{1.47}} \qquad 65\% \qquad (2\text{-}10)$$

where S_a is the ocean salinity.

This ratio is a measure of the division of the delta into distributaries. As the river discharge is greater so the flow is confined to fewer passages. Smaller velocities tend to increase the number of blockages which result in more distributaries. An increase in temperature has a strong influence on the width of land per distributary, as could be expected from arguments presented earlier. A decrease in river slope also increases this width, as it implies lower river velocities and easier deposition as bifurcation commences. An increase in salinity makes for a greater number of distributaries as the saline wedge penetrates well upstream and so causes blockages which result in new outlets.

Delta topography (D_T):

$$D_T <= R_s \qquad 74\% \qquad (2\text{-}11)$$

Delta slope is directly related to river slope as might be expected. This relationship is the least conclusive of all the equations as only 15 samples were available for analysis.

Side topography (S_T)

$$S_T < = \frac{C_s^{0.3}}{S_a^{1.7}} \qquad 45\% \tag{2-12}$$

This equation has very low determination compared with the previous relationships. The angular measure for side topography is closely associated with delta width and delta length and hence should indicate similar forces predominating. An increase in continental shelf slope increases this angle. This is to be expected when it had previously decreased apex length. A larger tidal range also increased the angle, since it also decreased the apex length. Salinity has a strong influence but appears to act in reverse to the trend expected. It could be expected that greater salinity, with its concomitant well-developed salt water wedge would cause accretion well upstream and hence a widening of the angle S_T. However, the bifurcation may be up beyond the reach of the saline wedge. Other factors not included explain more than 50% of side topography.

In the above equations the importance of temperature should be noted, as it influences many aspects of delta formation. It increases area, width and the width per distributary. High average temperatures can dry out material which has been deposited and so fix it for following floods and tidal cycles. Sand can be dried out to be blown into dunes, whilst silt cakes into a solid mass at around high-tide level. High temperatures also promote lush vegetation in such well-watered zones as deltas. High sea temperatures aid the formation of carbonates through precipitation and action of fauna. The accompanying high salinities also cause flocculation of the wash load issuing from the river, which is forced well upstream by the saline wedge. Any future analysis along the above lines should include latitude as an independent variable.

Discharge is directly related to sediment yield, an increase in either will expand the delta in some way. A decrease in river slope also strongly increases the size of the delta. Such mild slopes will be associated with larger catchments that contain an extensive river basin of sedimentary structure.

The variables associated with the sea itself appear to be less influential. However, quantifying some of these for regression analysis requires further thought. The input from the sea, in the form of waves, tides and perhaps ice, certainly control specific deltaic mouths and are of great engineering significance.

Besides the latitude, noted above, other variables that should be included in future research are sediment size, incidence of local winds (for their drying effect), quantifying of vegetation growth, fluctuations in river discharge, occurrence of cyclones and storm surges, rate of littoral drift towards or away from the delta and the duration of sea ice over the year.

There is little doubt that a concerted attack on the delta problems of the world would provide an enormous reward in the saving of life and property. It is not being suggested that man can cope with Nature at her worst, but if trends can be detected man may be able to emulate her and so have predictable tendencies.

PROBLEMS

1
Describe the sequence of events when a section of coast suffers a decreased replenishment of sediment being removed by an annual sequence of swell and storm waves.

2
Draw two headlands and the sedimentary beach in equilibrium between them, when waves approaching the headland alignment are at 45° to it. Trace a number of wave orthogonals from the upcoast headland onto the curved section of the bay periphery. Why can it be concluded that both diffraction and refraction are involved in the bay formation?

3
By using hydrographic charts determine the constant α in the logarithmic spiral curve of a number of bays in close proximity on a coast. Check the status of equilibrium and give reasons for the existing condition from an assessment of sediment supply to the bays. If not in equilibrium, draw the recessed beach line for such a condition.

4
From hydrographic charts or other sources find consecutive bays which are oriented at angles to each other, as depicted in Fig. 2-11A,B. Determine the offshore approach angle of the persistent swell and draw the wave orthogonals to assess β. Measure α and check the degree of stability. For bays not in equilibrium, draw the bay outline for this limit when sediment supply has been cut off.

5
From the upcoast ends of two physiographic units, each from separate continental margins, find bays which are in equilibrium or nearly so.

6
List various tests by which the direction of net sediment movement along a coast can be judged. Try and add to the number discussed in this chapter.

7 Select a specific portion of a continental margin and inspect the hydrographic charts covering same. From these determine the direction of net longshore drift, to verify or otherwise the results given in Fig. 2-16 to 2-21. Check where possible in the literature proof of this drift by erosion or siltation reported therein. Also inspect sea and swell data of the adjacent sea and note the correlation between wave energy vectors and the resultant sediment motion.

8
Choose a section of a continent or large island and study its geomorphology along similar lines to that carried out for Japan in this chapter. At specific points along the coast indicate future trends of erosion or siltation. What other benefits can accrue from such an analysis?

9
From an examination of a lengthy physiographic unit of a continent measure the width of the

PROBLEMS

continental shelf (to the 100-fathom contour should suffice) at intervals along it. Explain the reasons for the variations found in terms of the sediment supply and wave climate.

10

The loss of sediment from the island structure of Japan is a serious economic consequence. From the geomorphological study made in this chapter, indicate where structures might be considered to promote accretion onshore of reasonable magnitude.

11

Examine the technical literature available on deltas to ascertain in what manner engineers may improve the economic management of them, by reducing costs of navigation, water supply etc. and minimizing the effects of natural catastrophies which normally befall such areas.

12

Find references to regression analyses conducted on deltas, or river systems generally, in respect to sediment motion and effects therefrom. What engineering benefits can be effected from such statistical studies?

13

Examine the literature for comprehensive studies conducted on deltaic river systems. Choose one and summarize the conclusions reached. Discuss any factors, such as wave climate, that may have been omitted or not given sufficient weight.

REFERENCES

[1] A.A. Meyerhoff, 1970. Continental drift: implication of paleomagnetic studies, meteorology, physical oceanography, and climatology. *J. Geol.*, 78: 1–51.

[2] J.L. Davies, 1964. A morphogenic approach to world shorelines. *Ann. Geomorphol.*, 8: 127–142.

[3] R. Silvester, 1960. Stabilisation of sedimentary coastlines. *Nature (London)*, 188(4749): 467–469.

[4] J.N. Jennings, 1955. The influence of wave action on coastal outline in plan. *Austr. Geogr.*, 6: 36–44.

[5] J.L. Davies, 1958. Wave refraction and the evolution of shoreline curves. *Geogr. Stud. (London)*, 5: 1–14.

[6] R.K. Greswell, 1957. *The Physical Geography of Beaches and Coastlines*. Hulton Educ. Publ., 128 pp.

[7] W.E. Yasso, 1965. Plan geometry of headland-bay beaches. *J. Geol.*, 73: 702–714.

[8] R. Silvester, 1970. Growth of crenulate-shaped bays to equilibrium. *Proc. ASCE*, 96(WW2): 275–287.

[9] N. Vichetpan, 1971. *Development of Crenulate-shaped Bays*. Thesis no. 280, Asian Inst. Technol.

[10] R. Silvester, 1962. Sediment movement around the coastlines of the world. *Proc. Conf. Inst. Civil. Eng., (London), 1962*: 289–315.

[11] R. Silvester, 1968. Sediment transport, long-term net movement. In: R.W. Fairbridge (Editor), *Encyclopedia of Geomorphology*. Reinhold, London, pp. 985–988.

[12] *Monthly Meteorological Charts:* Indian Ocean, Mo.519, 1949; Western Pacific, Mo.484, 1956. H.M.S.O., London.

[13] *Atlas of Sea and Swell Charts:* South Atlantic, Publ.799B, 1948; Western Pacific, Publ.799CE, 1963; North Eastern Pacific, Publ.2990, 1963. U.S. Hydrogr. Off., Wash., D.C.

[14] *Oceanographic Atlas of the North Atlantic, Section IV Sea and Swell.* U.S. Naval Oceanogr. Off., H.O. Publ.700, 1963.

[15] J.T. McGill, 1958. Map of coastal landforms of the world. *Geogr. Rev.,* 48: 402–405.

[16] R. Silvester, 1963. Design waves for littoral drift models. *Proc. ASCE,* 89(WW3): 37–47.

[17] R. Silvester, 1965. Coastal sediment movement – some fundamental problems with discussion of research support. *J. Inst. Eng., Austr.,* 37: 311–323.

[18] D.R. Stoddart, 1969. World eosion and sedimentation. In: R.J. Chorley (Editor), *Water, Earth and Man.* Methuen, London 588 pp.

[19] F.P. Shepard, 1963. *Submarine Geology.* Harper and Row, London 348 pp.

[20] D.L. Leet and S. Judson, 1958. *Physical Geology.* Prentice Hall, Englewood Cliffs, N.J., 466 pp.

[21] H.W. Menard, 1961. Some rates of regional erosion. *J. Geol.,* 69: 154–161.

[22] M.D. Hayes, 1967. Relationship between coastal climate and bottom sediment type on the inner continental shelf. *Mar. Geol.,* 5: 111–132.

[23] P.H. Kuenen, 1950. *Marine Geology.* McGraw-Hill, New York, N.Y., 568 pp.

[24] R. Silvester, 1966. Sediment transport and accretion around the coastlines of Japan. *Proc. 11th Conf. Coastal Eng.,* 469–488.

[25] G.H. Smith and D. Good, 1943. Japan – a geographical review. *Am. Geogr. Soc., Spec. Publ.,* 28.

[26] M. Minato, M. Gorai and M. Hunahashi, 1965. *The Geologic Development of the Japanese Islands.* Tsukiji Sjokan, Tokyo.

[27] M. Hom-ma and K. Horikawa, 1960. Coastal protection works and related problems in Japan. *Proc. 7th Conf. Coastal Eng.,* 904–930.

[28] Tj.H. van Andel, 1967. The Orinoco delta. *J. Sed. Petrol.,* 37: 297–310.

[29] P.C. Scruton, 1956. Oceanography of Mississippi delta sedimentary environment. *Am. Assoc. Petrol. Geologists,* 40: 2864–2952.

[30] V.N. Mikhailov, 1966. Hydrology and formation of river mouth-bars. *Proc. UNESCO Delta Symp., Dacca, 1966:* 59–64.

[31] C.C. Bates, 1953. Rational theory of delta formation. *Am. Soc. Petrol. Geologists,* 37:2119–2162.

[32] V.N. Nagaraja, 1966. Hydrometeorological and tidal problems of the deltaic areas in India. *Proc. UNESCO Delta Symp., Dacca, 1966:* 115–119.

[33] A. Volker, 1966. Tentative classification and comparison with deltas of other climatic regions. *Proc. UNESCO Delta Symp., Dacca, 1966:* 175–179.

[34] I. Douglas, 1971. Man, vegetation and the sediment yield of rivers. *Nature,* 215: 925–928.

[35] R. Silvester and C.R. de la Cruz, 1970. Pattern forming forces in deltas. *Proc. ASCE* 96(WW2): 201–217.

[36] R.W. Fairbridge, 1947. The geology and geomorphology of Point Peron, Western Australia. *J. R. Soc. West. Austr.,* 34: 35–72.

Chapter 3

COASTAL DEFENSE

This chapter should be one of the most important sections of both *Coastal Engineering, 1* and *2*. The discussions of wave generation and propagation, of experimental observations regarding sediment transport, of forces on structures etc., all lead to the end result of some marine structure. In spite of wave action being ubiquitous, the actual wave climate can vary drastically. The task of the coastal engineer is to determine the capacity of the locally generated waves or swell to move material, and then to design structures that will impede it, expedite it, or maintain it in steady motion.

Some presentations on coastal defense consist, in the main, of detailed reports on certain structures, emphasis being given to the foundation conditions and the materials used. Papers are published soon after completion of the new structure, so that longer-term trends due to its presence cannot be reported. There is a need for follow-up articles, some years later, which evaluate honestly the success or otherwise of the installation. In this way mistakes will not be duplicated.

That the state of the art in respect to coastal defense is not well advanced can be gauged by the statement in "Effective Use of the Sea", by a panel on oceanography of the United Stated President's Science Advisory Committee. This was quoted in *Coastal Engineering I,* but warrants repetition here.

In the following discussion the general outline of various problems will be presented. Details of structures will be noted only in-so-far as they require dissipation or reflection of wave energy to serve a given purpose. Present practices will be critically examined, but specific failures will not be cited for obvious reasons.

The term "coastal defense" refers to protection of the coast from the action of the sea, or protection of a man-made facility, which may take the form of a dredged channel. The title implies defense against erosion or siltation, either of which can prevent a marine structure from fulfilling its purpose. Both tendencies emanate from the longshore transport of material, either within the surf zone or beyond it. The coastal engineer must solve this problem by either impeding or preventing the drift, or by maintaining its motion past his facilities. In accomplishing the former he must remain cognisant of the effects to adjacent downcoast areas, which may suffer erosion from the interception of sand supply, for example. To succeed with the latter he must prevent deposition until the sediment is downcoast of his structure or channel.

It may be thought that beach erosion is a modern problem, amplified by the pressure of public and commercial use of the seashore. It is at least 60 years old as gauged by the following quotation from Carey [1] in 1907:

"The first thought of the authorities in every district naturally is to safeguard the stability of the foreshore of that district; thus interminable complications arise, and almost every frontager or municipality is openly or covertly at war with the interests of his neighbour, who depends for the existence of his foreshore upon the regular travel of that circulating medium of defense provided by nature."

Utilisation of coastal strips, for commercial or other purposes, generally involves the stipulation of a static shoreline. The natural fluctuations of the beach, which were previously of little consequence, now become an economic hazard. Also, the structures installed for employment of the beach zone, or the waters adjacent to it, may produce their own erosive effects, through reflecting waves for example. Other human factors making for greater erosion could be dredging of navigation channels across the littoral zone, and the interception of sand supply from rivers being harnessed for flood control and irrigation.

Any deleterious effects of a remedial measure may not be noticeable for some years. This is because the wave climate can fluctuate from year to year and because the beach zone has reserves which cover immediate contingencies. The offshore zone must be deepened and steepened before the full demand for beach material is felt. This landward shift of the complete beach profile constitutes a vast quantity of material, which must be replaced fully if the original water line is to be maintained.

Unlike erosion, accretion against a man-made obstacle is very swift, since much of the action takes place within the surf zone. Perhaps it is this speedy reaction that gives engineers the impression of swift success with erosion-abatement structures. However, for the new water line to be stable, it is necessary for the offshore profile to be built up to, and maintained, at the equilibrium level commensurate with the inner zone. If the replenishment problem has not been solved "in depth", long-term stability may not be effected.

GROINS

A groin is a structure of moderate dimensions which runs from the beach into the sea; it is generally at right angles to the shoreline. The action of the groin is to intercept and hence accumulate material on the upcoast side, during periods of oblique persistent swell. Whilst material is being so accreted, the beach downcoast of the groin will suffer erosion, due to the non-replacement of sediment being removed by the swell. After the water line has built out to the toe of the groin, or nearly so, material will again be fed to the downcoast side, to maintain a slightly receded shoreline. (See Fig. 3-1A.)

During a subsequent storm sequence waves will arrive from a variety of direc-

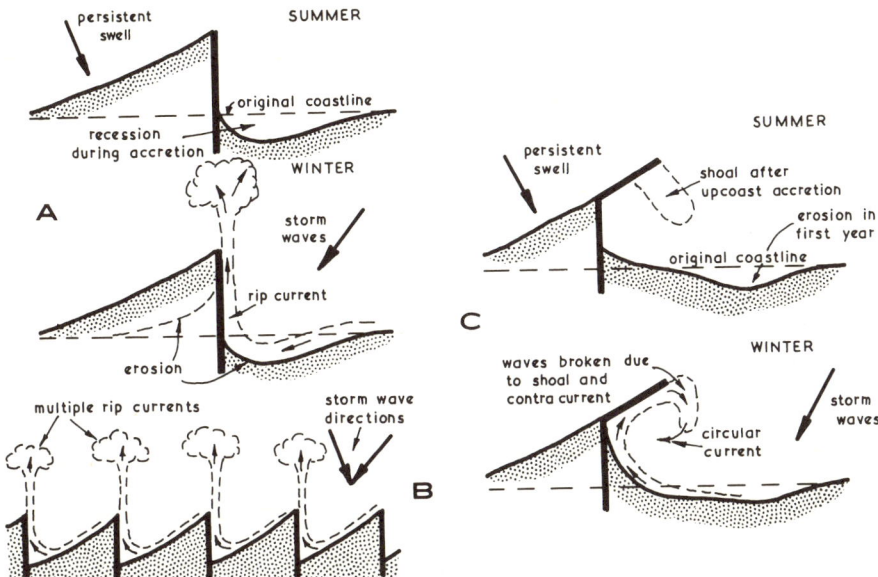

Fig. 3-1. A. Shoreline changes for groin normal to beach; B. Rip-current development in a groin field; C. Influence of Y addition to normal groin.

tions. As illustrated in the winter version of Fig. 3-1A, these storm waves could readily approach from the seaward quadrant alternate to that of the persistent swell. In this event a rip-current is triggered on the denuded side of the groin, which transports material much further seaward than the offshore bar constructed by the storm waves in the normal manner. When returning such sediment to shore, the oblique swell will transport it further downcoast than the closer offshore-bar material. In this way, the groin may expedite transmission of littoral drift when, in fact, it was installed to impede it. This acceleration of erosion can be multiplied by the use of a field of groins, as depicted in Fig. 3-1B. The many miles of Danish coast "protected" by groins at intervals of approximately 300 yards do not seem to be achieving the stabilisation desired according to Lehnfelt and Svendsen [2]. These authors conclude: "The groynes can, under favourable conditions, protect the bed of the sea from erosion as far as their extremities but not much further. Beyond the end of the groyne, erosion continues and it is evident that after a certain time this erosion will attack the groyne and later the land itself."

Suggestions have been made to angle the groin downcoast a little, or to add an oblique leg to its toe, in order to protect the heel of the structure and prevent the formation of rip-currents [3,4]. The use of the half Y shape, as illustrated in Fig. 3-1C, has the effect of directing any rip-current against the incoming storm waves producing it. This contra stream helps dissipate the waves, whilst a circular current is generated which produces an underwater shoal as indicated in the figure.

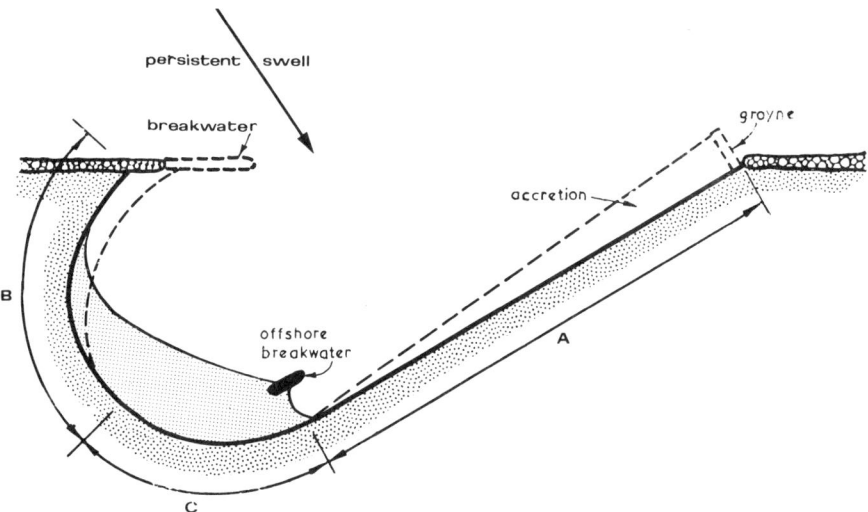

Fig. 3-2. Methods for beach accretion around a crenulate-shaped bay.

Subsequent swell during the summer season will sweep this sediment to the shore and help maintain a spit or shoal in the lee of the outer arm. All this action helps protect the heel of the groin. However, in an excessively erosive situation, even this design feature will not save the structure from being exposed and finally isolated from the shoreline.

The literature abounds with articles on the use of groins to prevent beach erosion. Many groins are installed as replacements for those previously found ineffective, for reasons which are directed more at the type of structure, rather than the suitability of the site for such "protective" measures [4]. Consider, for example, the crenulate-shaped bay of Fig. 3-2, remembering that this coastal feature is the rule rather than the exception where coasts are suffering denudation [5]. If the bay is proceeding to its equilibrium shape, from erosion caused by lack of replenishment, longshore drift will still be taking place around the periphery. In this event any groins placed in section A of Fig. 3-2 will cause the desired accretion, whereas any placed in the curved zones of B or C will not prevent coastal recession in the long run. In the short term they may maintain the beach line. But the offshore area will continue to erode and so produce an instability, which will result in a collapse of the shoreline and structure during a particularly fierce storm sequence.

As illustrated in Fig. 3-2, the best means of building up zone B is to extend the headland with a breakwater or reef. This displaces the point around which the waves must diffract and refract, so that their new crest alignment causes accretion. Sometimes this is an undesirable development when zone B is a harbour being protected by the breakwater [6]. Zone C is a more difficult area to protect, although the suggestion depicted in Fig. 3-2, of an offshore breakwater, could serve to break the bay into two bays, the equilibrium shapes of which could be determined by procedures already outlined.

There is no doubt that the simple decision to install a groin needs to be taken in the context of a large length of coastline. It needs to be taken by an engineer who is aware of the forces involved. This may appear axiomatic, but Carey's story of 1907 is still applicable in 1973 [1] : "Added to these considerations is the fact that every longshoreman considers himself an expert in coast defence. Every boatman considers himself aggrieved by the carrying out of works which interfere with the foreshore, and may thus add difficulty or danger to the exercise of his calling. Small-town and district councils can seldom be induced to look ahead and provide the means of defence against the inevitable forces of attack. When they do so they are often denounced by the indignant ratepayer for extravagance. I heard a case the other day of a local Board which had spent money in building groynes, 'and now,' said a sapient rateplayer, 'look at them, sir, they are covered with sand and shingle and completely buried, and our money is wasted'."

HEADLAND CONTROL

Since crenulate-shaped bays are formed by persistent natural forces, when sedimentary coasts exist in the presence of headlands, they would appear to serve as a good engineering example for man to follow. As noted in previous chapters, the bay in full equilibrium prevents longitudinal movement altogether, except for minor seasonal fluctuations. These transient transfers of material alongshore may be caused by batches of sediment being passed through the bay, or by annual changes in the direction of the resultant swell or storm-wave energy. Even without equilibrium the shape of the bay has been shown to be predictable and forecasts made possible in the event of changing wave or sediment supply. Should new headlands be established by man, the outline of the resulting bays can be determined for the limiting condition of non-replenishment of littoral drift.

Because of the limited influence of groins on the overall accretion or erosion of a coast, it would seem that greater widths of the beach profile must be influenced in order to stabilize it. Such a protection scheme involving man-made headlands, must encompass substantial lengths of coastline. An eroding straight coastline could be stabilized by offshore breakwaters. By displacement seawards of sections of coastline, so a series of bays can be formed as depicted in Fig. 3-3A. Since these pseudo-headlands are required to intercept longshore drift by the formation of a tombolo or spit, they would have to be extremely long if located offshore any distance. At such depths offshore, also, the provision of such headlands could prove uneconomical, unless a reef or shoal were already present to serve as a foundation. For these

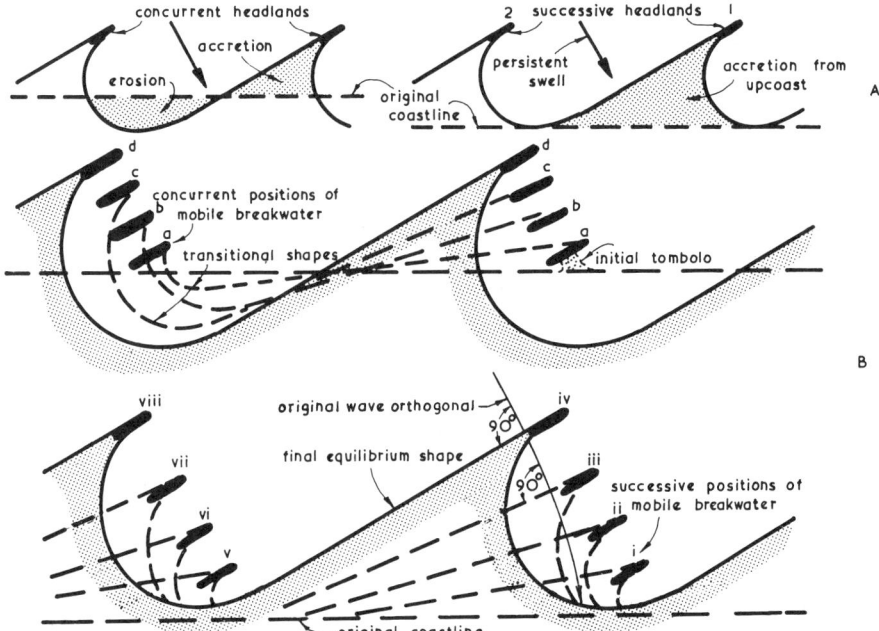

Fig. 3-3. A. Crenulate-shaped bays formed by offshore breakwaters; B. Progressive approach to stable bay by use of mobile breakwaters.

reasons a progressive approach is suggested as in Fig. 3-3B. Either fixed or mobile offshore breakwaters could be located successively at *a, b, c* and *d*, or at *i* through to *viii* if minimum erosion of the initial coastline were desired. Research is needed into a suitable form of breakwater that could be shifted periodically seawards as accretion took place [7].

The dimensions of the bays envisaged (half to two miles between "headlands") require a comprehensive coastal protection scheme and a substantial change in beach alignment. Such projects would require the cooperation of a number of administrative bodies, plus a sharing of costs. However, with the large losses incurred spasmodically, if not repeatedly, by coastal councils through beach erosion and property damage, the insular and haphazard approach to coastal defense must be replaced by projects of a scale to emulate Nature herself. Dunham [8] has suggested long impermeable groins to serve as headlands. Since these are intended essentially to prevent further longshore movement of sediment the legal implications are more important than for smaller and less effective structures [9]. This is an added reason for planning coastal defense measures on a wider financial and legislative base.

The headland approach has the great advantage of providing a more stable beach zone. Because the waves approach the equilibrium bay boundary normally, material remains sensibly in the same place. Any removed to an offshore bar will be returned directly onshore. Once the reserve of sand required for this purpose has been assessed from experience, the remaining beach may be utilized to the limit, in the knowledge that long-term erosion will not occur.

The above discussion relates particularly to sandy coasts where material can be readily accreted by moderate reductions in wave energy and where the seabed serves as a good foundation for even heavy structures. In the case of mud being the predominant content of the longshore drift, alternative means may have to be found to accrete it. Since the bed may not be capable of taking loads of rock-fill or other-heavy structures, consideration must be given to floating breakwaters.

One form of this could be a long floating groin extending from the shore to a long distance offshore. Because the mud makes for a very flat beach profile, such large distances need not entail great depths at the seaward limit. A structure which could prove economical is depicted in Fig. 3-4. It consists of a series of floats connected together by flexible couplings, which are also attached to piles on either side of the groin. Such piles are easy to install in shallow mud conditions. The floats, some 30 ft. in length, could consist of reject petroleum drums welded together and adequately protected with bitumen. They could be filled to near-capacity with fresh water so that they would be partly submerged. To each side

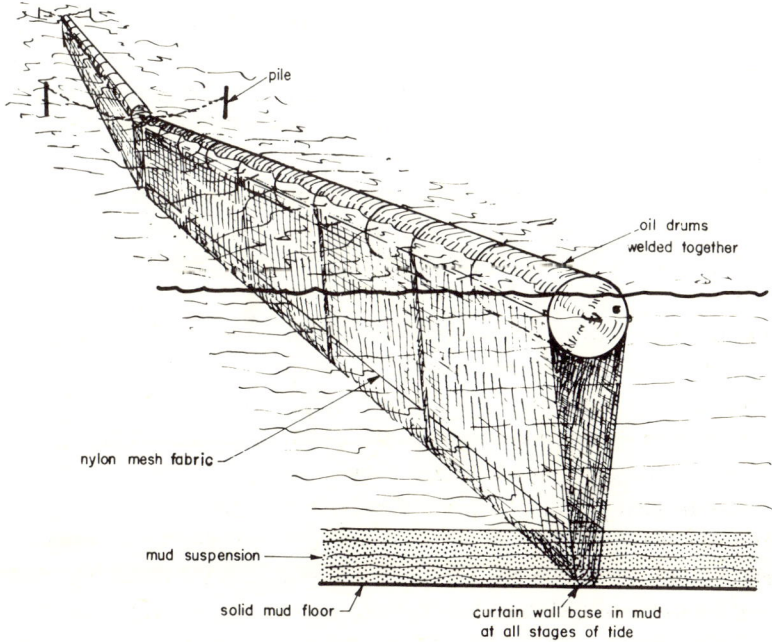

Fig. 3-4. Suggested floating groin for silty coastline.

could be attached the ends of suitable nylon mesh fabric, that extends down to the soft mud to form a double curtain wall. This could be of such a length to remain in the liquid mud at the bed even at high tide. It would sag into it at other stages of the tide. Any longshore current carrying suspended matter would be impeded by such a wall and so promote accretion.

An offshore breakwater facility which dissipates waves could consist of bamboos and logs, together with nylon mesh, depending upon the availability, transportability and type of labour force in an area. Another factor in the choice of material for floating structures is the economic value of it to local fisherman, who could not be continually policed in respect to removing desirable elements.

The references [10] to [22] should help the reader commence a survey of the problems attached to the design of floating breakwaters. The major problem is to attenuate the bulk of the wave energy in the lee of the structure. The waves then penetrating behind it from either end will sweep material towards the centre and so produce the desired shoal or even tombolo. It is not even necessary for such breakwaters in echelon to be placed closer than one wave length apart for the required action to be achieved.

INLET ENTRANCES

Inlet entrances will here refer to mouths of rivers or openings from lagoons to the sea. They may or may not have tidal inflow and outflow, but stratification is excluded from this discussion. The major problem for the coastal engineer is that of longshore drift [23], particularly when the tidal range is very small. For the entrances to be commercially useful, channels must be dredged through the bars or shoals which form across the mouth. Whilst fishing vessels find 10 ft. adequate, normal maritime traffic requires around 30 ft. depth, and large bulk carriers are demanding 60 ft. channels. The problem is to maintain these depths for the full width at all times.

Such a maintenance programme generally involves dredging, either on a spasmodic or continuous schedule. Material arriving from upcoast is discharged on the downcoast side of the opening to be carried away as littoral drift. This method is considered an economic cost that must be borne, but it can become excessive during storm seasons if equipment is damaged or lost.

An alternative to such pumping is a natural method of expediting the passage of material across the opening by means of the wave energy available. Since it will be oblique swell or storm waves that bring sediment to the mouth, the double application of this energy to the seabed should aid in its transfer across the deeper water of the opening. This suggests reflection of these oblique waves, so that angled wave trains or complete clapoti occur in the zone where sediment is required to be mobile.

INLET ENTRANCES

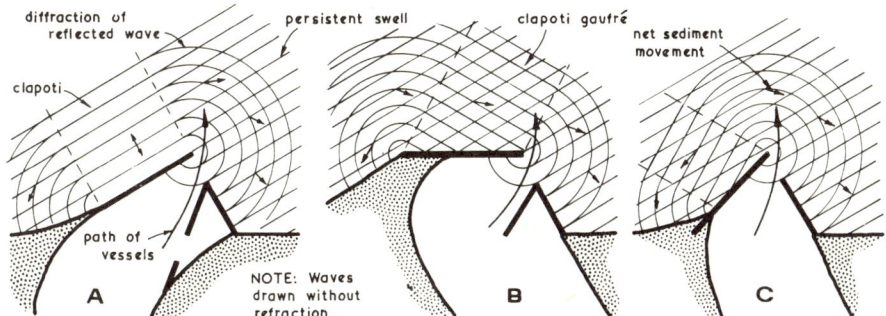

Fig. 3-5. Reflecting breakwaters for re-applying wave energy to sea-bed to expedite sediment transport.

Some suggestions for producing these conditions are depicted in Fig. 3-5. They consist of an upcoast groin which reflects the incoming oblique swell. As seen in the alternative schemes A, B or C, this reflection can produce complete clapoti, as in A, or clapoti gaufré (or angled wave trains) in B and C. At the boundaries of the wall the reflected waves diffract as they proceed seawards. The interaction of these diffracting waves and the incoming waves produces a short-crested system, which, as previously discussed, is extremely conducive to vortex generation and hence suspension of bed particles. The arrows indicate the vector sum of the net motion of sediment, based upon equal height of incoming and diffracted wave, with refraction omitted. Although this is a drastic simplification, the principle is self-evident. In practice the d/L ratio could greatly influence the result, as well as the littoral current passing across the mouth of the waterway.

The ideas presented above warrant investigation, since the greater the drift arriving the greater the wave energy available for transferring it across the mouth. Modelling of the process presents similar difficulties to any movable bed reproduction, but comparative studies would soon indicate the efficacy of the ideas.

The short-crested waves should not prove any more difficult for even small vessels to negotiate than the original long-crested swell. In fact, a small craft that can travel seawards at the speed of the reflected wave could sit in one of its troughs and experience smaller wave heights than if it has to proceed through the normal crest heights of the swell. The paths of vessels from the harbour refuge are suggested in the various layouts of Fig. 3-5.

In this context the usual location of river mouths within a bay should be noted. As discussed by Bascom [24], the curved zone of a bay receives the least wave energy, due to the diffraction and refraction suffered by the incoming swell. Hence, the beach berm is the lowest of the whole bay periphery. Any river in the vicinity that has been silted continually at the mouth will break through during flood at the weakest point and then continue to debouche in that zone, in the lee of the upcoast

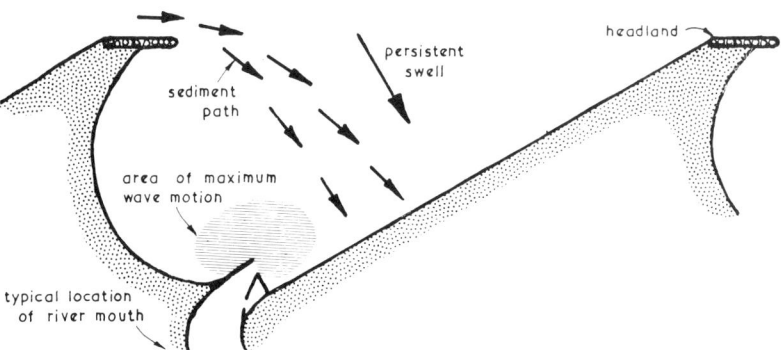

Fig. 3-6. Crenulate-shaped bay showing typical river outlet, zone of influence of reflected waves, and usual path of sediment across bay.

headland. Such a location is depicted in Fig. 3-6 where the offshore zone of maximum wave agitation is indicated for the reflecting groin of Fig. 3-5A. The path followed by a large proportion of the sediment across the bay is also illustrated. This has been proven in tracer tests [25], where a strong persistent swell was known to exist. Such a by-passing action depends greatly on the reflecting capacity of the upcoast headland and the general dimensions of the bay. However, when it is proved to exist, by tracer tests, for example, a great benefit ensues from the reduced sediment load to be transmitted past the river or lagoon mouth in the lee of the headland.

Where an inlet mouth has a strong tidal current in and out, the material feeding to it from the upcoast side is swept seawards or upstream. In the former case a crescentic bar is formed, as illustrated in Fig. 3-7, whose greatest width will be on

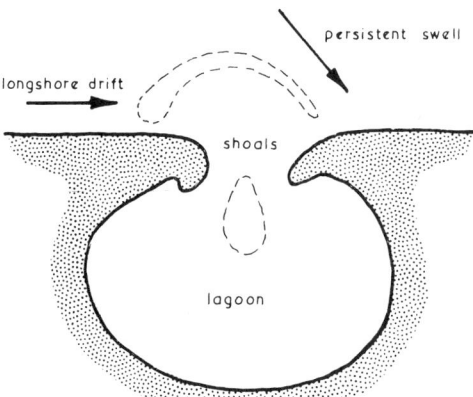

Fig. 3-7. Typical shoals formed at the mouth of a tidal inlet.

INLET ENTRANCES

the upcoast side because of the angle of the waves to its alignment. On the upcoast side the waves arrive almost normally to the elongated shoal, so that a wider expanse of shallow bed is required to transmit a given load. On the downcoast side the approach angle is more oblique, so that a given discharge of sediment can be effected over a narrower elevated path. Such a shoal occurs at the entrance to San Francisco Bay [26] and requires dredging to provide a channel for the large vessels at low tide.

For inlets which have a ready supply of sand that is flushed in and out by a tidal current, there is a relationship between the area of outlet opening and the tidal prism. O'Brien [27] has plotted these two variables for which the curve:

$$A = 1.62 \cdot 10^{-4} p^{0.905} \qquad (3\text{-}1)$$

appears to fit the data with an accuracy of ± 25%, where A = area of inlet for MSL (ft.^{2}), and p = volume of water in MHWS tidal prism (ft.^{3}).

Eq. 3-1 is depicted in Fig. 3-8, which applies to inlets with or without control jetties on one or both sides. Where such structures are used to limit the width, the width automatically increases the depth to maintain the appropriate area.

Assuming the duration of the ebb and flood are equal, the tidal prism can be expressed as:

$$p = \frac{AU_{max}T}{\pi} \qquad (3\text{-}2)$$

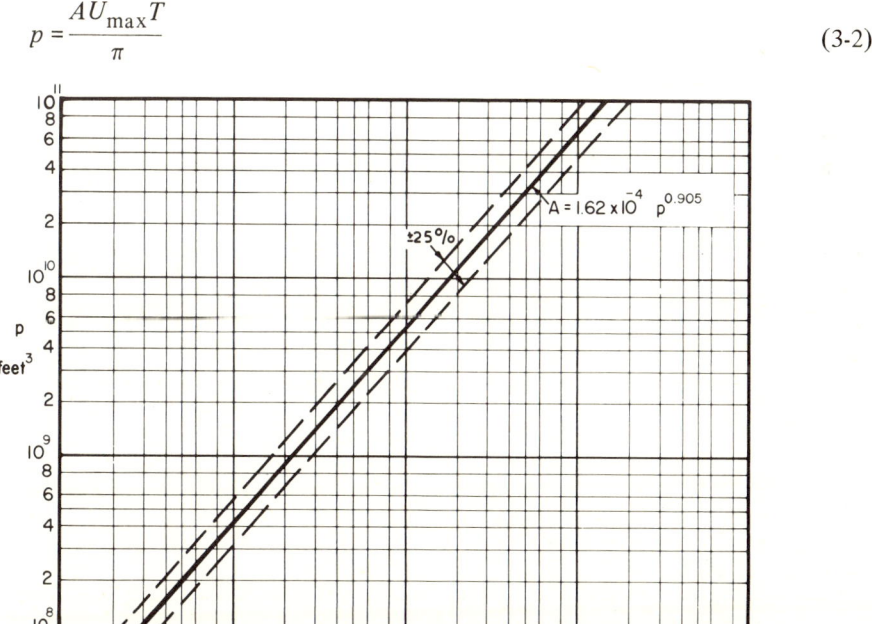

Fig. 3-8. Relationship for area of waterway through mouth of tidal inlet.

where U_{max} is the maximum mean velocity through the waterway, and T is the tidal period (= 44,700 sec).

Equating the areas in eq. 3-1 and 3-2 provides a value of expected velocities:

$$U_{max} = 0.434 p^{0.1} \qquad (3\text{-}3)$$

which, for the range of p in Fig. 3-8 gives U_{max} from 2.7–5.5 ft./sec, the larger velocities being associated with the larger prisms and hence the larger inlets.

For smaller inlets, where it may be assumed that the waterway section is a segment of a circle, a relationship can be derived as follows:

$$W^2 = 2.44 A^2/d^2 - 4d^2 \qquad (3\text{-}4)$$

where W = width in ft.; A = area in ft.^{2}; d = greatest depth in ft.

This is illustrated in Fig. 3-9 to cover areas up to 10,000 ft.^{2} and depths from 5 to 60 ft. As the width is decreased by some jetty structure, so the depth will increase for a given area, which is determined by the tidal prism as afore-mentioned.

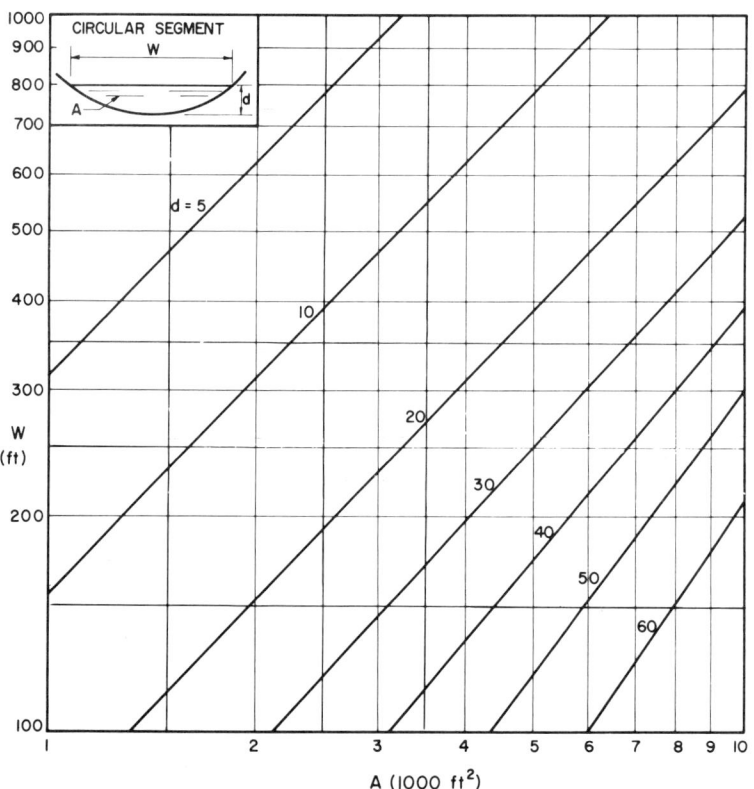

Fig. 3-9. Relationship between area (A), width (w), and depth (d) of inlet, assuming it to be the segment of a circle.

BY-PASSING

Means have just been discussed for by-passing material across inlet mouths with the same wave energy as is bringing the burden to the site. Similar principles will be enunciated later for delivering sediment downcoast of a harbour entrance. To date, most attempts at transmitting sand across such openings has been mechanical means. This process has been summarized by Watts [28].

Most installations consist of dredging units of either barge or pipeline disposal, or by fixed loading devices feeding a pipeline running under the entrance or around the harbour periphery. This latter method is the one for which the term 'by-passing' is generally reserved. The major difference between dredging and by-passing is that the former is removal of sediment after it has been deposited in a channel or harbour area, whilst the latter deals with material on the up-coast zone prior to it accumulating in an undesirable location. The exception to this procedure is the dredging of depressions on an upcoast beach, which can absorb littoral drift for a reasonable period of time.

Where the spoil from either method is used to nourish downcoast areas, care must be taken to see that it is spread along the beach by wave action. It is therefore necessary to dump it outside the shadow zone of any downcoast marine structure, such as a breakwater, but within the limit of the offshore bar formed during the winter season [29,30]. This limit of the storm beach-profile will vary from place to place. It may be computed from relationships already given or from observations over a number of years. A method for determining the location of the offshore bar is to take aerial photographs soon after a storm sequence. The outer line of breakers from the incoming swell indicates the seaward margin of the bar.

Generally associated with the maintenance of inlet entrances is the problem of dredging approach channels from the open sea [31]. With the greater depths demanded for larger and larger vessels, such channels have become a major cost in establishing harbour or loading facilities for oil and other bulk carriers. Their economic operation demands a tight schedule for ships using them, so that stability of the sides and siltation generally must be assessed beforehand. For this purpose the wave climate and tidal conditions are paramount.

If a predominant swell exists, no matter how small the wave height, more care must be taken than if only spasmodic storm sequences provide the wave energy. The critical section of the channel will be in the vicinity of the normal surf zone. This may migrate with the tide, but, fortunately, is close to the harbour where dredging equipment for maintenance operations can be located.

When storm waves only occur, as in some tropical regions, longshore drift from both directions can be expected in the surf zone. Further offshore the macro-turbulence generated by such short-crested systems is not associated with horizontal transport of great magnitude. Neither do tidal currents produce a significant net movement of the water mass near the seabed, even though it might be substantial at the surface. In assessing the influence of such currents it is their magnitude and constancy of direction near the bed that is of greatest importance.

The stability of the sides of dredged channels across the continental shelf depends greatly upon the nature of the bed material. In the tropical latitudes any shallow zone of sea is sufficiently heated for calcium carbonate precipitation and shell fish production to be magnified. Flat shell fragments can maintain almost vertical faces in dredged channels. The deposition of this calcareous powder and shell debris on the sea floor occurs over geologic time, together with a cementing action. The calcareous content of dredged shelf sands has reached 97% in some tropical areas. If such spoil is dumped above sea level the fresh water from rain can dissolve the carbonate and carry it down through the mound to redeposit it at a lower level. In time, limestone is formed through the filling of the voids with the carbonate. The mound thus becomes extremely stable and is not subject to spreading by the infrequent storm waves. A mound so constructed above high-water spring-tide level, parallel to a dredged channel, can prevent tidal cross-currents and so improve navigation in the dredged channel.

SHORELINE HARBOURS

A shoreline harbour is one created by the projection of either one or two breakwaters in a curved fashion (see Fig. 3-10). These virtually enclose a body of water that is sensibly free from wave action. In respect to longshore drift the breakwaters have the same effect as a groin, but this is magnified because of their greater protrusion across the beach profile. The upcoast accretion, and concurrent

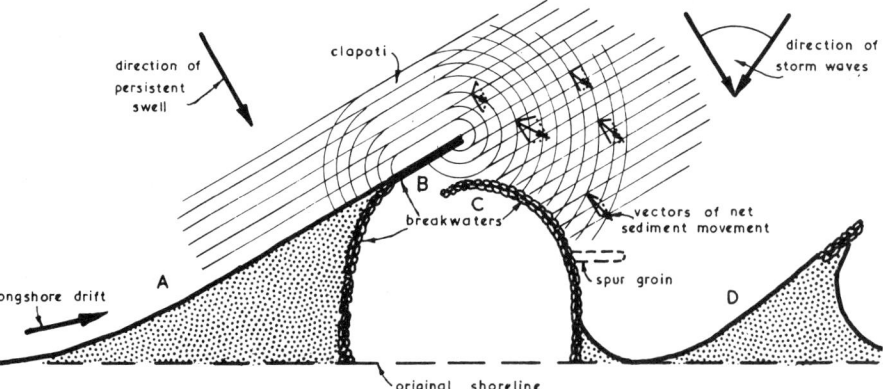

Fig. 3-10. Typical shoreline harbour showing use of reflected waves to prevent shoaling at the mouth.

downcoast erosion, thus present more impressive problems than a groin or groin field. Besides this, the ultimate by-passing of the sediment across the harbour mouth involves a siltation problem.

In establishing a harbour on a shoreline, which suffers a strong littoral drift, plans must be made for the progressive addition of land upcoast and depletion of the beach downcoast [32,33]. Some measures which can be taken are illustrated in Fig. 3-10. The direction of the persistent swell can be determined from the orientation of crenulate-shaped bays in the vicinity of the proposed harbour site. This will, with appropriate adjustments for local refraction, give the likely alignment of the beach at A in the figure. Because of the imminent protection of the breakwater arm leading to section B, it could be of relatively light construction.

Once the equilibrium beach-profile at A is established, and this will be similar to the profile existing upcoast from the site, the normal longshore drift will have to traverse section B of the breakwater. In fact, material will be arriving in this zone prior to saturation of the beach at A, due to the littoral current carrying it around the breakwater. If the profile at A is allowed to develop at B, sediment will be trapped in the harbour entrance. To obviate this, it is suggested that section B of the breakwater consist of steeply walled units which can reflect the persistent swell. This will produce almost perfect clapoti in front of section B and so expedite the passage of material past it, even though the depth is greater than along the beach zone A. Downcoast of B the reflected waves will diffract in curved arcs. This is illustrated in Fig. 3-10 without refraction, for the sake of simplicity. As noted in the discussion on inlet mouths, the combined crests will create a net movement of sediment in the individual directions of propagation. The resultants, as shown, together with the vortex generation and other macro-turbulence, will effect a swift transmission of material across the harbour entrance. A large zone seawards of the entrance will experience short-crested waves, which will impede the formation of shoals. As with the inlet entrances, navigation for large vessels will be unaffected by the angled waves and even small vessels should not suffer greatly under normal swell conditions.

Donnelly and MacInnis [34] have reported the self-maintaining character of two curved breakwaters with an opening at the extremity. This was on a shoreline subject to storm waves and swell with a moderate littoral drift (see location map Fig. 3-11). The cleansing action was put down to some inner walls which purported to dissipate the waves and generate "a periodic flushing current through the gap in the outer breakwaters" (see Fig. 3-11). It is believed that the modest reflection which took place from the exposed surfaces of these rock-fill structures was instrumental in creating clapoti and short-crested waves which expedited the removal of material across the opening and downcoast of it. As seen in Fig. 3-11 this action was accompanied by an accretion seawards where the 8-ft. contour has penetrated the 10-ft. channel dredged in 1968.

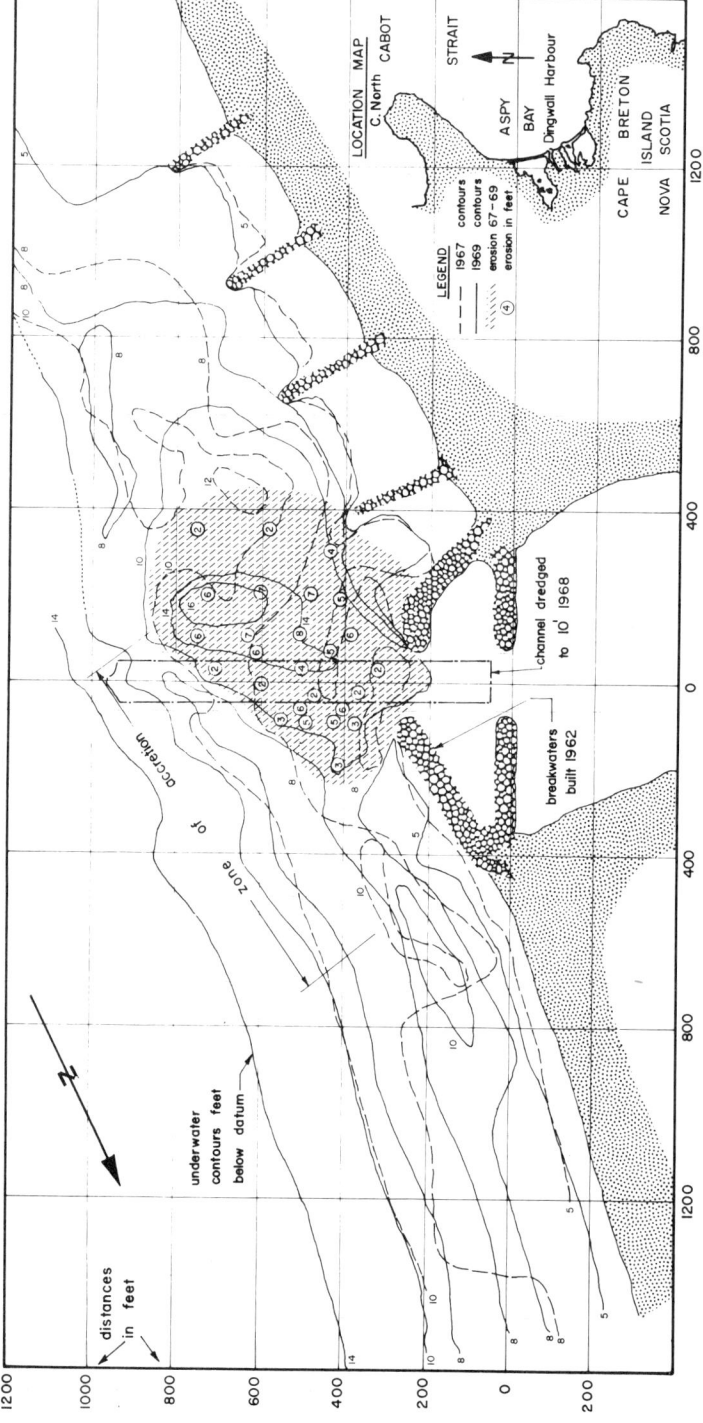

Fig. 3-11. Location and detailed hydrographic plan of Dingwall Harbour showing the zones of erosion and accretion over a two-year period. (Ref. [34].)

Section C of the downcoast breakwater (see Fig. 3-10) should be designed for dissipating both the persistent swell and any storm waves that can arrive from the wide range of directions indicated. This steepening and breaking should be accomplished solely by the breakwater, without assistance from the seabed through any shoaling. Particular attention, therefore, must be given to its design. To safeguard the heel of the breakwater C an offshore structure placed downcoast as in Fig. 3-10 could create a stable beach alignment D. If storm waves arriving obliquely into this bay were to generate a strong rip current, which carried material around to the harbour mouth, the spur groin indicated could readily prevent this. The headland structure forming beach D could be established progressively as discussed previously.

WALLS AND REVETMENTS

Walls of vertical or sloping character (revetments) have been used for many decades as a purported protection in an erosive situation. It is unfortunate that they have, in the main, promoted further erosion. It is of some concern to read accounts of bigger and better sea walls to replace those that have subsided into the sea.

Tests [35] have indicated that beaches in front of walls will recede to the point of being non-existent, due to the action of standing waves resulting from reflection. The influence of beach saturation in accelerating erosion has been discussed in the section on beach processes. The above action has been noted in wave flumes, where the sediment is necessarily retained in front of the wall. As noted already in this chapter, the three-dimensional aspects of sediment removal are much more serious. Obliquely reflected waves can create a short-crested system, with its concomitant turbulence and longshore mass-transport, which expedites the passage of material in front of cliffs or walls.

Where the tidal range is reasonably large, revetments have been used instead of walls. These may consist of cement slabs or blocks, stone pitching, or just plain bitumen sealing of a light soil pavement. The construction is carried down to the lowest tidal level. It is submitted that all such attempts at coast stabilisation are doomed to failure in the long term if not in the short run. This view is supported by Bruun [36].

Assuming that the defense program is occasioned by depleting supplies of sand in the area, the situation will deteriorate unless means are provided for rectifying this shortage. The waves will reflect from the relatively smooth sloping surface and will accelerate the removal as noted earlier. The seabed profile in front of the wall will steepen and deepen until subsidence of one section will occur during a particularly bad storm, and will cause the destruction along major lengths of the structure.

Not only does erosion in front of the so-called protective structure get acceler-

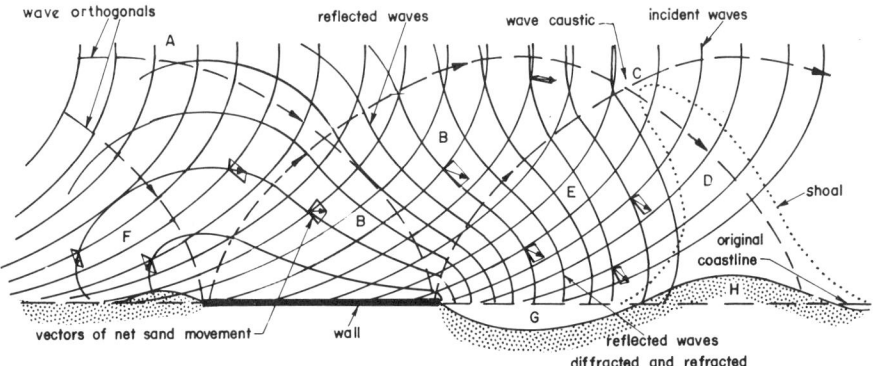

Fig. 3-12. Crest pattern of waves reflected from a sea-wall showing vectors of net sediment drift.

ated, but the shore just beyond the downcoast end is also eroded, even to the stage of shoreline recession. The reason for this is illustrated in Fig. 3-12, where a seawall exists along part of a sedimentary shoreline. Waves arriving obliquely in deep-water at A are refracted towards the shore. Those portions which are reflected from the wall propagate seawards through zone B. They are refracted, in turn, passing through zone C to zone D on their way back to shore. Area C is known as a "wave caustic", where the energy is concentrated and theoretically approaches infinity [37]. At least much dissipation, not only of the reflected but also of the incoming waves, will occur, so that zone D will be relatively calm. Outside the bounding orthogonals of zone B the reflected waves spread their energy through diffraction and refraction as illustrated by the crests in zones E and F.

Throughout all areas F, B, E and D, the reflected waves and incoming waves will interact, to create short-crested systems of various angles of propagation. As noted previously, each angled train has its own net effect on the particles at the bed, which for simplicity may be summed by the addition of the two vectors. The magnitude of the vector of each train will depend upon its wave height at that particular location. The arrows in Fig. 3-12 are only diagrammatic, in order to display the principle that a strong alongshore tendency exists in zones B and E. In zone D all waves will have parallel vectors, with a longshore component equal to the normal beach zones. However, the excessive longshore transport rate from E will overload the wave capacity in zone D, so that accretion will take place and a shoal built up. The shoreline at G facing zone E will suffer degradation and probably prompt the promoters to extend the wall downcoast. This will then shift the erosion and accretion zones along shore commensurately. In zone F, upcoast from the wall, the diffraction of the reflected waves causes the resultant sediment vector to oppose the downcoast component of the incoming swell. This could result in some accretion.

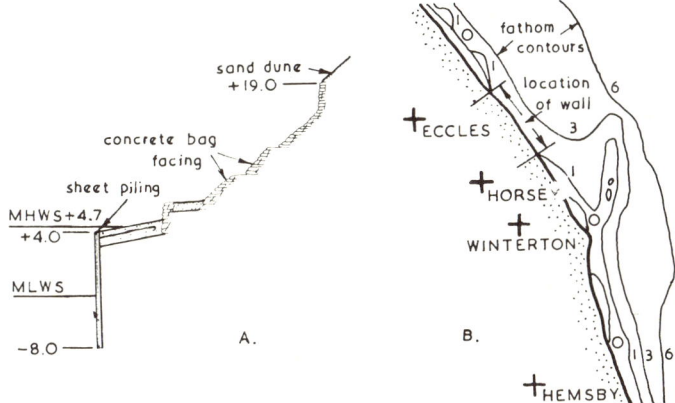

Fig. 3-13. Norfolk coast defenses. A. Details of revetments between Eccles and Horsey prior to 1953; B. Sea bed contours in 1960.

Hardy [38] has reported erosion and deposition similar to the above which has occurred on the Norfolk coast in England. Fig. 3-13B shows the extent of the coastal defenses depicted in Fig. 3-13A, which are virtually sea walls [48]. This "protection" was commenced in 1938, and completed in 1942. Extensions were made after the disastrous 1953 flood, with contemplated additions down to Winterton. The shoal as exhibited near Winterton, by the 1960 bed contours in Fig. 3-13B, would then be displaced further south, according to the explanation given above. This influence of clapotis gaufré appears also to be confirmed in the case of deepening at Dingwall Harbour previously discussed [34] and the associated shoal downcoast. Similar action can be observed at headlands where a shallow zone often occurs at the downcoast end, because the sediment accelerated along the cliff face must construct a shallower platform for the normal swell to cope with the longshore load.

Depending upon the angle of wave approach to a wall, secondary crests can be created which are normal to it. This is known as the Mach-stem effect [39], which is bound to increase the erosive effect of the reflected waves, and generate a stronger current along beach zone G of Fig. 3-10. The dissipation of waves in caustics (zone C) requires more study, as stated by Wiegel [40]. Previous work on this topic has dealt mainly with certain hypothetical situations [37].

BEACH NOURISHMENT

Schijf [41] has discussed the dynamic character of the coastline and the need to allow a buffer zone between the sea and buildings on the beach. Whilst deprecating the indiscriminate use of groins Schijf suggests renourishment of depleting sand supplies. At least such action can be stopped if the erosion happens to be transient, whereas a groin is a more-or-less permanent feature. This emphasizes the need to study the transient or more permanent nature of the problem, by successive surveys over a number of years. In assessing the sand reserves held by a beach, the possibility of high water levels from storm surge should not be forgotten.

Replacement of beach sand, and in some cases shingle, by reclamation from adjacent beach zones, or even offshore, has become a continuing economic cost in some places [42]. Where there is a known longshore drift such renourishment should be accompanied by construction which will impede its removal from zones requiring the greatest protection. The most successful scheme will be achieved between prominent headlands, where removal can only occur transverse to the beach. The loss to the offshore zone will depend upon the underwater slopes to be filled before a stable profile results.

To compute the volume of fill required, the equilibrium beach profile should be calculated. The limit of this profile will depend upon the period of the predominant swell in the area, plus the intensity of the storm waves that normally arrive. The renourishment should be spread over a number of years for the cyclic wave climate to effect a uniform distribution. In this respect, it should not be surprising to see a large proportion of the fill disappear during the first winter season.

Depending upon its source, the sediment used for renourishment may be larger or smaller in diameter than the natural material. If coarser the equilibrium slope will be steeper than the existing one, and vice versa for finer particles. Appropriate quantities can be computed on this basis. There is a possibility of wave reflection from a steeper beach slope, which may expedite the removal of finer material offshore, thus aggravating the problem whose solution is being attempted.

ESTUARINE PROBLEMS

Using the example of a large funnel-shaped indentation into which a large river debouches, as a typical estuarine problem, consider now some methods of reducing or dealing with its siltation. As already noted, these zones of the coast are extremely important for commerce, in spite of the navigational hazards they present. The first task of the coastal engineer is to ascertain the source of the bulk of the accretion.

If most material is being supplied by smaller rivers each side of the indentation, and being transported to the river mouth by the incoming swell, some kind of impedance must be supplied. Here again the use of crenulate-shaped bays through the use of headland structures suggest themselves, as in Fig. 3-14. This requires a comprehensive scheme, to which many authorities must lend their support. In spite

ESTUARINE PROBLEMS 147

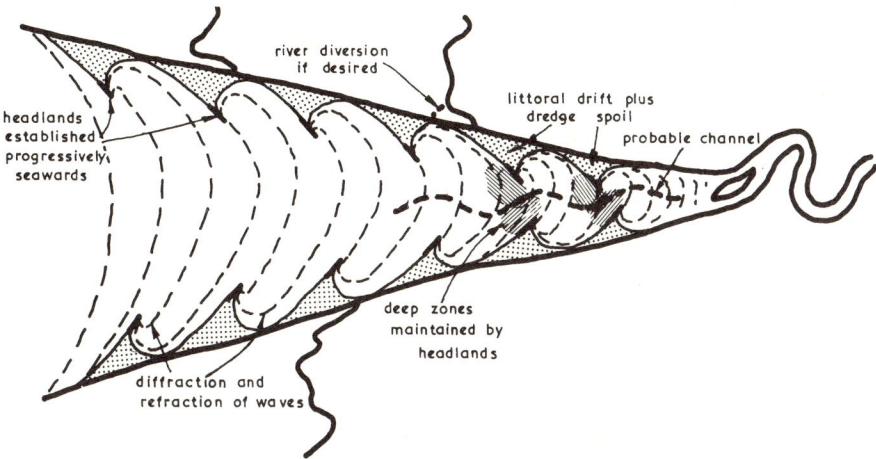

Fig. 3-14. Suggested use of headlands to control siltation in a typical estuary.

of its cost, which can be spread over decades, it would appear the only long-term solution to a seemingly perpetual problem. The bays depicted in Fig. 3-14 could be developed from the river mouth seawards, using the longshore drift to fill out the curved beaches, which can be valuable industrial sites. Any material issuing from the river mouth can be dredged to help fill these zones.

Also indicated in the figure is a possible use of wave reflection from the inner group of headlands, so as to keep sediment mobile, in order that deep channels can be maintained for shipping. Another desirable feature is to keep long ocean swell from penetrating to the river mouth. Location for the bays must be chosen with caution, cognisance being taken of stable outlets for the rivers supplying the sediment and perhaps furnishing havens for fishing and other small craft.

Such a solution as suggested above must take into account many complex issues, the strength of the tidal currents with respect to wave energy, for example. Such currents can be extremely erosive when meandering out of a river mouth. The resulting channel can migrate from side to side over a period of years. Cycles of fifty years or more have been recorded [43]. Such trends should be considered in the most effective use of a headland control scheme. The tidal-current pattern in the proposed final scalloped coastline should be studied in a large model, to check that vortices are not generated that could be a navigational hazard.

Each estuary or indentation must be treated on its merits. A research and development authority should be established for each major physiographic feature such as this. It could make a continuous study of the marine problems associated with the area and advise the many organisations who depend on the waterway for their multifarious activities. For example, port authorities, power station and refinery operators, public health engineers, beach recreational authorities, fisheries groups, transport organisations and many others, could be aided in their attempts to utilize the estuary for the overall benefit of the community. It is the prerogative of the coastal engineer to alert the relevant political circles to the need for such cooperation and wise expenditure.

DREDGING

Dredging is a term applied to removal of material from the bed of a river or the sea and placing it elsewhere. The object of the operation is usually to create a deeper channel for the passage of ships, although reclamation activities today employ similar types of equipment. When navigation channels are being created it is essential to see that the dredged spoil does not readily find its way back to the excavated zone. In estuarine situations this may mean cartage by barge for long distances out to sea, or deposition on reclaimed areas.

Besides deepening harbours, turning basins and approach channels, dredges are nowadays used for gaining minerals from underwater. These vary from sand in river shoals to diamonds on the outer zones of some continental shelves. The deeper the level of excavation, and the further offshore, the more sophisticated the equipment required. The discussion to follow will centre on normal coastal engineering requirements, where depressions from 5 to 12 fathoms are dredged for navigational purposes.

The components of a normal dredge [44] are as follows:

(*a*) Cutter − is rotated at the bed to break up soil or soft rock and throw it into suspension. A motor is attached at the top end of a ladder down which the shaft runs to activate the cutter (see Fig. 3-15A), or gears are used from a motor installed inside the dredge.

(*b*) Suction pipe − is slung beneath the ladder and is connected to the main pump suction by means of a flexible joint. The open end of the suction pipe is located close to the cutter so that suspended material is drawn into it.

(*c*) Main pump − is housed in the body of the dredge and may consist of one or several pumps in parallel or series. Horsepower can vary from 300 to 10,000 with speeds varying from 900 to 300, the latter speed for the larger horsepowers. These pumps are designed to transmit the sediment load.

(*d*) Discharge lines − consist of three major sections, on the dredge, floating, and on shore. The dredge section connects to the floating line through a swivel elbow joint with ability to turn through wide angles. The floating line consists of 30−50 ft. lengths supported on pontoons. Reasonably flexible joints are provided in the floating line so that it may be swung in a large arc, to allow for dredge advance without frequent changes in length. The shore section consists of pipes 10−15 ft. in length, temporarily fixed together for rapid changes in location of the

DREDGING

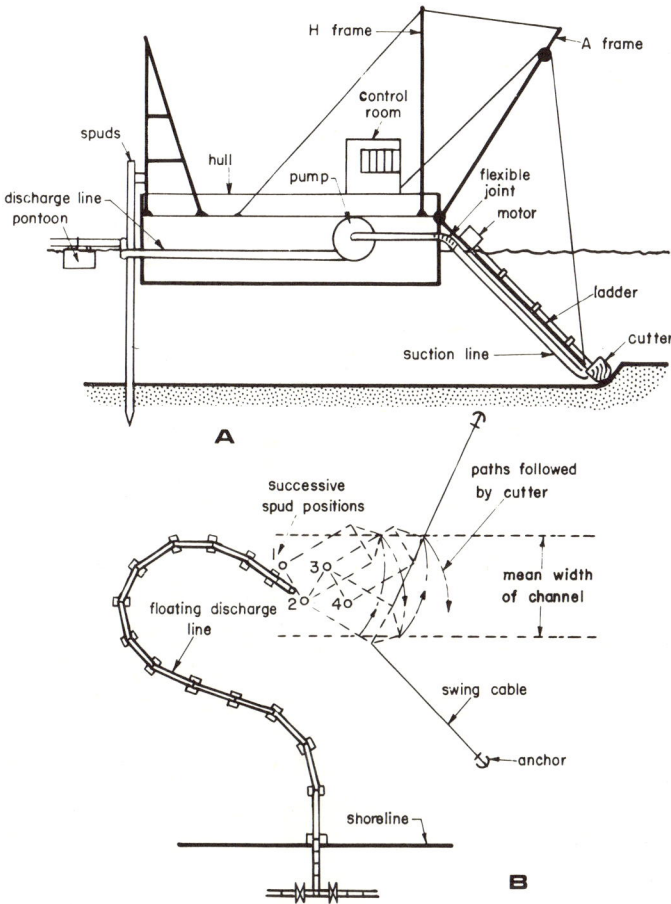

Fig. 3-15. A. Schematic diagram of dredge components; B. Method of "walking" dredge along channel

discharge end. For this purpose a T or Y joint with valves on each permits one of the two openings to be altered without closing down the pumps.

(e) Spuds — are vertical steel piles with pointed toes which are dropped or pressed into the bed, about which the dredge pivots when operating. There is one on either side of the stern. These are imbedded in turn so that the dredge virtually walks along the channel as the cutter swings from side to side (see Fig. 3-15B).

(f) Anchors — are fixed into the bed either side of the channel to be dredged, the cables from which are attached to synchronous drums on the dredge. The dredge is swung from side to side around a spud as a pivot. The anchors are set far enough ahead so as not to require shifting too often.

(g) Hoisting equipment — is required for raising and lowering the ladder with the cutter and suction pipe on it, for hoisting the spuds, and for swinging the dredge. This equipment may consist of mechanical lever systems or hydraulic actuators.

(*h*) Control room — is located at the bow end where the ladder can be observed closely. It contains controls for all elements of the operation as listed above. The pressures and suctions on the pumps are indicated on dials, by which the operator can judge the sediment load being carried.

The mode of dredging will depend greatly upon the material to be removed. Soft mud will readily slump so that a single cut can be made down to the desired level even where this is great. Harder soil or limestone rock will require smaller cuts, a number of swings being required to reach the bed level. A channel may be dredged to half depth and then re-run for short return lengths or even a full length. The width worked on each swing is determined by the dimensions of the dredge. If two runs are necessary these may be taken successively, or a whole half width may be completed to provide a minimal passage whilst the remaining half is being dredged.

In deciding minimum depths for channels, turning circles and harbour berths account must be taken of LLWS levels and the height of tide at which sailing is required. For economy's sake long channels generally cannot be dredged to take fully laden ships at all levels of the tide. It may be possible to utilize only the top half of a MHWS tide, but this may restrict sailings to three or four days each fourteen days. In channels some miles in length consideration has to be given to acceleration and top speed in shallow conditions for a fully laden vessel. The vessel must be clear of the seaward end before the lowering tide has reached the limit.

Besides a safety margin of a few feet between hull and bottom, allowance must be made for heaving in swell or storm conditions. This will depend upon the length of the vessel in respect to the length and height of the predominant waves. Depressions next to wharves must be able to float the vessel at all stages of the tide. In the case of bulk carriers, which must be moved fore and aft for loading, the berthing area must have a length up to twice that of the ship.

Dredged channels are generally designed for side slopes or trapezoidal form. However, they are cut to vertical faces when dredged, so that slumping provides the acceptable side slope. Some faces have been known to stand almost vertical for long periods, even where wave action is strong. This depends upon the depth of the ocean bed where the channel is made and on the bottom material. Where this consists of a large percentage of shell fragments which are relatively flat, intermittent layers of limestone, or complete limestone sections, virtually vertical walls will result. Caution should be exercised in designing for such slopes. The deeper zones of a berthing area can be planned to have steeper slopes than upper regions nearer the influence of waves and tidal currents.

Disposal of dredge spoil is a very important topic, since it can multiply the cost of a dredging commission many fold. To ensure that it will not be brought back to the channel, travel to other channels, or accumulate in a harbour entrance, an assessment must be made of net sediment movement in the area. The present economic maximum length for pipeline discharge is around 3500 ft., so that spoil-grounds must be chosen within this radius. The wind and wave climate, plus other indicators, should be used to determine the direction of discharge. It may be found useful to dump material in a dyke or mound that protrudes above MHWS level. This prevents storm waves from shifting a shoal towards a channel. If this mound runs parallel to the channel it could reduce cross-currents of tidal or other origin.

There are many types of dredge the choice of which depends upon the material to be removed, the distance for disposal, the size of the job and the time required for completion. Finally, the availability of units may determine the machine to be used. Contracts run for several months and even shifting from one site to another may take several weeks, when allowance is made for good weather. Sundry other factors can enter the choice of dredge, such as proximity to fuel and spare parts, or wave climate of the operating zone (pipe lines may be unworkable). It may be efficient to have one large machine and a second smaller machine working together, the former to complete the bulk of the work and the latter to remain continuously on site to complete the task and carry out maintenance dredging.

In spite of the high cost of operation and the need to lower the unit-cost of dredging, Huston [44] reports that no practical methods exist for determining accurately the instantaneous output of material. The difficulty arises from the variability in concentration of sediment in the pipeline. Output is generally determined by surveys of the cross-section dredged. Such information is normally available days or weeks after the operation.

Performance can be gauged roughly by the vacuum and discharge pressures of the pumps. A large batch of sediment in the suction line will increase the vacuum reading and when passed through the pump will increase pressure slightly. When the vacuum increases beyond a certain limit the cutter must be raised in order to induct water into the line. When pressures are recorded on a continuous chart these indicate hours of operation and the relative amount of water and sediment passing through.

Efficiency of dredging differs somewhat from efficiency of pumping. The latter relates to pumping a unit of material at the lowest power cost. The former is pumping at the lowest over-all cost. Although high velocities in pipelines incur higher head losses they may be more effective in transporting material. A range of velocities from 13 to 21 ft. per second is recommended, which spans low carrying capacity to uneconomically high friction losses [44].

Friction losses in pipe lines carrying mixtures of water and sediment can be computed from graphs supplied by the British Hydromechanics Research Association [45], from which the following relationship can be derived:

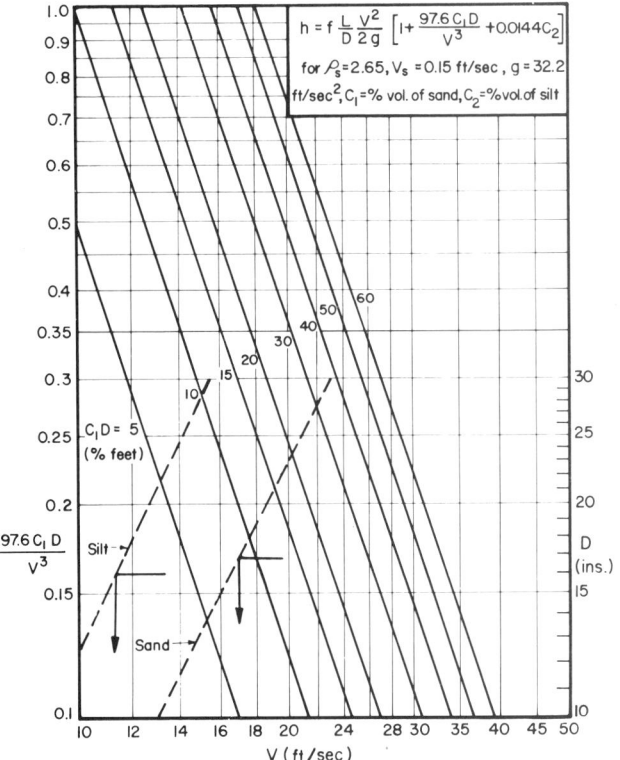

Fig. 3-16. Parameters for computing friction loss in pipelines carrying sand-silt mixtures.

$$h_t = f \frac{L}{D} \frac{V^2}{2g} \left(1 + \frac{97.6 C_1 D}{V^3} + 0.0144 C_2\right) \tag{3-5}$$

where h_t = head loss in ft. due to water and sediment with specific gravity = 2.65; f = friction factor for water alone; V = velocity of water in ft./sec; D = pipe diameter in ft.; L = length of pipeline in ft.; g = acceleration due to gravity (= 32.2 ft./sec^2); C_1 = % concentration by volume of sand with settling velocity of 0.15 ft./sec; C_2 = % concentration by volume of silt.

Fig. 3-16 should aid in computation, which also includes lines from which the minimum velocity required in given sizes of pipe to prevent settlement of either silt or sand. Friction loss in lifting a sediment mixture is that for water alone multiplied by the specific gravity of the mixture.

As material is dredged from deeper and deeper depths, so suction limitations of normal centrifugal pumps handling sediment exert their influence. These generally take the form of cavitation at the entrance to the pump. This has been overcome by boosting the suction flow with a water jet pump action [46]. A comparison has

PROBLEMS   153

been made of a complete pumping system incorporating the water jet pump [47] with that of the normal centrifugal system. Whilst the latter may utilize slightly less power for any given discharge of sediment, there are a number of advantages of the booster system employing jets that should be considered [47].

PROBLEMS

1
Discuss the problems that can result from the installation of a single groin, or a field of groins. Under what particular conditions can this type of structure be useful in protection of coasts?

2
On what basis would you determine the length and spacing of groins in order that an existing beach line be maintained?

3
What effect does an island or offshore breakwater, which is parallel to the shoreline, have on a mainland that consists of sedimentary material? Consider both cases of normal and oblique wave approach. Cite cases in the literature or on hydrographic charts where such influences are evident.

4
In stabilizing a section of coast by headlands progressively moved seawards, why is it advantageous to start at the downcoast end? In choosing such a location, what features or developments would be sought?

5
Suggest some economical designs for floating, semi-submersible or other type structures suitable for headland control measures. Consider cases for swell conditions only, storm waves limited by fetch, and oceanic margins where energy intensity is high.

6
It is desired to reclaim an area in the ocean which is of nearly circular shape. Swell approaches the area from a specific direction. Draw the headland controls necessary to prevent the erosion of this island. Sketch the probable beach line and underwater contours for the case of no control measures.

7
The coast where an inlet entrance exists suffers periods of storm and then periods of swell. Describe the conditions to be expected at the inlet during both of these sections of the year. If such an inlet suffers periodic blockage of its outflow what measures besides dredging would you suggest for maintaining the opening.

8
An inlet on a sandy coast connects a bay with an area of 5 sq. miles at MSL. The tides in the area are semi-diurnal in period and of 7-ft. range for the MHS condition. Compute the area of the inlet opening. Assuming this to be a segment of a circle, determine the ratio of width to depth. If it is desired to maintain a 10-ft. depth over the bar, what width should be planned for jetty-type structures either side of the opening?

9

What problems should be kept in mind when a deep approach channel is planned across a beach profile where an oblique persistent swell exists? What suggestions would you make in respect to disposal of dredge spoil and maintenance of the channel?

10

Discuss various types of by-passing installations and give examples from the literature. Note where each method is likely to be the most economical.

11

Sketch reflective groin orientations which will maintain the passage of sediment across a river mouth without the shoaling normally experienced when littoral drift is strong. Cite examples in the literature where this has occurred either as planned or fortuitously.

12

List references to shoreline harbours and the problems resulting from their construction. State whether these may have been obviated if means had been provided for by-passing the littoral drift either mechanically or by reflecting breakwaters.

13

An upcoast breakwater of a shoreline harbour consists of concrete boxes floated into posiiton and fixed onto a rubble mound base by filling them with soil. If the persistent swell arrives normal to this breakwater, what zone should be observed carefully to prevent subsidence? What measures would minimize this tendency?

14

List several factors which make it difficult for a beach to exist at the foot of a steep cliff.

15

Whilst walls and revetments give the appearance of stability above the water line, what could be happening below the water line to worsen the original condition?

16

A section of straight beach is in equilibrium, that is, the sediment being transported into the section by the oblique swell equals that being removed downcoast. A wall is then constructed along the beach front from which the waves are reflected. Describe the resultant action on the beach in front of and at both ends of the wall.

17

Cite examples in the literature where walls or revetments have failed and have been replaced by bigger and "better" structures of similar type. In the case of such a wall failing, what enquiries would you make when advising a remedial measure to be installed?

18

Before suggesting a renourishment plan for a section of coast, what information should you gather about the area? What particular conditions could make renourishment an economical proposition?

19

A sedimentary cape has an almost triangular form with the projecting angle approximating 90°. One side of the triangle is almost parallel to the incoming persistent swell. What direction will this physiographic feature move in respect to this wave system? Why? Cite two examples of such accretions on a coast.

20
It is planned to stabilize a section of coast with a strong littoral drift by a headland control measure. This is to be complemented by a renourishment scheme to expedite the equilibrium bays. When and where should such renourishment be effected?

21
What are some of the major problems associated with maritime activities in estuaries? Cite references in the literature where these have been overcome.

22
Give six examples from the literature where estuaries of the funnel shape, cited in the text, occur and are important from a commercial point of view. Show any benefit that would have accrued from the formation of an overall estuary control body.

23
Search the literature for articles in which costs of dredging large channels and basins are quoted. List the general geologic conditions and the costs per cubic yard. Give any reasons for apparently high or low figures, such as remoteness, danger, etc. The dates of valuations are important to note.

24
What are some of the problems encountered in dredging as the depth of excavation becomes greater? What methods are being used to overcome these?

25
In determining the optimum depth of channel to be dredged across a wide and mildly sloped continental slope, what traffic factors should be taken into account?

26
List the topics to be investigated before recommending the type and size of dredging units to be used on an excavation contract.

27
Compute the HP demand for pumping sand at 20% concentration by volume through a 24-inch diameter pipe line 3000 ft. in length, if a channel 40 ft. below MWL is to be excavated and the spoil deposited on a mound to a maximum height of 10 ft. above MWL. Compare this power demand with the same delivery of silt.

28
As a coastal engineering consultant you are to report on the best location for a refuge harbour for small boats along a specific length of coast. List the topics in an orderly fashion which you would have to investigate in the preparation of such a report.

29
An iron ore deposit could have a port outlet at three locations on the coast, one hundred miles each apart. It is required to assess the most economical site from the point of view of marine activities. List the topics of investigation and discussion for such a consulting task.

REFERENCES

[1] A.E. Carey, 1907. The protection of sea shores from erosion. *J. Soc. Arts*, 1907: 650–663.

[2] A. Lehnfelt and S.V. Svendsen, 1958. Thyboroen channel - difficult coast protection problem in Denmark. *Ingenioeren*, 2: 66–74.
[3] D.W. Berg and G.M. Watts, 1967. Variations in groin design. *Proc. ASCE* 93(WW2): 79–100.
[4] M. Petersen, 1963. Review of German experience on coastal protection by groins. *Beach Erosion Board Bull.*, 17: 38–54 (translated by O.W. Kabelae).
[5] R. Silvester, 1962. Sediment movement around the coastlines of the world. *Proc. Conf. Inst. Civil. Eng. London, 1962*: 289–315.
[6] J. Laurent, 1955. Le régime de la rade de Tanger. *Proc. 5th Conf. Coastal Eng.*, 364–378.
[7] R. Silvester, 1965. Coastal sediment movement – some fundamental problems with discussion of research support. *J. Inst. Eng., Austr.*, 37: 311–323.
[8] J.W. Dunham, 1966. Use of long groins as artificial headlands. *Proc. Santa Barbara Conf. Coastal Eng., 1965*: 755–762.
[9] O.J. Lillevang, 1966. Groins and effects – minimizing liabilities. *Proc. Santa Barbara Conf. Coastal Eng., 1965*: 749–754.
[10] B.S. Bulson, 1967. Transportable breakwaters. *Dock Harbour Authority*, 48: 41–46.
[11] J.H. Carr, 1950. Mobile breakwaters. *Proc. 2nd Conf. Coastal Eng.*, 281–295.
[12] R.Y. Hudson, 1945. Model tests of portable breakwaters for D-day invasion harbors. *Civil Eng.*, 15: 405–408.
[13] E.L. Bourodimos and A.T. Ippen, 1968-69. Characteristics of an open tube wave attenuation system. *Proc. ASCE*, 94(WW4): 465–488. (Discussion in: 95(WW2): 259; 95(WW3): 430–436; 95(WW4): 588–595.
[14] A.M. Kamel and D.D. Davidson, 1968. Hydraulic characteristics of mobile breakwaters. *U.S. Army Waterways Exp. Sta., Vicksburg, Miss., Tech. Rep.* H.68-2.
[15] J. Kato, S. Hagino and Y. Uekita, 1966. Damping effect of floating breakwater to which anti-rolling system is applied. *Proc. 10th Conf. Coastal Eng.*, 2: 1068–1078.
[16] R. Lochner, O. Faber and W. Penny, 1948. The "Bombardon" floating breakwater. *Civil Eng. War, Inst. Civil Eng.*, 2: 256–296.
[17] W. Marks, 1966. A perforated mobile breakwater for fixed and floating application. *Proc. Conf. Coast. Eng., 10th Conf. Coast. Eng.*, 2: 1079–1129.
[18] J.F. Ripken, 1960. An experimental study of flexible floating breakwaters. *St. Anthony Falls Hydraul. Lab., Univ. Minn., Tech. Pap., Ser. B.*, 31.
[19] K. Takano, 1960. Effets d'un obstacle parallélépipèdique sur la propagation de la houle. *Houille Blanche*, 15: 247–259.
[20] S. Tanaka, 1966. Researches on double curtain wall type breakwater. *Proc. 10th Conf. Coastal Eng.*, 2: 913–931.
[21] J.J. Vinjé, 1966. Increase of effective working-time during operations at sea by means of movable structures. *Delft Hydraul. Lab., Publ.* 42.
[22] R.L. Wiegel, H.W. Shen and J.D. Cumming, 1962. Hovering breakwater. *Proc. ASCE*, 88(WW2): 23–50.
[23] P. Bruun and F. Gerritsen, 1958. Stability of coastal inlets. *Proc. ASCE*, 84(WW3); Pap. 1644.
[24] W.N. Bascom, 1954. The control of stream outlets by wave refraction. *J. Geol.*, 62: 600–610.
[25] E.P.C. Hughs, 1957. The investigation and design for Portland Harbour, Victoria. *J. Inst. Eng. Austr.* 29: 55–68.
[26] J.W. Johnson, 1966. Near-shore sediment movement central California coast. *Proc. Santa Barbara Conf. Coastal Eng., 1965*: 537–560.
[27] M.P. O'Brien, 1966. Equilibrium flow areas of tidal inlets on sandy coasts. *Proc. 10th Conf. Coastal Eng.*, 1: 676–686.
[28] G.M. Watts, 1962. Mechanical by-passing of littoral drift at inlets. *Proc. ASCE*, 88(WW1): 83–99.

REFERENCES

[29] J.V. Hall Jr. and W.J. Herron, 1950. Test of nourishment of the shore by offshore deposition of sand. *Beach Erosion Board, Tech. Mem.*, 17.

[30] P. Bruun, 1967. By-passing and back-passing off Florida. *Proc. ASCE*, 93(WW2): 101–128.

[31] B. Blackman, 1951. Dredging at inlets on sandy coasts. *Proc. 1st Conf. Coastal Eng.*, 169–174.

[32] J.W. Johnson, 1959. Littoral drift problems at shoreline harbors. *Trans. ASCE*, 124: 525–546.

[33] N. Shiraishi, 1959. The measurement of littoral drift at shoreline harbors. *Coastal Eng. Japan*, 2: 59–72.

[34] P. Donnelly and I. MacInnis, 1968. Experience with self-dredging harbour entrance. *Proc. 11th Conf. Coastal Eng.*, 2: 1283–1294.

[35] J.B. Herbich and S.C. Ko, 1968. Scour of sand beaches in front of seawalls. *Proc. 11th Conf. Coastal Eng.*, 1: 622–643.

[36] P. Bruun, 1953. Coastal protection – review of methods for defence. *Dock Harbour Authority*, 34: 217–222; 233–237.

[37] W.J. Pierson Jr., 1951. The interpretation of crossed orthogonals in wave refraction phenomena. *Beach Erosion Board, Tech. Rep.*, 21.

[38] J.R. Hardy, 1966. An ebb–flood channel system and coastal changes near Winterton, Norfolk. *East Midland Geogr.*, 4: 24–30.

[39] T.C. Chen, 1961. Experimental study of the solitary wave reflection along a straight sloped wall at oblique angle of incidence. *Beach Erosion Board, Tech. Mem.*, 124.

[40] R.L. Wiegel, 1964. *Oceanographical Engineering*. Prentice Hall, Englewood Cliffs, N.J., 532 pp.

[41] J.B. Schijf, 1959. Generalities on coastal processes and protection. *Proc. ASCE*, 85(WW1): 1–12.

[42] J.V. Hall Jr., 1952. Artificially nourished and constructed beaches. *Beach Erosion Board, Tech. Mem.*, 29.

[43] S.C. Roy, 1969. Hydraulic investigations on behalf of Hooghly estuary. *Mitt. Franzius-Inst. Grund- Wasserbau Tech. Univ. Hannover*, 32: 45–149.

[44] J. Huston, 1968. Dredging fundamentals. *Proc. ASCE*, 93(WW3): 45–69.

[45] A. Linford and G. Fish, 1968. The hydraulic transport of silt and sand. *BHRA Rep.*, RR948.

[46] M.S. Gorton, 1971. Comparative full scale tests of a jet pump on the suction of a 22" hydraulic dredge. *Proc. World Dredging Conf., 3rd, 1971:* 119–139.

[47] R. Silvester and S. Vongvisessomjai, 1971. Characteristics of the jet-pump with liquids of different density. *Proc. World Dredging Conf., 3rd, 1971:* 293–315.

[48] K.E. Cotton, 1954. Flood damage in Norfolk and Suffolk. *Proc. Conf. North Sea Floods, Inst. Civil Eng., 1954:* 200–211.

Chapter 4

LONG-PERIOD WAVES

Wind-generated waves range in period from 1 to 30 sec. At each end of this range the energy content is relatively small so that from the engineering point of view the range is 4—20 sec. But besides these there are longer-period waves with which the coastal engineer must be concerned. These include: *surf beat*, with periods from 50 to 300 sec; *storm surge*, with periods from 1000 to 2000 sec; *tsunamis*, with periods from 1000 to 2000 sec; *tides*, with periods from 45,000 to 90,000 sec.

Surf beat is caused by the interaction of two-swell-wave trains of almost equal period. As they approach the coast together they form zones where the crests are cumulative, whilst in between the trough of one cancels the crest of the other. Thus periods of calm are interspersed with short sequences of high waves, which build up the water level at the beach temporarily due to wave set-up and other factors [1]. This transient increase in mean water level is then transmitted seawards again as a low long-period wave. In *Coastal Engineering, I* it has been discussed how these can generate resonant surges or seiches in harbours.

Storm surges are built up by the "suction" created over an area of the ocean in a low-pressure cyclonic centre and by the stress exerted on the water surface by the accompanying wind. Such surges are particularly strong when generated within tropical cyclones, although extratropical cyclones can cause them from wind stress alone. Storm waves will necessarily be generated at the same time, but large surface undulations are not needed for the wind to effect the frictional drag of this phenomenon, only small, short waves can supply the required roughness. Another feature important in storm surge is a wide zone of shallow water, such as inland waterways or the continental shelf in the ocean basins.

Tsunami is the Japanese term for a wave, or a series of waves, produced by a sudden movement of the earth's crust, such as occurs in earthquakes. These are sometimes termed "tidal waves", but this is erroneous since they have nothing to do with the generation or propagation of tides. However, their wave length is long enough for them to be shallow-water waves, in which case their celerity is dictated by the depth of the ocean where they are travelling (i.e. $= \sqrt{gd}$). As tsunamis traverse the continental shelf they are slowed up greatly and hence transfer their kinetic energy to potential energy. The resultant increase in height causes great havoc to facilities in certain coastal zones whose offshore depths concentrate this energy over a limited length of shoreline.

Tides are produced in the waters of the globe from the attraction of the moon and the sun, as well as other planets. Because the periodicity of this energy input is known, so tides around the world can be predicted for 2 or 3 years ahead. However, since the tidal range slowly varies over a 19-year cycle and other local geomorphological factors affect it, a record should be made over a period at intervals of a few years, from which the various spectral components can be assessed and used in a further prediction. Tide recording, analysis and prediction is a highly organized business today and is becoming more important as larger vessels must operate by the tide to traverse portions of the continental shelf. Like other long waves the tidal oscillation is magnified across the shelf whose width is a predominant factor in determining the height and time of arrival of the crest to a shore station.

In this chapter some methods will be provided for calculating characteristics of long waves as they propagate across shallow zones of the coast, or into restricted areas, which amplify the height. The graphs and equations can then be applied to surges, tsunamis and tides alike, so long as the periods of each are known or can be measured. Relationships are given for the computation of storm surge from known wind fields and some background information on tides around the world should prove useful.

TIDES

The forces involved in the attraction of the moon and the sun on the waters of the globe are well described by Dean [2], to which the reader is referred. Suffice it to say that the earth and the moon make a single rotating system, the centre of which is located within the radius of the earth. Whilst water nearest the moon is attracted towards it, that on the opposite side of the earth is also forced outward by the increased centrifugal force and the reduced influence of the moon. Thus two "bulges" occur on opposite sides of the globe which are associated with high tides. Because of the 24-hour rotation of the earth and the 27.3-day rotation of the moon around the earth, the periodicity of high tide at points on the great circle which is coplanar with the moon's rotation is 12 hours 25 minutes (44,700 sec). This is termed the *semi-diurnal* (half-daily) tide. Places more distant from this plane of greatest attraction (closer to the poles) experience a lesser tidal range and one which fluctuates once per day or *diurnally*. At intermediate latitudes various combinations of diurnal and semi-diurnal tides occur, in which a high tide is followed by a not-so-high tide, the difference in the sequential peak levels being termed the *diurnal inequality*.

The great circle co-planar with the moon's orbit is angled to the equator an amount of 28°30', known as the *declination*. Because of this, points on it are located in both the Northern and Southern Hemisphere. As the moon rotates about

TIDES

the earth, so the zones nearest to and furthest from the moon vary in location over a period of about 28 days. This results in any one place on the same great circle having periods of high tide and periods of low tide successively. These are known respectively as *spring tides* and *neap tides* and they are each repeated about every 14 days.

The sun also attracts the waters of the oceans and so induces a tidal cycle with a period of 24 hours. Its influence is about half that of the moon. The plane of the earth's rotation about the sun is known as the *ecliptic circle* and this is inclined at 5°9′ to the moon's orbit about the earth. Where the sun and the moon are more or less aligned with the earth (either on the same side or on opposite sides), so will their attractive forces combine to produce extra high tides, known as *equinoctial spring tides*).

Thus at periods of new moon and full moon spring tides occur, the high-water level of which varies throughout the year. For most engineering purposes it is sufficient to average these over 12 months and so obtain a mean high-water spring level (MHWS), as also a mean low-water spring level (MLWS). On the same basis, averages of *neap* tides can be computed (MHWN and MLWN) which occur when the sun and the moon are 90° apart. With such a continuous variation in water level it is difficult to establish a fixed datum for the purpose of hydrographic surveys. One method is to average the low low-water for each year (MLLW) over 19 years. In 1929 the United States adopted a mean sea level as averaged over 19 years.

Although the tide has a wave length of half the earth's circumference and period of about half a day, its speed across the oceans is dictated by their depths. The Atlantic has an average depth of around 4000 metres (see Fig. 4-1A), which gives a wave celerity $\sqrt{gd}$ = 200 m/sec (383 knots). However, the earth's speed of rotation at the equator is 450 m/sec and 225 m/sec at 60° latitude. Thus the tidal oscillation lags behind the point of greatest moon attraction and is, in fact, about 180° out of phase with it. Besides this the major ocean basins have their own natural period of oscillation which, fortunately, are not the same as the semi-diurnal forcing function. The natural period for the Atlantic Ocean is 19 hours and of the Pacific Ocean 42 hours, for dimensions existing at the equator.

Tidal ranges in the deep ocean are difficult to measure, but values at islands with little or no shelf area can give a good approximation. Pekeris and Accad [3] have computed the ranges and relative times of high water for the major M tide, taking friction into account by a linear law only. Fig. 4-1B has been prepared from their maps, which have major coastlines outlined in a 1° latitude-longitude grid. The figure contains *co-tidal lines* (joining points with concurrent HW) and *co-range lines* (joining points with similar tidal range). Co-tidal lines marked with a zero are synchronous with time of lunar transit at Greenwich meridian, whilst range is in contours of 0.25 m. It is seen that tidal crests rotate about certain centres in the oceans, termed *amphidromes*.

For most land boundaries the deep-water oscillations range from 0.25 to 1.0 m,

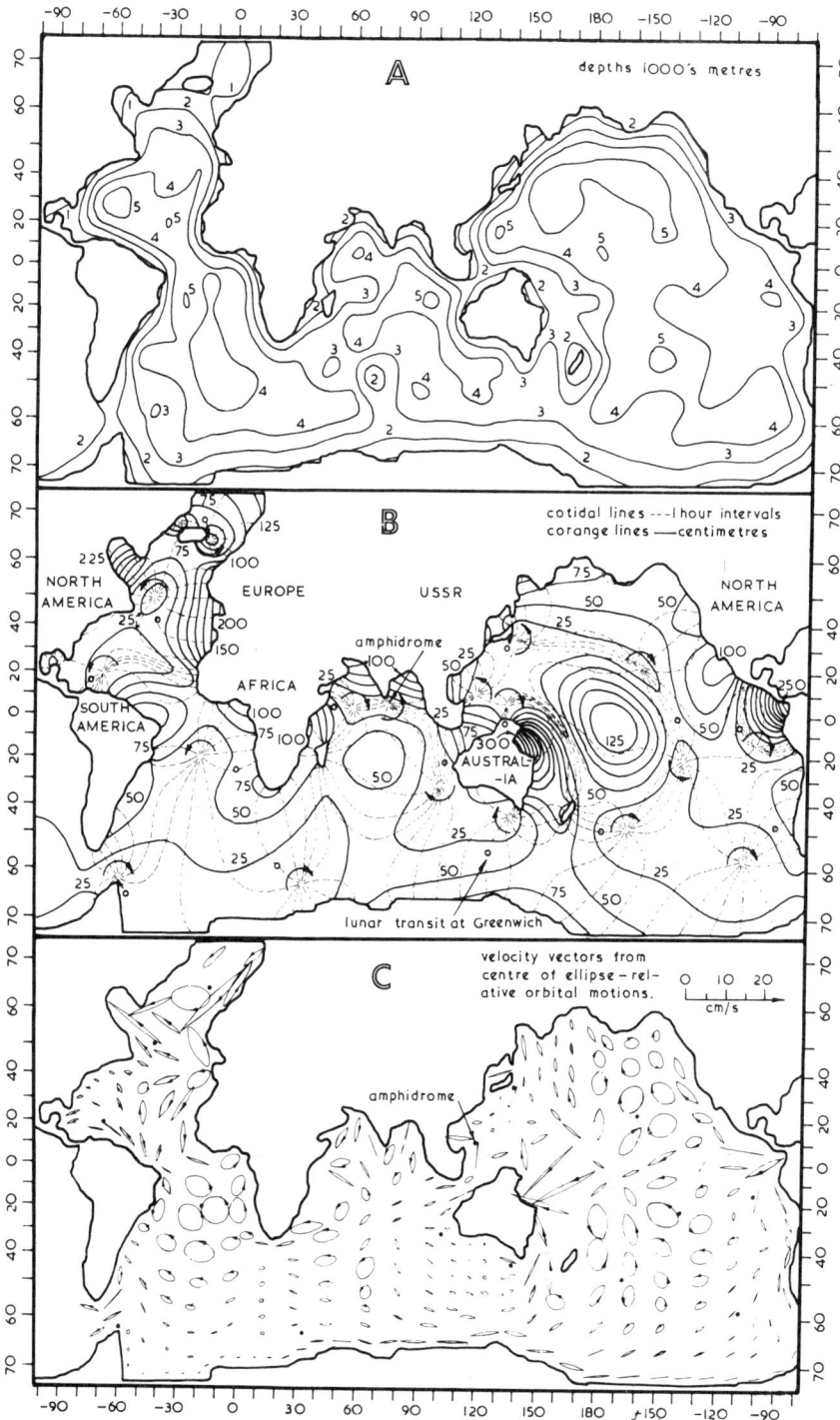

Fig. 4-1. Data on world oceans as presented in Ref. [3]. A. Depth contours; B. Cotidal and corange lines for M_2 tide only; C. Velocity vectors or relative orbital motions of water particles.

although in four locations the offshore tide exceeds 2 m. Comparison of observed island tides with the theoretical values shows that the former are about half the latter, indicating the need for some further refinement in the analysis. Even so, Fig. 4-1B should serve the purpose of supplying the input value for computations of tidal range at the shoreline if 50% is added for the influence of the solar component with almost the same period of oscillation.

Tidal currents are, in essence, the orbital velocities of the water particles. The difference between those for tides and those for waves being the much larger magnitude of the former, which introduces the influence of the earth's rotation or the *Coriolis force*. The currents as computed by Pekeris and Accad [3] are depicted as ellipses of velocity vectors in Fig. 4-1C. These also give the relative magnitudes of water-particle excursion from the centre of the ellipse. It is seen that some places have almost a linear oscillation of water, whilst others have almost a circular orbit. The amphidromic centres are marked by dots, where it is noted that currents have values commensurate with the surroundings. The authors emphasize the preliminary nature of the velocity vector graph from which Fig. 4-1C has been prepared.

Influence of continental shelf

Whilst the negligible ranges and currents produced by tidal action in the open ocean are of little concern to engineers, it is this minimal input that creates the large oscillations at the shoreline, due to the massive amplification as these long waves traverse the shallow continental margins. These margins comprise a gently sloping section termed the *continental shelf*. At a certain depth, which is reasonably uniform around oceanic basins, the slope increases substantially and runs down to the abyssal depths, this section being termed the *continental slope*. Tides can be amplified when travelling up this slope and crossing the shelf to shore.

They can be amplified further by being contained in funnel-shaped indentations, or resonating in near-circular embayments. As tides propagate up rivers of decreasing depth they can also be amplified, but friction at this stage will be taking its toll of tidal height. The main parameter in this increase of tidal range or wave height is the ratio slope length/entry wave length. The wave length may vary as the tide travels over a shoaling bottom.

Transitions – no friction

Considering the overall problem of waves propagating through transitions, either of varying depth or varying width or both, two solutions can be applied in combination. The first assumes an infinitely long transition (or $B \geqslant L$, where B is the length of the transition and L is the wave length) where no reflection takes place. The second assumes a finite length of transition ($B \leqslant L$) where reflection is of importance. Either of these solutions can include or exclude frictional effects.

(A) d_1/d_2 and $Z_A = \dfrac{4\pi d_1^{1/2}}{g^{1/2}TS_v} = \dfrac{4\pi d_1}{T\sqrt{gd_1}\, S_v} = \dfrac{4\pi B}{L_1}$ (4-1)

(B) $(B_1/B_2)^2$ and $Z_B = \dfrac{\pi W_1}{g^{1/2}d_1^{1/2}TS_H} = \dfrac{2\pi B}{L_1}$ (4-2)

(C) $(d_1/d_2) = (W_1/W_2)$ and $Z_c = \dfrac{4\pi^2 d_1}{gT^2 S_v^2} = 4\pi^2 \left(\dfrac{B}{L_1}\right)^2$ (4-3)

or $Z_c = \dfrac{2\pi^2 W_1}{gT^2 S_v S_H} = 4\pi^2 \left(\dfrac{B}{L_i}\right)^2$ (4-4)

Note that in these equations the length B is taken to the zero depth or zero width point (see inset of Fig. 4-2) and that the wave length L_1 refers to the depth d_1 only.

Thus in all cases the ratio B/L_1 is the determining factor, besides width and depth ratios, for the height amplification ratio H_2/H_1. Limiting cases arise where:

(a) $Z_A = 0$ (i.e., $B = 0$), which is an abrupt step giving $\dfrac{H_2}{H_1} = \dfrac{2\sqrt{d_1}}{\sqrt{d_1} + \sqrt{d_2}}$ so that as $d_2 \to 0, H_2/H_1 \to 2$.

(b) $Z_A = \infty$ (i.e., $B = \infty$) which is an infinitely long channel giving $\dfrac{H_2}{H_1} = \left(\dfrac{d_1}{d_2}\right)^{1/4}$ for shallow-water conditions (i.e., $C = \sqrt{gd}$), so that as $d_2 \to 0, H_2/H_1 \to \infty$, limited only by friction and breaking.

(c) $Z_B = 0$ (i.e., $B = 0$) the abrupt reduction in width gives $\dfrac{H_2}{H_1} = \dfrac{2W_1}{W_1 + W_2}$ so that as $W_2 \to 0, H_2/H_1 \to 2$, the same as for an abrupt reduction of depth in (a) above.

(d) $Z_B = \infty$ (i.e., $B = \infty$) which is an infinitely long channel giving $\dfrac{H_2}{H_1} = \left(\dfrac{b_1}{b_2}\right)^{1/2}$ for shallow-water conditions (i.e., $C = \sqrt{gd}$), so that as $W_2 \to 0, H_2/H_1 \to \infty$, not limited as much as in b above because of the constant depth d_1.

In conditions a and c above pure reflection is accepted (due to the long wave length), so that:

$$H_t = H_i + H_r = 2H_i \quad \text{or} \quad \dfrac{H_t}{H_i} = 2 \qquad (4\text{-}5)$$

The general solution for a swift reduction in width and depth (i.e., $Z_C = 0$) is:

TIDES

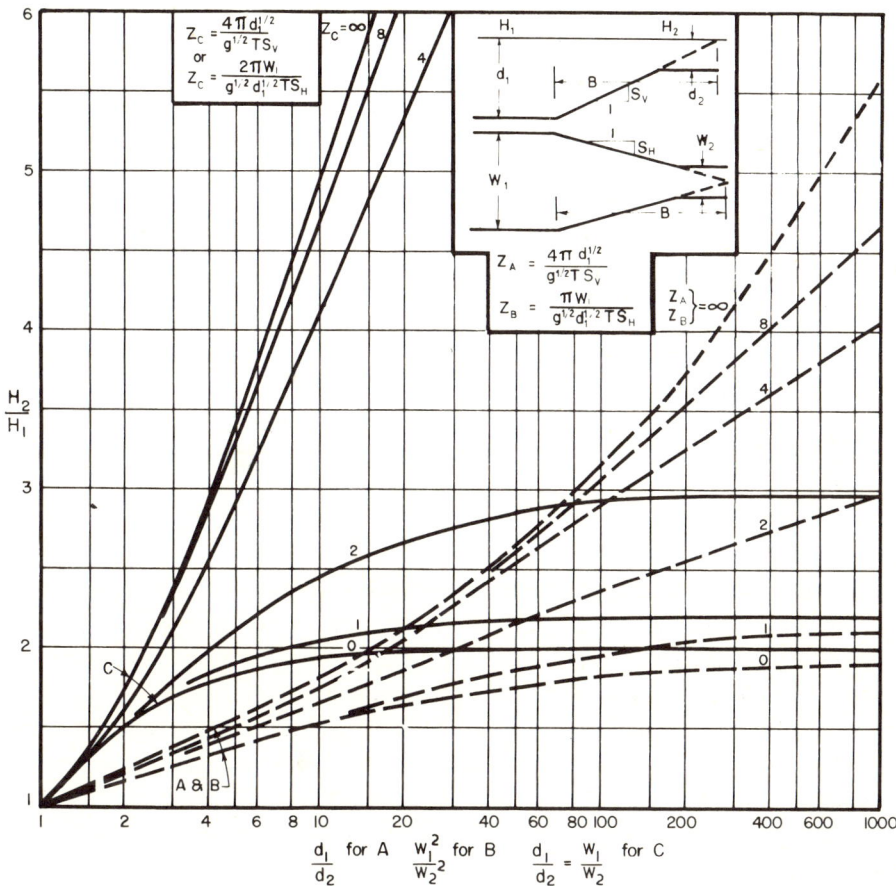

Fig. 4-2. Amplification of waves in channel transitions.

Of the latter Dean [4] has provided an analysis which can be readily applied. From his comprehensive coverage Fig. 4-2 has been prepared, which also contains the definition sketch where the suffix *1* refers to the inlet end of the transition and suffix *2* to any intermediate inside plane, or at the outlet end of the transition. The ratio H_2/H_1 is given for conditions: (*A*) uniform width W, reduction in depth d_1 to d_2; (*B*) uniform depth d_1, reduction in width W_1 to W_2; (*C*) reduction in width and depth.

For each of these conditions two sets of dimensionless ratios are employed, namely:

$$\frac{H_2}{H_1} = \frac{2L_1 W_1 n_1}{L_1 W_1 n_1 + L_2 W_2 n_2} \tag{4-6}$$

where $n = C_g/C$ as in eq. 4-35 of *Coastal Engineering, I*, which for shallow-water conditions gives: $n_1 = n_2 = 1$, $L_1 = T\sqrt{gd_1}$, $L_2 = T\sqrt{gd_2}$ and $d_2 = W_2 = 0$ reduces to $H_2/H_1 = 2$, the same as eq. 4-5 and the result for conditions *a* and *c*.

In conditions *b* and *d* above no reflection is considered, but all energy is conserved. This is the application of Green's law, in which the energy per unit area of water surface is constant at all points along the transition (i.e., $Z_C = \infty$), so that:

$$E = L_1 W_1 n_1 \, wH_1^2 = L_2 W_2 n_2 \, wH_2^2 \tag{4-7}$$

which for shallow-water conditions, in which $n_1 = n_2 = 1$, $L_1 = T\sqrt{gd_1}$ and $L_2 = T\sqrt{gd_2}$, reduces to:

$$\frac{H_2}{H_1} = \left(\frac{W_1}{W_2}\right)^{1/2} \left(\frac{d_1}{d_2}\right)^{1/4} \tag{4-8}$$

which is the same result as for conditions *b* and *d*.

For intermediate values of Z_A, Z_B and Z_C Fig. 4-2 provides values of H_2/H_1. For values beyond the limits of this figure appropriate relationships can be chosen from the above. Dorrestein [5] has shown that for varying depth only (condition *A*) a resonant length ratio of $B/L_1 = 0.192$ exists where H_2/H_1 becomes large. From eq. 4-1 this is equivalent to $Z_A = 2.4$, which is the point in Fig. 4-2 where the Z_A curves commence to curve upwards. For the varying width case the critical $B/L_1 = 0.383$, which from eq. 4-2 gives $Z_B = 2.4$, the same condition as for *A* in Fig. 4-2.

Where a change in depth is involved the wave is decelerated so that it takes longer to arrive at plane 2 than if depth d_1 existed throughout. This lag time can be determined for a uniform change in depth by:

$$\tan \frac{2\pi t}{T} = \left(\tan \frac{2\pi B}{L_1}\right) \frac{2\pi d_2}{B} \tag{4-9}$$

Transitions – with friction – progressive wave

For the purpose of analyzing changing depths it is convenient to assume such a transition consisting of a number of steps, each being a channel of uniform depth, with or without width variation. In each elemental channel attenuation due to friction can be considered and at each step the influence of reflection taken into account, in order to derive the wave height transmitted to the next section with a differing uniform depth.

The assumptions regarding a friction factor applied to wave action are:

(a) It is considered constant regardless of the varying orbital velocity (u).

(b) Harmonic motion exists i.e., $u = u_{max} \cos 2\pi t/T$ and varies with distance along the channel.

(c) Friction loss is based upon work done over a tidal cycle, the same for linear approximation as for the quadratic law of open channels.

(d) Friction factor is taken for u_{max}.

Using Darcy's resistance coefficient f_w for the wave, Ippen and Harleman [6] have shown that:

$$M = \frac{f_w u_{max}}{3\pi gd} = \frac{f_w H}{6\pi g^{1/2} d^{3/2}} \quad \text{(see eq. 4-22 in } Coastal\ Engineering,\ I\text{)} \quad (4\text{-}10)$$

Factor M was also equated as follows:

$$M = \frac{2\pi \tan 2\alpha}{Tg} = \frac{2\pi}{Tg}\left(\frac{2\tan\alpha}{1-\tan^2\alpha}\right) \quad (4\text{-}11)$$

where $\tan\alpha = \mu L/2\pi$ where μ = damping coefficient which is considered constant for the channel length, and L = wave length at any point inside the transition.

Equating M in eq. 4-10 and 4-11 gives:

$$f_w HT\sqrt{gd} = 24\pi^2 d^2 \left[\frac{\mu L/2\pi}{1-(\mu L/2\pi)^2}\right] \quad (4\text{-}12)$$

Using the suffix 1 to denote conditions at the entrance of the uniform channel, we have:

$$(2\pi/L_1)^2 = (2\pi/L)^2 - \mu^2 \quad (4\text{-}13)$$

so that:

$$\mu = \frac{2\pi}{LL_1}\sqrt{L_1^2 - L^2} \quad (4\text{-}14)$$

which on substitution into eq. 4-12 gives:

$$\frac{f_w H L_1}{d^2} = 24\pi^2 \frac{\sqrt{1-(L/L_1)^2}}{(L/L_1)^2} \quad (4\text{-}15)$$

In eq. 4-15 H is a representative wave height in the channel, taken to be the height on entry for a uniform channel or a mean height where the sides are converging. The depth d is also a mean value for the depth at entry and exit for the elemental length. Eq. 4-15 is graphed in Fig. 4-3, where the reduction in wave length ($L_1 - L$) due to friction can be ascertained. It is seen that increased friction reduces the ratio L/L_1.

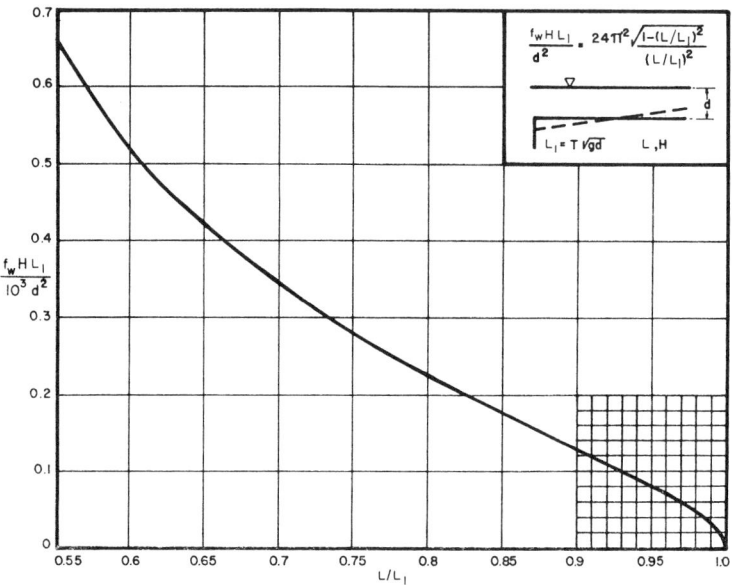

Fig. 4-3. Relationship between friction factor and wave length reduction.

To find values of f_w to be used in eq. 4-15, portion of the graphs derived by Jonsson [7] are modified and represented in Fig. 4-4 for shallow-water conditions in which $C = \sqrt{gd}$. Jonsson's abscissa used a wave Reynolds number $R_w = u_{max} x_{max}/\nu$

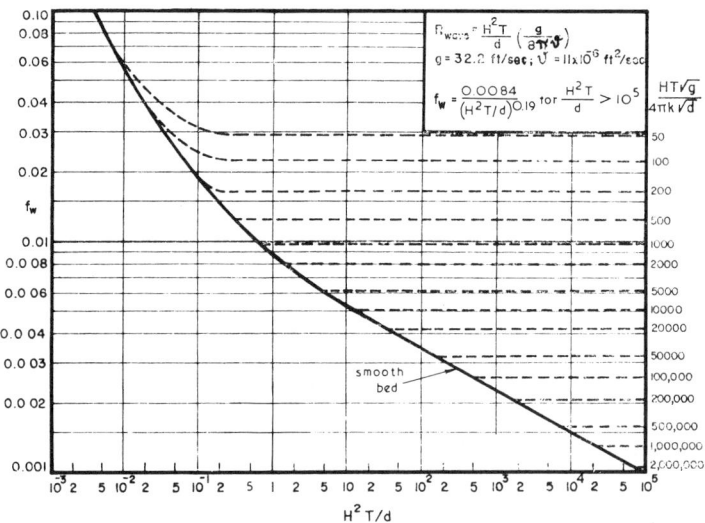

Fig. 4-4. Wave friction factor f_w for long waves where $2\pi d/gT^2 < 0.01$.

where x_{max} is the maximum horizontal excursion of a water particle from its mean position. The ordinate was defined by $\tau_{0_{max}} = f_w \frac{1}{2}\rho u_{max}^2$. The Reynolds number can be expressed from linear theory for shallow water conditions as:

$$R_w = \frac{H^2 T}{d}\left(\frac{g}{8\pi\nu}\right) \tag{4-16}$$

By substituting an appropriate value of $\nu = 11 \cdot 10^{-6}$ ft.^{2}/sec for seawater at ambient temperature, an appropriate conversion has been made in Fig. 4-4 for R_w in f.p.s units. The bed roughness ratio used by Jonsson for waves was x_{max}/k where k was the Nikuradse sand roughness on a smooth surface. In the context of long-period waves, this should be replaced by dune or ripple height k in which:

$$\frac{x_{max}}{k} = \frac{HTg^{1/2}}{4\pi k d^{1/2}} \tag{4-17}$$

which parameter is identified in Fig. 4-4. As shown by Kajiura [8] for sufficiently large values of d/k the friction factor varies only with this parameter. As noted in Fig. 4-4 for values of H^2T/d exceeding 10^5 on the smooth bed curve f_w is given by:

$$f_w = 0.0084/(H^2T/d)^{0.19} \tag{4-18}$$

Note that Fig. 4-4 is applicable only to shallow-water waves (i.e., $2\pi d/T^2 g \leqslant 0.01$).

Assuming an infinitely long channel, or no reflection, the wave progressing along the elemental length of channel can be considered progressive in character. In this case the height attenuation [6] is given by:

$$H_B/H = \exp(-\mu B) \tag{4-19}$$

where H_B is the height at a distance B from the entrance of the elemental section of channel. Thus:

$$\frac{H_B}{H} = \exp\left(-\frac{2\pi B}{L_1}\sqrt{(L_1/L)^2 - 1}\right) \tag{4-20}$$

where $L_1 = T\sqrt{gd}$ and L is the wave length at distance B from entrance. Eq. 4-20 has been graphed in Fig. 4-5 for a range of L/L_1 and $B/T\sqrt{gd}$. It is seen that as L/L_1 decreases or $B/T\sqrt{gd}$ increases so H_B/H decreases. In practice, if H, H_B and B are measured μ can be determined from eq. 4-19. This could also be checked by measuring celerity C in the channel and computing the frictionless value $C_1 = \sqrt{gd}$. Then from $C/C_1 = L/L_1$ in eq. 4-14, μ can be computed. By use of Fig. 4-3 and 4-4 the roughness dimension k could be ascertained, for further calculations on the one stretch of channel, or for applications in channels of similar size with similar tidal velocities. Values of L/L_1 have been found to vary [6] in rivers and canals from 0.7 to 0.95.

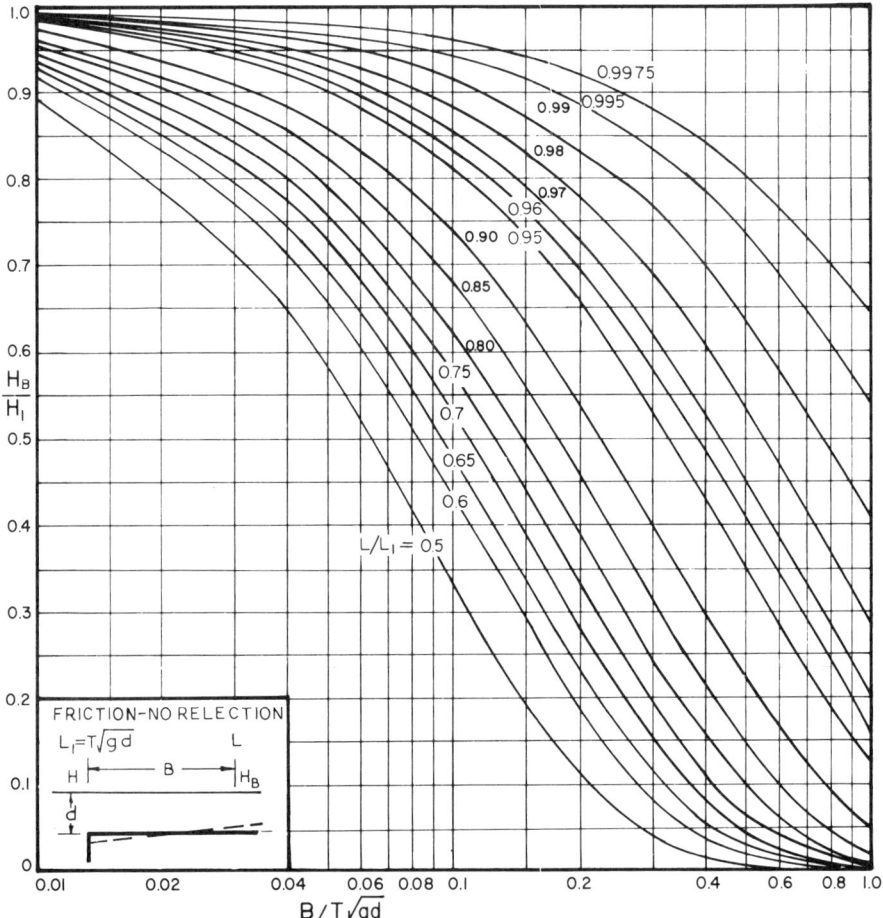

Fig. 4-5. Wave height attenuation along a channel due to friction.

The tidal currents (from the water-particle orbital velocities in shallow water) are attenuated along the channel, both due to the reduction in wave height and to the smaller wave length. The maximum tidal current at distance B from the entrance ($U_{B\,max}$) is given by [6]:

$$\left(\frac{U_B}{U_1}\right)_{max} = \frac{\exp\left(-\frac{2\pi B}{L_1}\sqrt{(L_1/L)^2 - 1}\right)}{\sqrt{2(L_1/L)^2 - 1}} \qquad (4\text{-}21)$$

In Fig. 4-6 $(U_B/U_1)_{max}$ has been graphed against L/L_1 for a range of B/L_1. It is seen that as L/L_1 reduces from 1.0 to 0.8 the velocity ratio reduces swiftly. The lower

TIDES 171

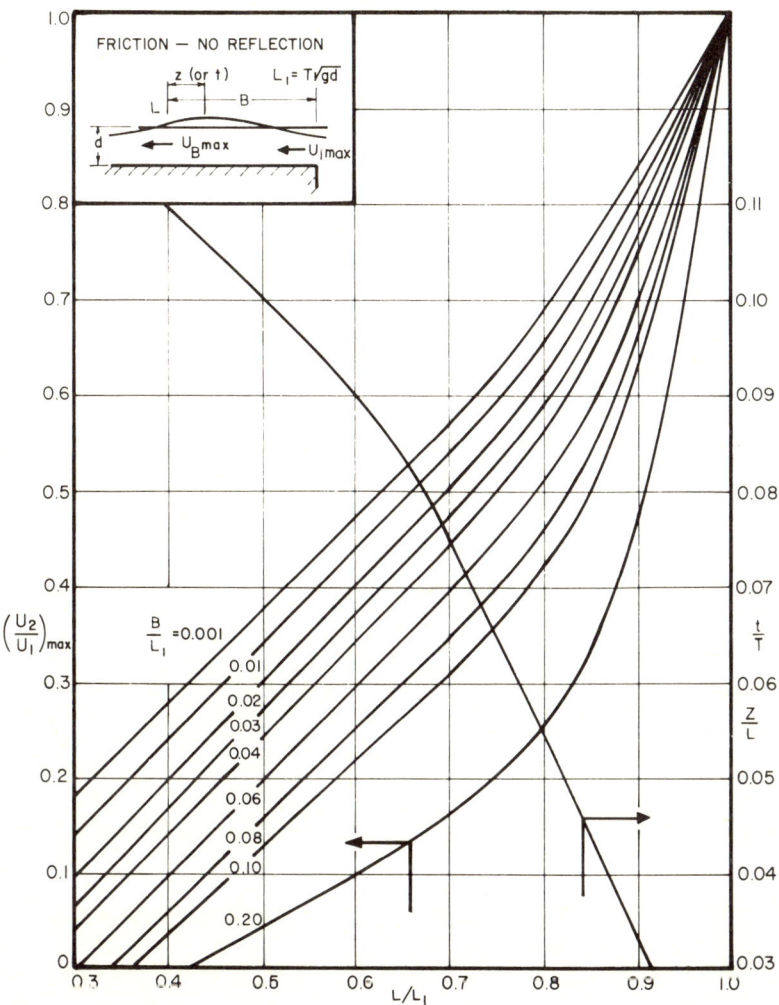

Fig. 4-6. Tidal current attenuation along a channel due to friction, and phase difference of maximum velocity and high water.

limit of B/L_1 contained in the figure (= 0.2) gives, for a 30 ft. deep channel and a semidiurnal tide, $B = 0.2 \times 44700\sqrt{32.2 \times 30}/5280 = 52.6$ statute miles. Comparing tidal ranges and tidal currents for $B/L_1 = 0.1$ and $L/L_1 = 0.8$, results in $(H_B/H_1)_{max}$ = 0.625 and $(U_B/U_1)_{max}$ = 0.425, which indicates a swifter reduction of velocity than of tidal range.

The velocity $(U_1)_{max}$ coincides with H_{max} at the entrance to the channel, but at distance B along the channel $(U_B)_{max}$ precedes high-tide level by time t (or distance

Z) given by:

$$\tan(2\pi t/T) = \sqrt{1 - (L/L_1)^2} \tag{4-22}$$

which is also graphed in Fig. 4-6, either as t/T or Z/L against L/L_1.

If in the above cases of progressive tidal waves the channel is converging in width, Green's law should be applied to height or range and the mean height used for the calculations. Where the depth is changing the friction factor should be based on wave characteristics at mid-depth in the elemental length of channel.

Transitions – with friction – total reflection

Where a channel is of finite length and has a reasonable reflecting surface at its head (such as a dam), then total reflection of the wave existing as a progressive wave to that point can be accepted. This doubles the height and this crest is

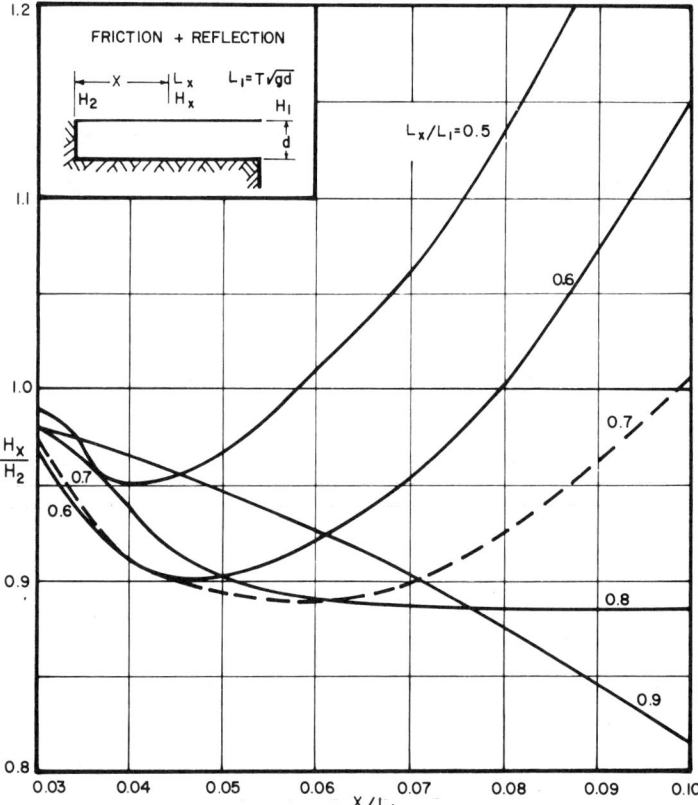

Fig. 4-7. Wave height attenuation from head of a channel for a standing shallow-water wave.

TIDES

progressively returned back to the entrance. The maximum tidal range $(H_X)_{max}$ at a point X from the channel end is given by [6]:

$$(H_X/H_2)_{max} = \sqrt{\cos 4\pi X/L_1 + \cosh 2\mu X/2} \qquad (4\text{-}23)$$

where $(H_2)_{max}$ is the maximum range experienced at the head of the channel. Eq. 4-23 is graphed in Fig. 4-7 where it is seen that for certain distances and certain L_X/L_1 the intermediate ranges are smaller than that at the channel head.

The time of high water at any point distance X from the closed end, in respect to high water at the end, is given by:

$$\tan \frac{2\pi t}{T} = \tan 2\pi \frac{X}{L_1} \frac{L_1}{L_x} \tanh \frac{2\pi X}{L_1} \sqrt{(L_1/L_x)^2 - 1} \qquad (4\text{-}24)$$

which is graphed in Fig. 4-8. In the case of no friction ($L_x/L_1 = 1$), then $t/T = 0$ and all stations along the channel have *HW* simultaneously.

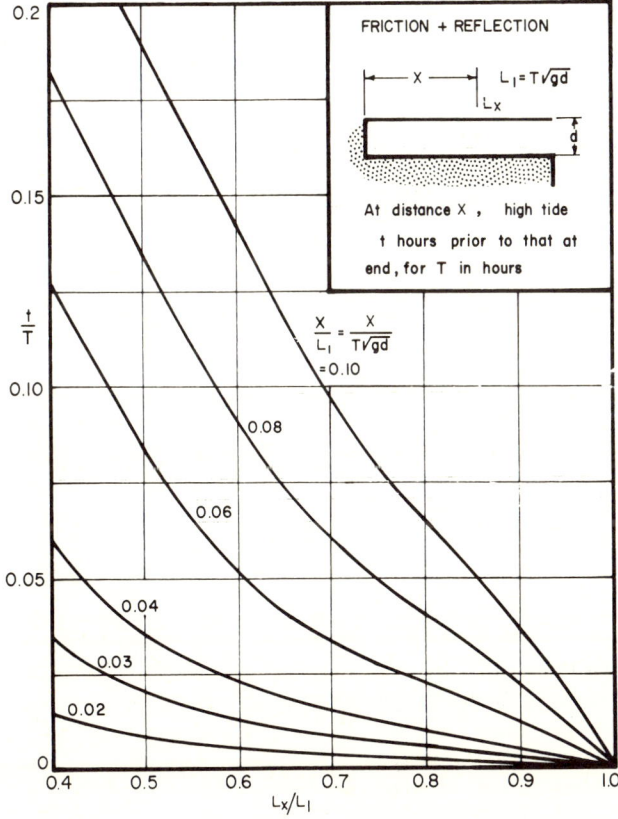

Fig. 4-8. Time differential of *HW* at the head of a channel and at points along it when friction is present.

An analysis is available [53], similar to the above, for a negative reflection where a lake exists at the head of an estuary.

Typical tides

Shepard [9] has computed averages from profiles measured around continental shelves of the world. He differentiates between the average of all and that of those shelves that are completely sedimentary. These are depicted in Fig. 4-9 where it is seen that the shelf edge could be taken as a depth of 65 to 72 fathoms. If the parabolic profile close to shore of a sedimentary coast, as proposed by Sitarz [10] is accepted, the 35-fathom mildly sloped zone would commence 7 NM's from shore in an overall width of 30 NM. The 5-fathom contour would appear 1/5 NM from shore. The equivalent uniform slope of the shelf averages 0.002 and that of the slope 0.06.

Such averages are not particularly useful to the coastal engineer who is concerned with a specific area of coast where the continental shelf could vary from 1 to 200 NM. To compute possible tides or variations of tide from one palce to another he must gather detailed information of the offshore hydrography. He should know the depth to the shelf edge (where the slope steepens appreciably), the distance from shore to the edge, the general profile, the depth just offshore, any major undulations of the shelf and the shoreline features such as embayments or river outlets.

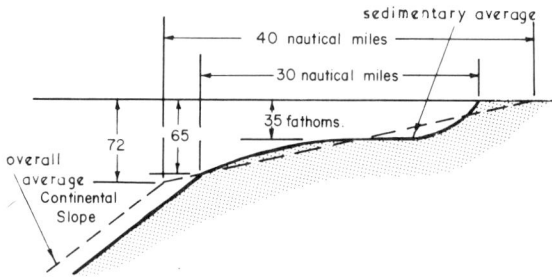

Fig. 4-9. Average profiles of continental shelves. (Ref. [9].)

But taking the overall average profile as in Fig. 4-9, with amplification factors as calculated in examples at the end of the chapter, the following tidal range results:

moon tide in deep ocean = (Fig. 4-1B)	0.625 m
solar addition (× 0.457) = 0.286	0.911 m
continental slope (× 1.65) $Z_A = 0$	1.5 m
continental shelf (× 1.6) $Z_A = 0.5$	2.41 m

Thus it would appear that 2.5 m would be the average tide from a deep-ocean oscillation of 0.625 m. This appears to be borne out by the tidal distribution depicted in Fig. 4-10, which has been modified from a presentation by Dietrich [11] of the mean spring tide $[2(M_2+S_2)]$. In making such a calculation as above it is seen from Fig. 4-2 that the choice of near-shore depth d_2 is very critical, especially when $Z_A \geqslant 2.4$ and the curve concaves upwards. If d_2 is assumed small, because of tidal flats etc., $d_1/d_2 \to \infty$ making for large H_2/H_1. With a normal parabolic profile it is wise to take the depth a short distance offshore, say 0.2 NM where the depth is likely to be 5 fathoms; because the volume of water shoreward of this line would not greatly affect the tidal action.

Fig. 4-10 displays a number of areas where tides are exceptionally large, up to 17 m. These are caused not only by a large width of shelf because of a substantial indentation, but also because these indentations are funnel-shaped. Such reductions in width of the incoming tidal crest concentrates energy and so amplifies the range. The Gulf of Kutch is treated as an example at the end of the chapter, and illustrates this point.

To illustrate the influence of shelf width and shoreline features on tidal range Fig. 4-11 compares these two variables in the three major ocean basins. Profiles on east and west coasts at various latitudes have been measured together with the tidal information at chosen port sites along them. Even this rough comparison shows a close correlation of tide and shelf characteristics. Some anomalies can be explained by the large deep-water tides as noted in Fig. 4-1, for example, the excessive tide on the Pacific side of Panama.

Tidal currents on the continental shelf are much greater than in the deep ocean and are, naturally, more influenced by the boundaries. In a progressive wave the maximum velocity occurs at the crest. With standing waves the largest horizontal velocities are at nodes or trough locations, or 90° out of phase with the crest. In partial standing waves the phase difference is between 0° and 90°. Thus tidal currents cannot be predicted readily from tidal range, so they should be recorded at intervals over a tidal cycle, with these intervals related to time of *HW*. When taking measurements for a coastal engineering project such velocities should be recorded at all depths, together with current directions during the interval. It will normally be found that as the bed is approached so the fluctuations in velocity and direction become greater. This is due to the vortices generated around obstacles or from undulations of the bed. These have a great influence on sedimentary processes and may cause difficulties in navigation of deep-draft vessels in channels dredged across the continental shelf.

For a prediction to be made of future tides at any location a record must be available over a portion, or preferably over the whole of a year. This can then be analyzed harmonically to find the constituents present. Each constituent has a period, a phase angle and an amplitude, and can be caused by the moon, the sun,

176   LONG-PERIOD WAVES

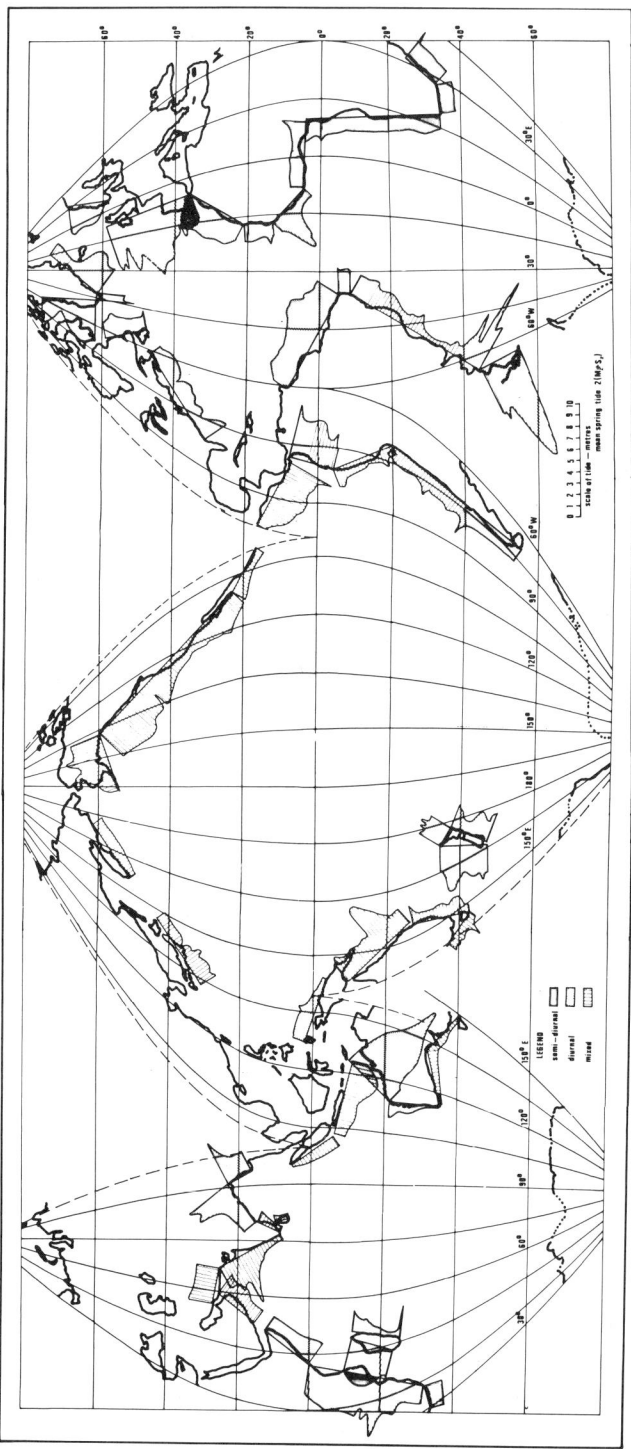

Fig. 4-10. Distribution of mean spring tide around the coastlines of the world.

TIDES 177

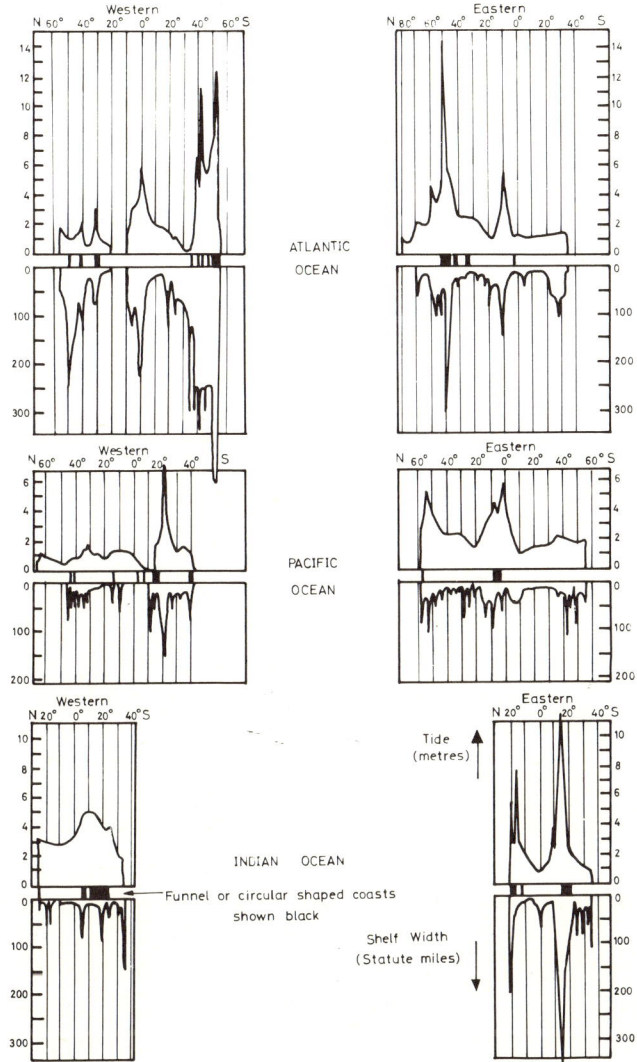

Fig. 4-11. Comparison of tidal range with width and other features of the continental shelf.

the ocean basin, the shelf, the coast and any adjoining waterways. Substantial silting or scouring can alter the tidal response at a port. It is therefore wise to have a new analysis carried out every few years. Once the constituents are determined they can be fed into a computer and future oscillations can be predicted in respect to date and time of day. These are published for the major ports of the world by several maritime institutes in the form of "tide tables".

As stated already, tidal information has reached a highly developed stage, but there are still many avenues for research [12,13]. The references supplied should assist the reader closely concerned with the subject to have an initial insight into the literature [14–26].

STORM SURGE

Storm surges are sometimes called meteorological tides since they are caused by the low barometric pressure and associated high wind velocities of cyclonic centres. As such centres travel across or along the continental shelf, or over enclosed bodies of water, so is a suction and a stress applied to the surface which results in a relatively swift build-up in level at specific points on the boundary. When the stress is released so the water will return to normality, but not without the despatch of a long wave to the surrounding zone of ocean. In the case of a lake or small sea, such an initial surge may set up a seiche or oscillation which is attenuated with time by friction. The main concern in this presentation is to compute the maximum surge height for a given cyclonic condition.

Wind stress

Generally the largest component of a storm surge is supplied by the wind stress on the water surface, which is a function of its roughness. However, this roughness is not directly associated with the concomitant storm waves generated, since it is the smaller elements of roughness which provide the "grip" for the wind on the sea surface. This can be likened to the sand-grain roughness rather than the bed dunes providing resistance to water flow in channels. Wu [27] has illustrated how waves in the order of 0.7 sec period and 10 cm height are the main roughness element. The stress thus varies with the wind velocity and the fetch available. For this reason it is understandable that optimum values can be reached which are different for enclosed bodies of water of limited size and the open sea. The former contain a larger proportion of short waves, which are approaching or are at their limiting steepness. When the latter arrive at their FAS condition they contain less of the short-period components and more of the longer-period waves [28]. The sea is slightly smoother, therefore, as far as wind stress is concerned, than a limited zone of inland waterway. It could also be presumed that the stress developed will vary with the duration and fetch of the wind.

The shear stress τ is given by:

$$\tau = \rho_a U_*^2 \tag{4-25}$$

where ρ_a is the density of air and U_* is the shear velocity near the water surface. This can be expressed as:

$$\tau = \rho_a C_y U_y^2 \tag{4-26}$$

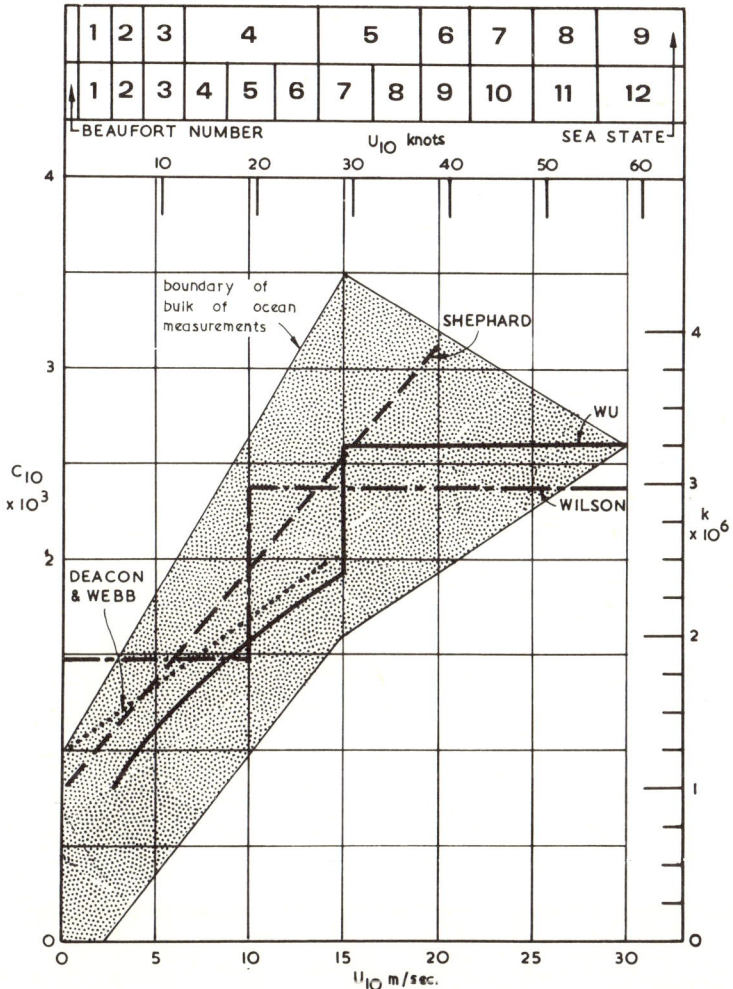

Fig. 4-12. Wind stress coefficient for various wind velocities, sea states or Beaufort numbers.

where C_y is the resistance coefficient which varies with U_y and U_y is the wind velocity at y metres above the water surface.

Many workers [29–31] have evolved relationships for C_y and U_y from measurements over lakes and the sea. Wu [32] has summarized these data and discussed their significance, from which Fig. 4-12 has been prepared. This shows C_{10} for a range of U_{10} (as most wind measurements have been made at the 10 m height). Within the boundary of the experimental values of the figure, specific relationships are identified.

It is now generally accepted that a limiting value of $C_{10} = 2.6 \cdot 10^{-3}$ can be applied to limited bodies of water when $U_{10} \geqslant 15$ m/sec, 30 knots, sea state no.5

or Beaufort no.7. The optimum for the open ocean is $C_{10} = 2.4 \cdot 10^{-3}$. For winds less than this, a relationship of:

$$C_{10} = 0.65 \cdot 10^{-3} \, U_{10}^{1/2} \tag{4-27}$$

(when U_{10} is expressed in m/sec) would appear from Fig. 4-12 to be acceptable, curving into the aforesaid limits at $U_{10} = 15$ m/sec. Wind velocities at other levels can be converted to values at 10 m height by use of Fig. 3-8 of *Coastal Engineering, I*.

In storm surge calculations it is more appropriate to use the relationship:

$$\tau = k \rho U_{10}^2 \tag{4-28}$$

where ρ is the density of water.

From eq. 4-26 it is seen that $k = \rho_a C_y / \rho$, which for seawater $= C_{10}/800$. The limiting values of $k = 3.3 \cdot 10^{-6}$ and $3.0 \cdot 10^{-6}$ are applicable to lakes and oceans respectively, as seen in Fig. 4-12.

Enclosed bodies of water

Considering a lake of rectangular proportions in plan, its longitudinal cross-section can be equated to a rectangular shape of equal area, giving an equivalent depth d and a length or fetch L. The equation for the water surface profile [33–35], when a steady wind velocity (U_{10}) is applied to it, is:

$$\frac{S}{d} = \frac{k_{10} U_{10}^2 L}{2gd(d+S)} = \frac{k_{10} U_{10}^2 L}{2gd^2} \tag{4-29}$$

where S is surge height above SWL at the downwind end. The second form of the equation assumes S to be small in respect to d. Eq. 4-29 also approximates the nodal point of the water slope at the centre of the length L.

The variables in this dimensionless relationship must be chosen consistently. The values of the fall (h_u) and rise (h_d) of the water surface at the upwind and downwind ends of the lake are presented in graphical form in Fig. 4-13. When the upwind bed is exposed, Fig. 4-14 should be used to evaluate the downwind surge height (h_d). Note that in these two figures h_d has replaced S in eq. 4-29. The surface passes through the mean depth plane around the centre of the water body [33] even when part of the bed is exposed. Lakes of irregular depth can be treated similarly by deriving an equivalent rectangular section.

In the case of narrow bodies of water, such as canals, the same value of $k_{10} = 3.3 \cdot 10^{-6}$ is applicable, since the optimum roughness is reached when waves of about 1-sec period reach their maximum steepness of around 1/7 [27].

For non-rectangular shapes in plan, of uniform depth or sloping bottoms,

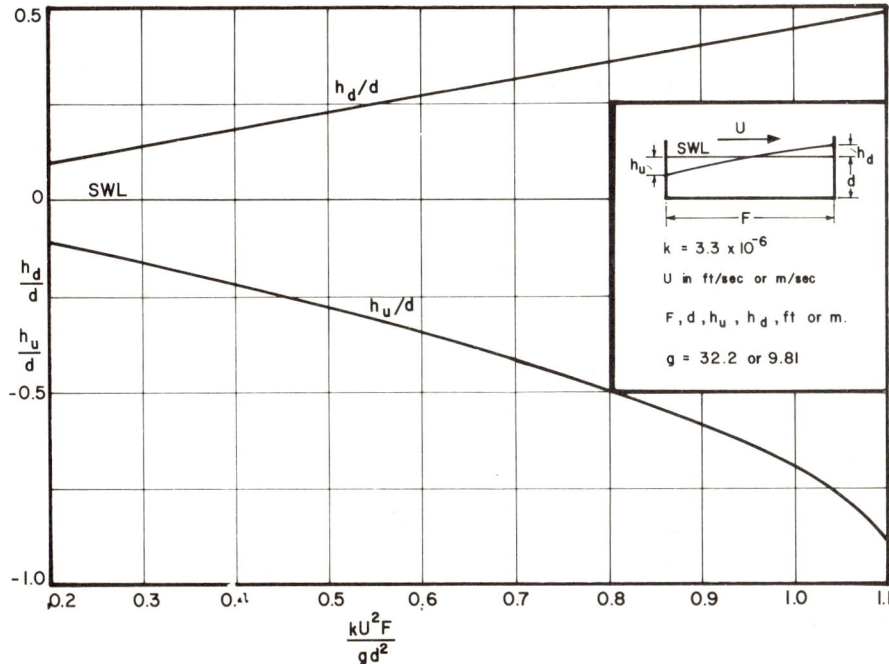

Fig. 4-13. Upwind and downwind surge levels in a lake of uniform depth and equivalent rectangular plan.

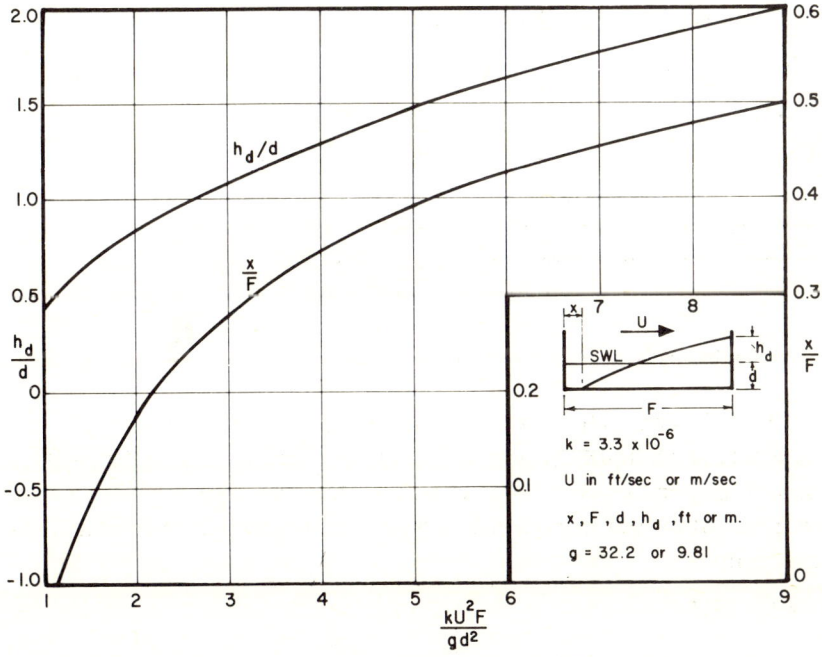

Fig. 4-14. Surge levels in a lake when the upwind bed is exposed.

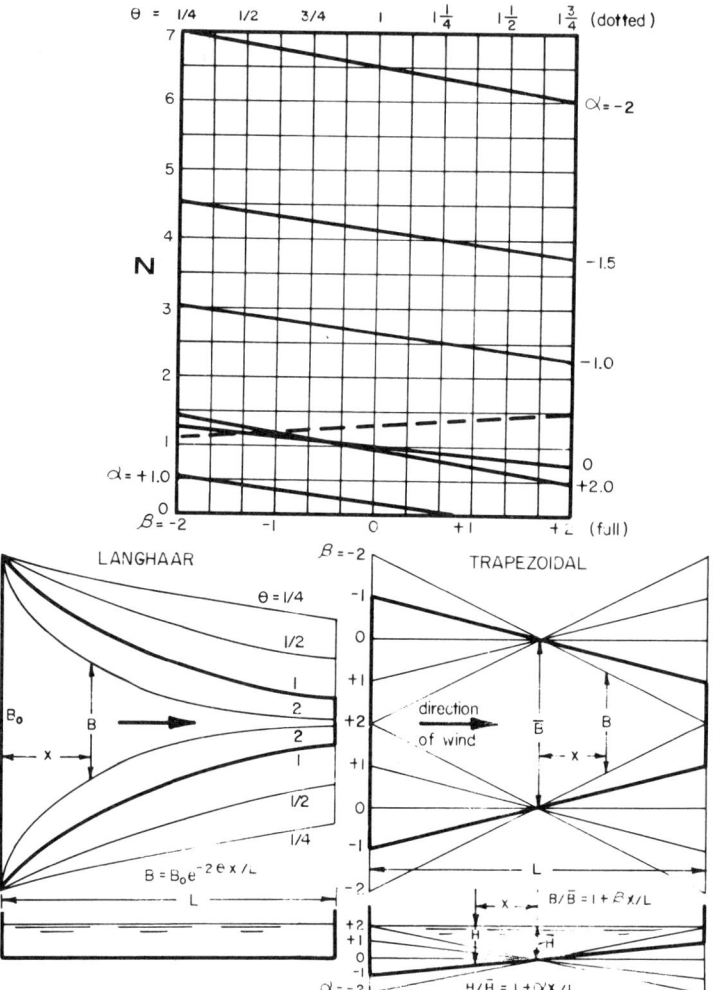

Fig. 4-15. Amplification factor N for Langhaar and trapezoidal lakes.

Keulegan [36] has derived form factors N, by which the values of S/d (i.e., h_d/d) in eq. 4-29 should be multiplied. These are given in the upper portion of Fig. 4-15 for two major shapes and cross-sections. The trapezoidal plan lakes are defined from their mean breadth $\bar{B}$ and their change in breadth in the up- and down-wind directions such that:

$$B/\bar{B} = 1 + \beta X/L \tag{4-30}$$

where B is the breadth at distance X from the mean and β is a factor varying from -2 to $+2$. Similarly the cross-section of these trapezoidal lakes can vary in depth H, so that:

$$H/\bar{H} = 1 + \alpha X/L \tag{4-31}$$

where α is a factor varying from -2 to $+2$.

The Langhaar lake is of uniform depth but exponential in plan, where the breadth B is given by:

$$B = B_o \exp(-2\theta X/L) \tag{4-32}$$

where B_o is the upwind breadth; X is distance measured downwind from B_o; θ is a factor varying from 1/4 to 2 for the shapes illustrated.

The high values of N in Fig. 4-15 should be noted for the cases where the downwind depth decreases. The same order of magnification will be observed for similar shoaling on the continental shelf.

Continental shelf

In the ocean situation the shallow area of water surface affected by wind stress can be as great or greater than the wind field, whereas for lakes the limiting factor is the area of the water body. Also the motion of the wind field assumes importance. Surges over the continental shelf will be divided into two classes, namely, static and moving wind fields.

Static wind field. Where a wind is blowing shorewards the shelf width can be taken as half the length of the lake considered previously. The water feeding the surge comes from the deep ocean. The major difference from the previous situation is the decreasing depth towards the shore. For the purposes of surge calculation it is reasonable to assume a uniform variation from a depth d_1 at the shelf edge to d_2 near the coast (see inset of Fig. 4-16).

From the average profiles in Fig. 4-9 it is reasonable to assume $d_1 = 400$ ft. (120 m) if no other information is available on the shelf profile. Also the shoreline depth d_2 could be assumed to approximate 30 ft., which depth exists within 0.2 NM of the beach generally. Where a large tidal range is associated with the inundation of extensive sand or mud flats, a smaller value of d_2 should be assumed.

The width of the shelf will be designated as L, which can be different from the fetch length F of the wind field. As seen in Fig. 4-16:

$$d_1/d_2 = (L/x) + 1 \tag{4-33}$$

where x is the horizontal distance from the beach to where the offshore bed profile meets the SWL. Where a storm zone has a fetch (F) in excess of the shelf width (L), only that portion across the relatively shallow zone is effective in producing surge, thus $F = L$. This is likely to be the case in extra-tropical cyclones, where wide

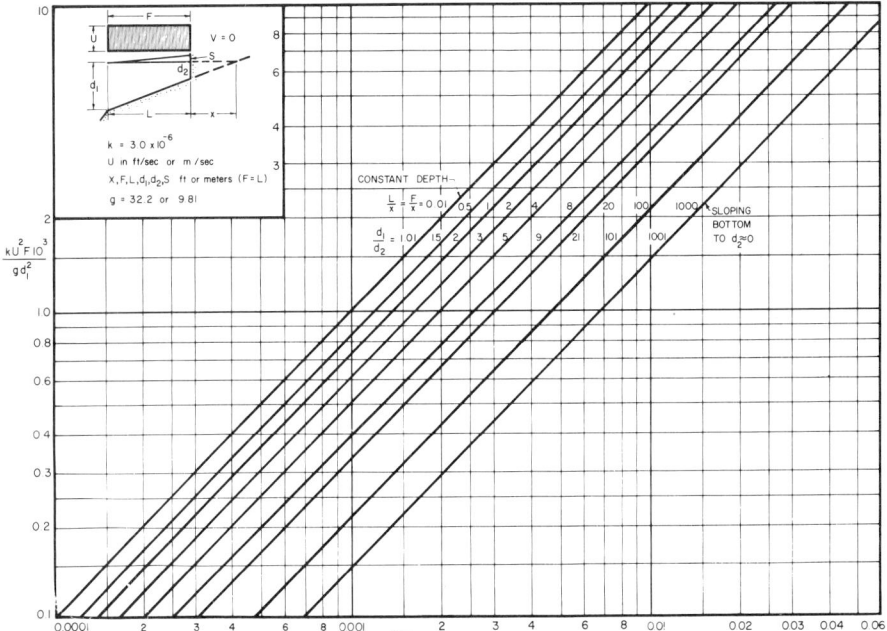

Fig. 4-16. Surge produced by a static uniform wind field extending across the continental shelf.

expanses of ocean can suffer winds of relatively uniform speed and direction. In tropical cyclones the fetch lengths are smaller and are more likely to be less than the shelf width.

(a) Uniform wind velocity ($V = 0$). For a wind of steady and uniform speed applied to a continental shelf the following formula has been derived [37]:

$$S = \frac{kU^2L}{g(d_1 - d_2 - S)} \ln\left(\frac{d_1}{d_2 + S}\right) \quad (4\text{-}34)$$

Since S is small in respect to d_2 eq. 4-34 can be rewritten:

$$\frac{S}{d_1} = \frac{kU^2}{gd_1}\left(\frac{L}{d_1 - d_2}\right)\ln\frac{d_1}{d_2} \quad (4\text{-}35)$$

so that:

$$\frac{S}{d_1} = \frac{kU^2L}{gd_1^2(1 - d_2/d_1)} \ln\frac{d_1}{d_2} \quad (4\text{-}36)$$

Values of S/d_1 versus kU^2L/gd_1^2 have been graphed in Fig. 4-16 for a range of $L/x = F/x$ or d_1/d_2. The curve for $d_1/d_2 = 1001$ can approximate for $d_2 = 0$.

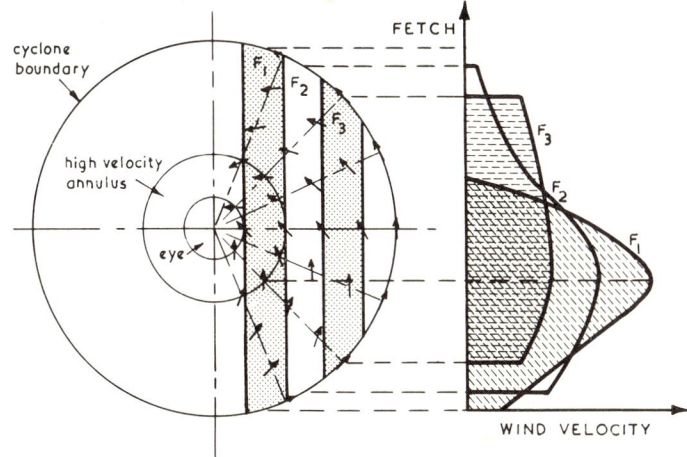

Fig. 4-17. Typical surface-wind structure inside a tropical cyclone.

(*b*) Triangular wind velocity ($V = 0$). The wind in a tropical cyclone is circular in character, but is deflected towards the centre such that it is around 45° to the radii [38]. This is depicted in Fig. 4-17, where the circulation for the Northern Hemisphere is indicated. As far as surge is concerned the component wind velocities along a fetch strip, such as F_1, F_2 or F_3, determine the overall stress applied to the sea surface. Because of the angled approach the lengths of these fetches are limited, as illustrated in the velocity-fetch graph in Fig. 4-17. Also the magnitudes of the peak velocities vary from fetch to fetch, the most intense winds being experienced adjacent to the eye of the cyclone. As illustrated in Fig. 2-24 and 2-25 of *Coastal Engineering, I*, the distribution of surface wind speeds in hurricanes, typhoons and other tropical cyclones can vary immensely, so that the near-triangular distribution along such a fetch as F_1 in Fig. 4-17 could vary substantially.

For a triangular wind distribution of more-or-less isosceles shape Reid [39] has derived the following surge formula:

$$S = \frac{kU^2 T}{C_1} \left(\frac{d_1}{d_2}\right)^{1/4} Z \qquad (4\text{-}37)$$

where U is the maximum velocity in the triangular distribution; T is the mean time for the surge wave to traverse the shelf; C_1 is the celerity of the surge wave at the shelf edge ($= \sqrt{gd_1}$); and Z is a factor ($= 0.56$ for $F = L$ and $V = 0$).

Substitution of $T = \dfrac{L}{\frac{1}{2}(\sqrt{gd_1} + \sqrt{gd_2})}$ into eq. 4-37 gives:

$$\frac{S}{d_1} = \frac{kU^2 L}{gd_1^2} \left[\frac{1.12}{1 + \sqrt{d_2/d_1}} \left(\frac{d_1}{d_2}\right)^{1/4}\right] \qquad (4\text{-}38)$$

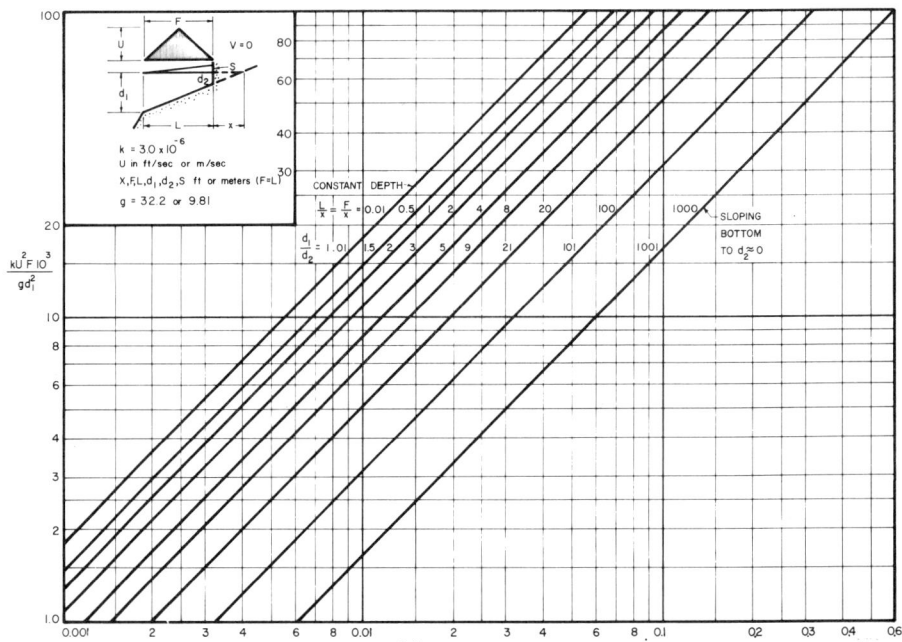

Fig. 4-18. Surge produced by a static triangular wind field extending across the continental shelf.

Eq. 4-38 has been graphed in Fig. 4-18 in a similar manner to Fig. 4-16 for a similar range of d_1/d_2 or L/x and $V = 0$.

Moving wind field. When a cyclone is in motion the critical zone as far as surge or wave generation is concerned is in that quadrant where the centre and wind vectors are in the same direction.

In Fig. 4-17 it is seen to be the right rear quadrant. In the Southern Hemisphere it is the left rear quadrant as viewed in the direction of cyclone advance. The worst affected zone will be the shoreline in the path of fetch F_1 in Fig. 4-17, although surges of more moderate magnitude will be experienced at points along the coast to the limit of the boundary of the cyclone.

When a storm centre is travelling towards the coast the initial surge wave is being reflected as later portions are still arriving. The interaction of these establishes a new surge system which Reid [39] has analyzed for various ratios of F/L and $V/\bar{C}$, where V is the velocity of the centre and:

$$\bar{C} = L/T = \tfrac{1}{2}(\sqrt{gd_1} + \sqrt{gd_2}) \tag{4-39}$$

From the graphs so presented [39] the ratio (R) of maximum surge (S_{max}) to that

STORM SURGE

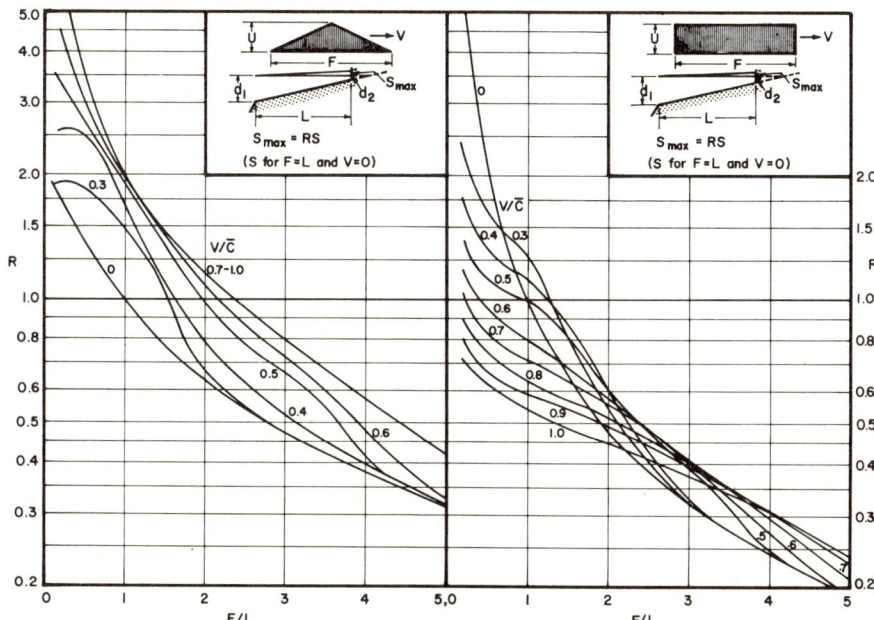

Fig. 4-19. Optimum surge resulting from various fetch lengths and velocities of advance.

for static storm conditions (S), in which $F = L$ and $V = 0$, has been plotted in Fig. 4-19, for both uniform and triangular wind distributions.

To find S_{max} it is first necessary to compute *S for a hypothetical shelf whose width L equals the fetch length F* (not the reverse) but with the same d_1 and d_2 values; R is then obtained by deriving the ratios F/L (actual F over actual L) and $V/\bar{C}$. From R obtain $S_{max} = R \times S$. It is seen in Fig. 4-19 that the influence of the storm-centre speed differs in the two wind distributions. Reid's analysis also provides information on the timing of the maximum surge in respect to the location of the fetch, but this is not included herein.

Barometric pressure. Besides the wind stress forcing water shorewards the reduction of atmospheric pressure at the centre will also cause a rise in the water surface. For these to be cumulative the latter should strictly be determined near the eye of the cyclone and not directly at its centre. Since these values are difficult to assess at any time, the weather reports of possible values should be used. The maximum barometric surge may be concurrent with the wind stress surge or it may precede or follow it by a small margin. For engineering purposes it is wise to consider them as synchronous.

The surge due to pressure deficiency (S_a) is given by:

$$S_a = \left(\frac{1013-p_c}{34}\right)\left(\frac{13.59}{12\times 1.003}\right) = (1013-p_c)0.033 \tag{4-40}$$

where S_a is the storm surge in feet of water; p_c is the pressure at the cyclone centre in millibars (34 millibar = 1 inch mercury).

Eq. 4-40 assumes a normal atmospheric pressure of 1013 mbar = 29.8 inches of mercury.

Examples of Hong Kong surges

Since 1962 the Department of Public Works in Hong Kong has collated some valuable data on typhoons approaching the vicinity of the island colony. These have been summarized in a data sheet for each event, one of which has been modified to be reproduced in Fig. 2-27 of *Coastal Engineering, I*. The original sheets contain data on rainfall, wind speed, wind direction, atmospheric pressure and sea level, all plotted against time in hours. As seen in that figure, the sea-level graph contains the predicted tide curve, from which the surge level has been extracted. The chronological relationship between the variables provides sufficient information to draw a plan of the cyclone and determine the diameters of its several parts. A geographic plan of the ocean area, with the typhoon path traced on it plus time markings, permits the speed of travel to be assessed. The distance across the continental shelf that the typhoon has traversed can also be measured. The wind speed trace indicates whether a triangular or rectangular distribution has produced the surge.

Data as above were available for seven typhoons, as listed in Table 4-I [40] where relevant parameters leading to the final surge ($S_T = S_{max} + S_a$) are included. Comparison of the computed and measured values indicates the accuracy of the procedure presented.

To determine a surge just prior to its arrival, the probable path in respect to the site must be predicted. Also the maximum wind speed at the boundary of the eye must be assessed, for example from a known relationship between this velocity and the atmospheric depression at the centre. Kraft [41] provides an equation for Atlantic hurricanes of:

$$U_{max} = 12.2\sqrt{1013-p_c} \tag{4-41}$$

where U_{max} is expressed in knots for p_c in mbar.

This relationship may not apply to typhoons or more modest tropical cyclones. From Table 4-I the only typhoon which was centred directly on Hong Kong was Wanda, when U_{max} = 80 knots for 1013-952 = 61 mbar. This would indicate that:

$$U_{max} = 10.3\sqrt{1013-p_c} \tag{4-42}$$

which is slightly smaller than indicated by eq. 4-41. This appears reasonable when

TABLE 4-I

Comparison of computed and measured surges at North Point, Hong Kong Harbour

	Name Date	Wanda Sep.62	Faye Sep.63	Viola May 64	Ida Aug.64	Ruby Sep.64	Sally Sep.64	Shirley Aug.68
Δp (mbar)		61	17	21	38	45	24	44
S_a (ft.)		2.02	0.57	0.70	1.27	1.50	0.80	1.47
U_{max} (knots)		80	25	30	45	60	35	42
Distribution		△	□	□	△	△	△	△
F (NM)		55	220	110	210	166	107	90
d_1 (ft.)		400	400	400	400	270 [1]	400	400
$kU^2F\,10^3/gd_1^2$		3.55	1.4	1.0	4.3	13.7	1.32	1.6
$S/d_1 \cdot 10^3$ (Fig.4-16,4-18)		5.5	3.8	2.7	6.5	20.0	2.0	2.5
S (ft.)		2.2	1.52	1.08	2.6	5.4	0.8	1.0
L (NM)		180 [3]	240 [4]	120	120	80 [1]	120	120
F/L		0.3	0.92	0.92	1.75	2.08	0.89	0.75
V (knots)		12	10	8	15	12.8	13.3	9.1
$V/\bar{C}$		0.280	0.234	0.187	0.350	0.348	0.312	0.213
R (Fig. 4-19)		1.9	1.3	1.4	0.9	0.7	1.6	1.5
S_{max} (= RS)		4.18	1.98	1.51	2.34	3.78	1.28	1.5
S_t (= $S_a + S_{max}$)		6.20	2.55	2.21	3.61	5.28	2.08	2.97
S (measured)		4.6 [5]	2.8	3.0	4.0	5.0	1.8	3.0

(1) Typhoon changed direction part way across shelf where d_1 = 270 ft. and L = 80 NM.
(2) Assuming d_2 = 30 ft.
(3) Approach was from ESE, equivalent L = 180 NM.
(4) Approach was from east, equivalent L = 240 NM.
(5) Tide gauge out, water mark observations only.

the smaller diameters of typhoons are considered in respect to hurricanes. Such data should be compiled for any region under study so that optimum surges can be computed.

The maximum water level to be expected can be calculated from a knowledge of the most severe tropical cyclones known to have occurred in an area. The fetch for this should then be moved across the continental shelf at the critical speed, to hit the coast just to the left (Northern Hemisphere) or right (Southern Hemisphere) of the port under study. This maximum surge should then be added to MHWS level which could occur during the cyclone period of the year. Should the harbour be located near the head of a funnel-shaped inlet, or one that has a shoaling approach channel, an amplification factor would have to be applied as outlined in the first part of this chapter. The response of an embayment to a storm surge can be quite complex as indicated by Unoki and Isozaki [42] and Heaps [43,44]. It lends itself to computer solution [45,46].

On coasts where storm surges are known to occur a coastal engineer should compile as much information as possible, in order to have data readily available on minimum dike heights and other facets of coast and facility protection. Such a comprehensive survey has been carried out for Japan by Isozaki [47], whilst another has been conducted for the Florida coast by Bruun et al. [48]. These should be examined to see the type of information that should be gathered, in order that worthwhile recommendations can be made respecting these sporadic high water levels. Marinos and Woodward [49] have devised a method for arriving at a "Standard Project Hurricane" for a particular region, which should be applicable to other geographic areas subject to tropical cyclones. Kishi et al. [50] have examined historic data on water levels at meteorological information and so derived a relation between surge and its return period.

EXAMPLES

1

Compute the tidal range at the coast for the overall average shelf cross-section depicted in Fig. 4-9, for an oscillation of 2 ft. in 12,000 ft. beyond the continental slope. Make no allowance for friction.

It is necessary to trace the "wave" over the slope and then over the shelf. Since a per unit length of coastline is assumed the A curves of Fig. 4-2 should be employed (i.e., constant width channel).

(1) Continental slope. $d_1 = 12,000$ ft; $d_2 = 40 \times 6080 \times 0.002 = 436$ ft.

$$Z_A = \frac{4\pi d_1^{1/2}}{g^{1/2} TS_v} = \frac{4\pi (12000)^{1/2}}{g^{1/2} 44700 \times 0.06} = 0.09 \text{ (say 0)}.$$

This is equivalent to an abrupt step, since $L \gg B$. From Fig. 4-2, for $d_1/d_2 = 12000/436 = 27.5$, $H_2/H_1 = 1.7$.

(2) Continental shelf. $d_1 = 436$ ft, $d_2 = 0$.

$$Z_A = \frac{4\pi (436)^{1/2}}{g^{1/2} 44700 \times 0.002} = 0.52, d_1/d_2 = \infty.$$

From Fig. 4-2 for $d_1/d_2 = 1000$, $H_2/H_1 = 2$. Thus tidal range at shoreline $= 2 \times 1.7 \times 2 = 6.8$ ft. If the sedimentary shelf, 30 NM wide and 400 ft. at the edge, were used and d_2 for the shelf was taken as 30 ft. close to shore:

$$Z_A = \frac{4\pi d_1^{1/2}}{g^{1/2} TS_v} = \frac{4\pi (400)^{1/2}}{g^{1/2} 44700 \times 400/(30 \times 6080)} = 0.45$$

From Fig. 4-2, for $d_1/d_2 = 400/30 = 13.3$, $H_2/H_1 = 1.6$. Since amplification across the slope is the same as before the tidal range at the shoreline is $2 \times 1.7 \times 1.6 = 5.5$ ft. Thus the majority of coasts have a tide around 6 ft. or 2 m.

The above calculation shows an amplification of 2.75 to 3.4 for these two shelf profiles. Munk et al. [51] found that across the shelf directly out from La Jolla an amplification from deep-water to the shore was in the order of 2.7 for the semidiurnal tide. The shelf there is narrower than 30 NM.

2

The Gulf of Kutch on the west coast of India has a profile and longitudinal plan as in Fig. 4-20.

EXAMPLES 191

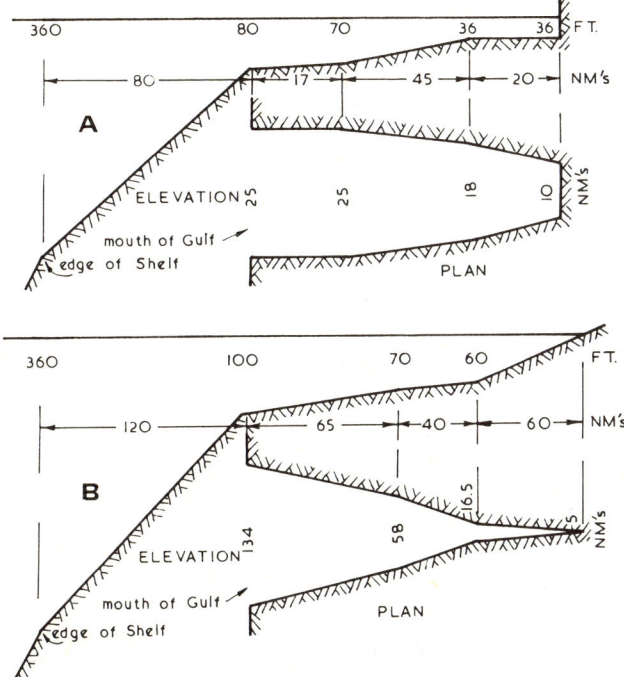

Fig. 4-20. Plan and section of: A. the Gulf of Kutch, and B. the Gulf of Cambay.

Tide tables indicate spring tides at the entrance and at the head of the Gulf as 10 ft. and 27 ft., respectively. It is required to check these values with the analysis available. Continental slope: for normal conditions the value of $H_2/H_1 = 1.7$ is acceptable.

(1) Continental shelf to Gulf entrance.

$$Z_A = \frac{4\pi d_1^{1/2}}{g^{1/2} TS_V} = \frac{4\pi\, 360^{1/2}}{g^{1/2}\, 44700 \times 280/(80 \times 6080)} = 1.64, \, d_1/d_2 = 360/80 = 4.5.$$

Fig. 4-2 gives $H_2/H_1 = 1.35$. Thus the deep ocean tidal range = $10/(1.35 \times 1.7) = 4.35$ ft. Compare this with the theoretical M_2 value in Fig. 4-1B of about 1 m, to which S_2 adds 0.46×4.35 giving range of 5.1 ft.

(2) Gulf 17 NM section.

$$Z_A = \frac{4\pi\, 80^{1/2}}{g^{1/2}\, 44700 \times 10/(17 \times 6080)} = 4.58, \, d_1/d_2 = 80/70 = 1.14.$$

Fig. 4-2 gives $H_2/H_1 = 1.05$.

It will be seen in the next section that friction can be ignored at this stage.

(3) Gulf 45 NM section.

$$Z_A = \frac{4\pi\, 70^{1/2}}{g^{1/2}\, 44700 \times 34/(45 \times 6080)} = 3.35, \, d_1/d_2 = 2.$$

Fig. 4-2 gives $H_2/H_1 = 1.2$.

$$Z_B = \frac{\pi\, 25 \times 6080}{g^{1/2} 70^{1/2} 44700(25-18)/(2\times 45)} = 2.9,\ (W_1/W_2)^2 = (25/18)^2 = 1.93.$$

Fig. 4-2 gives $H_2/H_1 = 1.2$.

The mean range (H) for purposes of friction assessment is given by $H = (10 \times 1.05) \times \left(\dfrac{1 + 1.2 \times 1.2}{2}\right) = 12.8$, $d = 58$ ft. Assume a dune height on the bed $k = 2$ ft.

$H^2 T/d = 12.8^2 \times 44700/58 = 127{,}000;$

$\dfrac{HT g^{1/2}}{4\pi k d^{1/2}} = \dfrac{12.8 \times 44700\, g^{1/2}}{4\pi \times 2 \times 58^{1/2}} = 17{,}000$ Fig. 4-4 gives $f = 0.0043$.

$\dfrac{fHT \sqrt{gd}}{10^3\, d^2} = \dfrac{0.0043 \times 12.8 \times 44700\sqrt{g\,58}}{10^3\, 58^2} = 0.0312;$ Fig. 4-3 gives $\dfrac{L}{L_1} = 0.99$.

$\dfrac{x}{T\sqrt{gd}} = \dfrac{45 \times 6080}{44700\sqrt{g\,58}} = 0.14;$ Fig. 4-5 gives $H_X/H_1 = 0.875$.

(4) Gulf 20 NM section.

$$Z_B = \frac{\pi\, 18 \times 6080}{g^{1/2} 36^{1/2} 44700 \times 8/(20 \times 2)} = 1.13,\ (b_1/b_2)^2 = (1.8)^2 = 3.25;$$

Fig. 4-2 gives $H_2/H_1 = 1.3$.

Reflection at end could double wave, but omit this aspect for present.

$H = (10 \times 1.05 \times 1.2 \times 1.2 \times 0.875)\left(\dfrac{1 + 1.3}{2}\right) = 15.2,\ d = 36$ ft.

Assume the dune height (k) to increase to 3 ft. since higher velocities are present.

$H^2 T/d = 15.2^2\, 44700/36 = 288{,}000,$

$\dfrac{HT g^{1/2}}{4\pi k d^{1/2}} = \dfrac{15.2 \times 44700\, g^{1/2}}{4\pi\, 3 \times 36^{1/2}} = 17{,}000;$ Fig. 4-4 gives $f = 0.0043$.

$\dfrac{fHT\sqrt{gd}}{10^3 d^2} = \dfrac{0.0043 \times 15.2 \times 44700\sqrt{g\,36}}{10^3\, 36^2} = 0.077;$ Fig. 4-3 gives $L/L_1 = 0.95$,

$\dfrac{x}{T\sqrt{gd}} = \dfrac{20 \times 6080}{44700\sqrt{g\,36}} = 0.08;$ Fig. 4-5 gives $H_X/H_1 = 0.85$.

Final $H = 10 \times 1.05 \times 1.2 \times 1.2 \times 0.875 \times 1.3 \times 0.85 = 14.65$ ft. If reflection doubles this a range of 29.3 ft. ensues. Actual tide is 27 ft.

3

The Gulf of Cambay, which is almost adjacent to the Gulf of Kutch on the west coast of India, has a profile and longitudinal plan as shown in Fig. 4-20. Tide tables show spring tides at entrance and head of this Gulf as 13 and 39 ft., respectively. It is required to check this using the deep-water range of 4.35 ft. computed in example 2.

(1) Continental slope. Use value previously derived, of $H_2/H_1 = 1.7$.

(2) Continental shelf to Gulf entrance.

$$Z_A = \frac{4\pi d_1^{1/2}}{g^{1/2} TS_V} = \frac{4\pi\, 360^{1/2}}{g^{1/2} 44700\, (360 - 100)/(120 \times 6080)} = 2.64,\ d_1/d_2 = 3.6;$$

Fig. 4-2 gives $H_2/H_1 = 1.35$.

EXAMPLES 193

Thus $H_2 = 4.35 \times 1.7 \times 1.35 = 10$ ft., which is not quite the same as 13 ft., but the actual location of this tidal range is not known. Use 10 ft. for calculations.

(3) Gulf 65 NM section.

$$Z_A = \frac{4\pi \, 100^{1/2}}{g^{1/2} 44700 \times 30/(65 \times 6080)} = 6.55, \, d_1/d_2 = 100/70 = 1.43; \qquad \text{Fig. 4-2 gives } H_2/H_1 = 1.1.$$

$$Z_B = \frac{\pi b_1}{g^{1/2} d_1^{1/2} TS_H} = \frac{\pi \, 134 \times 6080}{g^{1/2} 100^{1/2} 44700(134-58)/(2 \times 65)} = 1.74, \, (b_1/b_2)^2 = (134/58)^2 = 5.32$$

Fig. 4-2 gives $H_2/H_1 = 1.45$.

$$H = 10 \left(\frac{1 + 1.1 \times 1.45}{2}\right) = 13 \text{ ft}; \, d = 85 \text{ ft}; \text{ assume } k = 2.$$

$H^2 T/d = 13^2 \, 44700/85 = 89{,}000$

$$\frac{HT g^{1/2}}{4\pi k d^{1/2}} = \frac{13 \times 44700 \, g^{1/2}}{4\pi \, 2 \times 85^{1/2}} = 14{,}200;$$

Fig. 4-4 gives $f = 0.0045$.

$$\frac{fHT\sqrt{gd}}{10^3 d^2} = \frac{0.0045 \times 13 \times 44700 \sqrt{g85}}{10^3 85^2} = 0.019; \qquad \text{Fig. 4-3 gives } L/L_1 = 0.995.$$

$$\frac{x}{T\sqrt{gd}} = \frac{65 \times 6080}{44700 \sqrt{g85}} = 0.17; \qquad \text{Fig. 4-5 gives } H_x/H_1 = 0.89.$$

(4) Gulf 40 NM section

$$Z_A = \frac{4\pi \, 70^{1/2}}{g^{1/2} 44700 \times 10/(44 \times 6080)} = 11.1, \, d_1/d_2 = 70/60 = 1.17;$$

Fig. 4-2 gives $H_2/H_1 = 1.05$.

$$Z_B = \frac{\pi \, 58 \times 6080}{g^{1/2} 70^{1/2} 44700(58-16.5)/(2 \times 44)} = 1.11, \, (b_1/b_2)^2 = (58/16.5)^2 = 12.4;$$

Fig. 4-2 gives $H_2/H_1 = 1.6$.

$$H = (10 \times 1.1 \times 1.45 \times 0.89) \left(\frac{1 + 1.05 \times 1.6}{2}\right) = 19 \text{ ft}, \, d = 65 \text{ ft}, \, k = 3$$

$H^2 T/d = 19^2 \times 44700/65 = 250{,}000$

$$\frac{HT g^{1/2}}{4\pi k d^{1/2}} = \frac{19 \times 44700 \, g^{1/2}}{4\pi \, 3 \times 65^{1/2}} = 15{,}900;$$

Fig. 4-4 gives $f = 0.0044$.

$$\frac{fHT\sqrt{gd}}{10^3 d^2} = \frac{0.0044 \times 19 \times 44700 \sqrt{g65}}{10^3 65^2} = 0.0405; \qquad \text{Fig. 4-3 gives } L/L_1 = 0.985.$$

$$\frac{x}{T\sqrt{gd}} = \frac{40 \times 6080}{44700 \sqrt{g65}} = 0.119; \qquad \text{Fig. 4-5 gives } H_x/H_1 = 0.875.$$

(5) Gulf 60 NM section. Shoals appear at low tide but water may rise 39 ft. according to measured tide so that a mean depth of 20 ft. could be used for calculations.

$$Z_A = \frac{4\pi \, 60^{1/2}}{g^{1/2} 44700 \times 40/(60 \times 6080)} = 3.5, \, d_1/d_2 = 60/20 = 3;$$

Fig. 4-2 gives $H_2/H_1 = 1.3$.

$$Z_B = \frac{\pi 16.5 \times 6080}{g^{1/2} 60^{1/2} 44700(16.5-5)/(2 \times 60)} = 1.68, \quad (b_1/b_2)^2 = (16.5/5)^2 = 11;$$

Fig. 4-2 gives $H_2/H_1 = 1.55$.

$$H = (10 \times 1.1 \times 1.45 \times 0.89 \times 1.05 \times 1.6 \times 0.875) \left(\frac{1 + 1.5 \times 1.55}{2}\right) = 31.3 \text{ ft}; d = 40 \text{ ft.},$$

$k = 3$.

$$H^2 T/d = 31.3^2 \times 44700/40 = 1,100,000$$
$$\frac{HTg^{1/2}}{4\pi k d^{1/2}} = \frac{31.3 \times 44700 \, g^{1/2}}{4\pi 3 \times 40^{1/2}} = 33,400;$$

Fig. 4-4 gives $f = 0.0038$.

$$\frac{fHT\sqrt{gd}}{10^3 d^2} = \frac{0.0038 \times 33.6 \times 44700 \sqrt{g40}}{10^3 \cdot 40^2} = 0.12; \quad \text{Fig. 4-3 gives } L/L_1 = 0.91.$$

$$\frac{x}{T\sqrt{gd}} = \frac{60 \times 6080}{44700 \sqrt{g40}} = 0.228; \quad \text{Fig. 4-5 gives } H_x/H_1 = 0.525.$$

Final $H = 10 \times 1.1 \times 1.45 \times 0.89 \times 1.05 \times 1.6 \times 0.875 \times 1.3 \times 1.55 \times 0.525 = 24.4$ ft.

This differs from the 39 ft. quoted, but if some reflection takes place once the water rises up the banks this figure could be doubled — actually only about 50% has to be added. Also the actual range given for the mouth is 13 ft. instead of 10 used to derive 24.4 ft. at the head.

It is seen from examples 2 and 3 that the friction factor f is determined mainly from the roughness factor k when this has values of 2–3 ft. (which is not unreasonable). In these circumstances $f = 0.0045$ most of the time and such a value might be accepted throughout to save calculations. The order of f should be computed at the outset as its value is determined by x_{max}/k, in which x_{max} is a function of wave period T, amongst other things.

4

Test whether friction is likely to affect the tidal range onshore with a typical coastal profile. Accept a roughness dimension of 2 ft. Check shallowest section of any significance, say 30 ft. depth, with a tidal range of 10 ft.

$$\frac{HTg^{1/2}}{4\pi k d^{1/2}} = \frac{10 \times 44700 \, g^{1/2}}{4\pi 2 \times 30^{1/2}} = 18,500; \quad H^2 T/d = 10^2 \times 44700/30 = 150,000.$$

From Fig. 4-4, f is dictated by roughness = 0.0043.

$$\frac{fHT\sqrt{gd}}{10^3 d^2} = \frac{0.0043 \times 10 \times 44700 \sqrt{g30}}{10^3 30^2} = 0.066;$$

From Fig. 4-3, $L/L_1 = 0.965$.

$x/T\sqrt{gd} = 6080/44700 \sqrt{g30} = 0.0044$ assuming 30 ft. depth is 1 NM from shore; from Fig. 4-5, $H_x/H_1 = 1.0$.

Even if $x = 10$ miles $H_2/H_1 = 0.95$, so that for normal continental shelf profile friction can be ignored.

5

A fetch in an extra-tropical cyclone can be considered to be 200 NM in length with a mean wind speed at 19.5 m height of 38 knots. It approaches normally to a coast where the continental shelf is 100 NM wide, the shelf edge is at 400 ft. depth and the inshore zone can be considered as 30 ft. deep. Compute the maximum storm surge assuming: (a) steady conditions with the fetch equal to the shelf width; (b) steady conditions with the fetch twice the shelf width; (c) an approach velocity of the fetch in b equal to the wind velocity (i.e., speed of cold front); (d) an approach velocity half that specified in c above; (e) that $F/L = 1.0$ and $V/\bar{C} = 0.3$.

EXAMPLES

$U_{19.5}$ = 38 knots. From Fig. 3-8 of *Coastal Engineering, I*, $U_{19.5}/U_{10}$ = 1.08, U_{10} = 38/1.08 = 35 knots.

(a) $\dfrac{kU_{10}^2 F 10^3}{gd_1^2} = \dfrac{3(35 \times 6080/3600)^2 \, 100 \times 6080 \times 10^3}{10^6 \, g \, 400^2} = 1.24.$

d_1/d_2 = 400/30 = 13.3. Fig. 4-16 gives S/d_1 = 0.0035, S = 0.0035 × 400 = 1.4 ft.

(b) F is doubled, so that $\dfrac{kU^2 F 10^3}{gd_1^2} = 2.48.$

Fig. 4-16 gives 0.007, S = 2.8 ft.; from Fig. 4-19, for F/L = 2, and V = 0, R = 0.475. $S_{max} = R \times S = 0.475 \times 2.8 = 1.33$ ft.

(c) For $V = U_{10}$, $V/\bar{C} = V/\frac{1}{2}(\sqrt{gd_1} + \sqrt{gd_2}) = \dfrac{35 \times 6080/3600}{\frac{1}{2}\sqrt{g}(\sqrt{400} + \sqrt{30})} = 0.82;$

from Fig. 4-19, for F/L = 2 and $V/\bar{C}$ = 0.82, R = 0.52; S_{max} = 0.52 × 2.8 = 1.45 ft.
(d) For $V = U_{10}/2$, $V/\bar{C}$ = 0.41, R = 0.6; S_{max} = 1.68 ft.
(e) For F/L = 1.0 and $V/\bar{C}$ = 0.3, R = 1.25; $S_{max} = R \times S = 1.25 \times 1.4 = 1.75$ ft.

6

A fetch in a typhoon has a maximum wind velocity of 70 knots and the triangular wind distribution can be considered to have a 650 NM fetch. This typhoon approaches normally across a continental shelf whose width is 120 NM, edge depth 400 ft. and inshore depth 30 ft. If its average speed of approach is 10 knots, compute the maximum surge and determine what length of coast is affected by such high water.

For $V = 0$ and $L = F$:

$$\dfrac{kU^2 F 10^3}{gd_1^2} = \dfrac{3(70 \times 6080/3600)^2 \, 650 \times 6080 \times 10^3}{10^6 \, g \, 400^2} = 32.2.$$

From Fig. 4-18, for d_1/d_2 = 400/30 = 13.3, S/d_1 = 0.53, S = 21.2 ft.

For $V \neq 0$ and $L \neq F$,

$$F/L = \dfrac{650}{120} = 5.42, \quad \dfrac{V}{\bar{C}} = \dfrac{10 \times 6080/3600}{\frac{1}{2}\sqrt{g}(\sqrt{400} + \sqrt{30})} = 0.234.$$

From Fig. 4-19, R = 0.29, S_{max} = 21.2 × 0.29 = 6.2 ft. From eq. 4-42, for a typhoon, U_{max} = $10.3 \sqrt{1013 - p_c}$, so that 70 = $10.3 \sqrt{1013 - p_c}$, Δp = 46.3 mbar. $S_a = \dfrac{46.3}{34} \cdot \dfrac{13.59}{12} = 1.55$ ft., S_T = 6.2 + 1.55 = 7.75 ft.

Since fetch is 650 NM long, equivalent to F_1 in Fig. 4-17, this should be about 5/8 of cyclone diameter, which is then 1000 NM. From Fig. 2-24 of *Coastal Engineering, I*, it is seen that this is a larger than normal typhoon and the high velocity annulus is likely to be 1000 × 20/400 = 50 NM wide. Hence this is the length of coast to suffer the largest surge. Other points on the coast, (F_2 and F_3 of Fig. 4-17), to a distance of probably 250 NM will experience some surge from this typhoon. The coast on the other side of the cyclone centre-line will have depressed water levels as the typhoon approaches the shoreline.

PROBLEMS

1
A section of coast has a continental shelf which is 100 NM wide and 400 ft. deep at its edge. The adjacent ocean is 12,000 ft. deep. Compute the diurnal tidal height at the shore for a deep-ocean range of 3 ft., with and without friction. The outer slope of the shelf is 0.06. (It is suggested that the shelf be divided into sections, with depths of each half that of its seaward neighbour).

2
A tsunami, of 20 min period and height 2 ft. in 12,000 ft. depth, approaches a shoreline whose continental shelf is 390 ft. deep at the edge, 80 NM wide, and has an outer slope of 0.06. At 50 NM from the edge a 40 NM wide funnel is 150 ft. deep and at its head has a river 1 mile wide and 30 ft. deep running into it. Determine the height of the tsunami at the shelf edge, at the funnel mouth and the river mouth. (It is possible that a bore may be formed at some stage so determine its location).

3
It is proposed to build a new port at an equal distance between two other ports, which have semi-diurnal spring tides of 10 and 15 ft., respectively. If the edge of the continental shelf (depth 400 ft.) is a straight line, as is the coast, calculate the tide for the new port site. The width of the shelf at this site is 80 NM. What is the difference in time of high water at the three ports?

4
A river is very long, extending from the mouth, which is 2 NM wide and 40 ft. deep, to a distance of 100 NM inland, where the width is 1/2 NM and the depth is 20 ft. Surveys indicate a nearly linear variation of both depth and width. Compute the semi-diurnal tidal range and maximum stream velocity 30 NM upstream from the mouth, where the range is 10 ft. How long before high tide is the maximum velocity experienced? If a dam were constructed at this point, what would be the effect on tidal ranges at the dam site and mid-way between it and the mouth? At this mid-point what is the difference in time between HW and that at the dam site. Use a roughness dimension of 2 ft. for the bed.

5
A lake can be considered to be rectangular in shape, with its longest dimension 20 miles. The undulating bottom can be equated to a uniform depth of 12 ft. Compute the rise in level at the downwind end when a 40-knot wind (as measured at 25 m height) blows down the length of the lake. What occurs at the upwind end? Determine the frictionless period of oscillation for the lake once the wind stress is released.

6
A continental shelf is 100 NM wide and is uniformly sloped from the outer depth of 400 ft. to a near-shore value of 40 ft. An extra-tropical cyclone causes a uniform wind of 40 knots ($U_{19.5}$) to be applied over a 200 miles width of the ocean, normal and towards the shore. Determine the surge created at the shoreline when this wind system travels across the coast at 20 knots. If this fetch were to have its downwind edge 50 NM from shore as it travelled parallel to the coast, what is the maximum surge experienced during its passage. Compare the original surge with a triangular distribution of wind velocity which covers the whole shelf and has a peak wind speed of 80 knots. The advance of this tropical cyclone to shore is at the rate of 10 knots.

7
Fig. 4-21 contains relevant dimensions of embayments where large tides exist. From the ranges

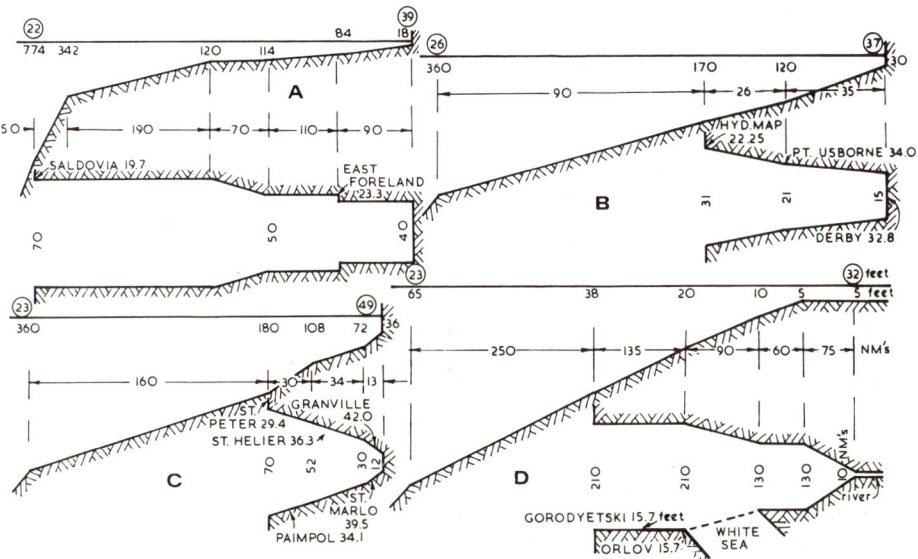

Fig. 4-21. Plans and sections of embayments where large tidal ranges exist, namely: A. Corks Inlet (Alaska); B. King's Sound (Western Australia); C. St. Marlo (France); D. Gulf of Mezen (Finland). Circled ranges in ft. are taken from the original of Fig. 4-10. Other tides are mean spring values taken from hydrographic charts or tide tables. Check these values from the deep-ocean tides (multiplied by 1.5) displayed in Fig. 4-1B using the normal continental slope amplification factor of 1.7.

as given, check the upstream and deep-ocean values, and compare the latter with those as given in Fig. 4-1B. (Note that mean spring tides are about 1.5 times the M_2 tide recorded in this figure.) Use roughness dimensions of 2–3 ft.-depending upon the relative ranges and depths.

8
A straight canal which has lock-gates at each end is 20 ft. deep and 10 miles in length. It is essential under all conditions of weather to maintain 18 ft. depth, in order to provide navigational safety. What velocity of wind in knots, blowing directly along the canal, could cause a hazard at the upwind end?

9
A *long* river has a depth of 20 ft. and it is dredged to a depth of 30 ft. for a distance of 5 miles from the mouth for purposes of navigation. Determine the effect on the maximum tidal velocities at the station 5 miles upstream. The tidal wave length was found to be 0.7 that of the value at the mouth ($= T\sqrt{gd}$) for the original condition of 20 ft. depth. For the same point, assess the change in time lag between maximum velocity and maximum tidal stage caused by the dredging. (Consider that the whole length of river has been deepened 10 ft. but indicate the nature of the error so introduced.)

10
Why do the tides at any location vary in range over a two-weekly cycle? What causes the slower variation throughout the year? What is done statistically to obtain ranges to serve engineering and navigational purposes? Draw the probable daily oscillation for ports at latitudes of 0°, 30° and 60°.

11
The deep-ocean tide varies from ocean to ocean, why is this? From an inspection of Fig. 4-1 comment on the distribution of corange lines in terms of ocean depths and/or land masses.

12
Discuss the factors which determine the shoreline tide from the deep-water oscillation beyond the continental shelf. What is the order of this tide for a typical shelf profile?

13
How would you obtain tidal predictions for a section of coast where none are available at present?

14
What variables are important in computing the surge produced from cyclonic centres? Indicate the relationship of any parameters by means of sketches or graphs.

15
The Hooghly River, which is part of the Ganges complex, and on which Calcutta is sited some miles upstream, has the following conditions, as reported in the literature [52]: "A notable feature in the Hooghly is the formation of 'bore-tides' characterized by a sudden and almost vertical rise of water level immediately after low-water. 'Bores' now occur practically on every spring tide. The maximum 'bore' height is in the order of 8 ft. and the speed of travel about 10−20 miles per hour". Verify these conditions by calculating the tidal range at the river mouth if the shelf is 80 NM wide. Assume the shoreline depth is negligible i.e., $d_1/d_2 = 1000$.

16
The calculations of storm surge over the continental shelf are generally based upon a normal approach to the coast. Discuss the likely variations from the surge so computed when the cyclone approaches at an angle of 45° either side of this normal. Consider in turn the tropical and extra-tropical centres and also stationary and moving centres.

17
The propagation of a tide up a river is somewhat similar to that of a shallow water wave. There is one major difference. Discuss this and outline means of measuring it.

18
What conditions produce large storm surges? Name three places in the world where these are important and note the predominant factor in each case.

19
In the Adriatic Sea east of Italy a major storm surge was reported at the northern end in 1970, when the low lands in the vicinity of Venice were inundated. What meteorological situation could produce this? Why could such a surge be more devastating than, say, a similar surge on the coast of The Netherlands or of East Pakistan?

20
Narragansett Bay on the east coast of the United States suffers hurricanes periodically, which result in storm surges at the beach. Table 4-II gives wind information for three such occasions. Compute the surge due to wind stress, from the knowledge that the continental shelf is 84 NM wide, and is 360 ft. and 36 ft. deep at edge and near shore respectively.

TABLE 4-II

Data for Problem 20

Hurricane	F (NM)	U_{max} (mph)	V (mph)
Sept. 1938	148	88	42
Sept. 1944	124	84	33
Aug. 1954	87	95	32

21
A tropical cyclone approaches normally to a coast that has a 120 NM wide continental shelf of normal depth proportions. The triangular wind velocity distribution has a maximum value of 80 knots and the base is the same as the shelf width. If the velocity of approach is 12 knots, compute the storm surge.

22
Using the general proportions of tropical cyclones as in Fig. 2-24 of *Coastal Engineering, I*, compare the surges developed by a hurricane, a typhoon and an Australian tropical cyclone for the same maximum wind speed at the eye edge of 90 knots. Carry out the test on a shelf width of 80 miles with normal depth values. Besides the difference in the maximum surge so calculated, what other major differences in the phenomenon should you note, for these three classes of cyclone?

REFERENCES

[1] M.J. Tucker, 1963. Long waves in the sea. *Sci. Progress*, 51: 413–424.
[2] R.G. Dean, 1966. Tides and harmonic analysis. In: A.T. Ippen (Editor), *Estuary and Coastline Hydrodynamics*. McGraw-Hill, New York, N.Y., pp.197–230.
[3] C.L. Pekeris and Y. Accad, 1969. Solution of Laplace's equations for the M_2 tide in the world oceans. *Phil. Trans. R. Soc.*, A265: 413–436.
[4] R.G. Dean, 1964. Long wave modification by linear transitions. *Proc. ASCE*, 90(WW1): 1–29.
[5] R. Dorrestein, 1961. Amplification of long waves in bays. *Florida Exp. Station, Tech. Pap.*, 213.
[6] A.T. Ippen and R.D.F. Harleman, 1966. Tidal dynamics in estuaries. In: A.T. Ippen (Editor), *Estuary and Coastline Hydrodynamics*. McGraw-Hill, New York, N.Y., pp.493–545.
[7] I.G. Jonsson, 1966. Wave boundary layers and friction factors. *Proc. 10th Conf. Coastal Eng.*, 1: 127–148.
[8] K. Kajiura, 1964. On the bottom friction in an oscillatory current. *Bull. Earthquake Res. Inst. (Japan)*, 42: 147–174.
[9] F.P. Shepard, 1963. *Submarine Geology*. Harper and Row, New York, N.Y., 2nd ed. 348 pp.
[10] J.A. Sitarz, 1963. Contribution à l'étude de l'évolution des plages à partir de la consistance des profiles d'équilibre. *Trav. Centre Etud. Rech. Océanogr.*, 1963: 10–20.
[11] G. Dietrich, 1943. Über ozeanische Gezeitenerscheinungen in geographischer Betrachtungsweise. *Ann. Hydrogr. Mar. Meteorol.*, 71: 123–127.
[12] J.M. Caldwell, 1969. New developments in the field of tidal hydraulics. *Proc. ASCE*, 95(HY1): 1–8.
[13] C.F. Wicker, 1969. New horizons in the field of tidal hydraulics. *Proc. ASCE*, 95(HY1): 147–160.

[14] K.F. Bowden and L.A. Fairbairn, 1952. A determination of the frictional forces in a tidal current. *Proc. R. Soc.,* A214: 371–392.

[15] A.T. Doodson, 1956. Tides and storm surges in a long uniform gulf. *Proc. R. Soc.,* A237: 325–343.

[16] J.J. Dronkers, 1969. Tidal computations for rivers, coastal areas and seas. *Proc. ASCE,* 95(HY1): 29–77.

[17] N.G. Heaps, 1969. Some notes on tidal theory and its possible relevance to a program of deep-sea tidal measurement. *Dtsch. Hydrogr. Z.,* 22: 11–25.

[18] O.H. Sheridan and R.M. Forney, 1970. Tidal motion in bays. *Proc. 12th Conf. Coastal Eng.,* 3: 2225–2242.

[19] G. Evangelisti, 1955. On tidal waves in a canal with variable cross section. *Proc. Congr. IAHR, 6th, 1955,* Pap. A-10.

[20] J.N. Hunt, 1964. Tidal oscillations in estuaries. *Geophys. J.,* 8 : 440–455.

[21] B. Johns, 1967. Tidal flow and mass transport in a slowly converging esturay. *Geophys. J.,* 13: 377–386.

[22] W.H. Munk, G.R. Miller and F.E. Snodgrass, 1969. Long period waves over Californian continental shelf. The decay of tsunamis and the dissipation of tidal energy. *J. Mar. Res.* 20: 119–120.

[23] A. Mysak, 1967. On the theory of continental shelf waves. *J. Mar. Res.,* 25: 205–227.

[24] P. Perroud, 1959. The propagation of tidal waves into channels of gradually varying cross-section. *B.E.B. Tech. Mem.,* 112.

[25] J. Proudman, 1957. Oscillations of tide and surge in an estuary of finite length. *J. Fluid Mech.,* 2: 371–382.

[26] A.C. Redfield, 1958. The influence of the continental shelf on the tides of the Atlantic coast of the United States. *J. Mar. Res.,* 17: 432–448.

[27] J Wu, 1969. A criterion for determining air-flow separation from wind waves. *Tellus,* 21: 707–714.

[28] T.P. Barrett and A.J. Sutherland, 1968. A note on an overshoot effect in wind-generated waves. *J. Geophys. Res.,* 73: 6879–6885.

[29] P.A. Shepard, 1958. Transfer across the earth's surface and through the air above. *Qu. J. R. Meteorol. Soc.,* 84: 205–224.

[30] E.L. Deacon and E.K. Webb, 1962. Small-scale interaction. In: M.N. Hill (Editor), *The Sea.* Interscience, New York, N.Y., pp.43–87.

[31] B.W. Wilson, 1960. Note on surface wind stress over water at low and high wind speeds. *J. Geophys. Res.,* 65: 3377–3382.

[32] J. Wu, 1969. Wind stress and surface roughness at sea interface. *J. Geophys. Res.,* 74: 444–455.

[33] B. Hellstrom, 1941. Wind effect on lakes and rivers. *R. Inst. Tech. Stockholm, Bull.,* 41.

[34] H.L. Langhaar, 1951. Wind tides in inland waters. *Proc. Mid Western Conf. Fluid Mech., 1st, 1951:* 278.

[35] G.H. Keulegan, 1951. Wind tides in small closed channels. *J. Res. Natl. Bur. Stand.,* 46: 358–381.

[36] G.H. Keulegan, 1952. The form factor in wind-tide formulas. *Natl. Bur. Stand., Rep.,* 1835.

[37] C.L. Bretschneider, 1966. Engineering aspects of hurricane surge. In: A.T. Ippen (Editor), *Estuary and Coastline Hydrodynamics.* McGraw-Hill, New York, N.Y., pp.231–256.

[38] S. Syono, 1963. Structure of typhoons. *Proc. Inter-reg. Seminar Trop. Cyclones, Tokyo, 1962:* 121–131.

[39] R.O. Reid, 1956. Approximate response of water level on a sloping shelf to a wind fetch which moves towards shore. *Beach Erosion Board, Tech. Mem.* 83.

[40] R. Silvester, 1970. Computation of storm surge. *Proc. 12th Conf. Coastal Eng.,* 3: 1995–2010.

[41] R.H. Kraft, 1961. The hurricane's central pressure and highest winds. *Mar. Weather Log,* 5(5).

REFERENCES

[42] S. Unoki and I. Isozaki, 1965. Mean sea level in bays, with special reference to the mean slope of sea surface due to the standing oscillation of tide. *Oceanogr. Mag. (Japan)*, 17: 11–33.

[43] N.S. Heaps, 1965. Storm surges on a continental shelf. *Phil. Trans. R. Soc.*, A257: 351–383.

[44] N.S. Heaps, 1967. Storm surges. *Oceanogr. Mar. Biol. Ann. Rev.*, 5 : 11–47.

[45] H.R. Davis, 1962. Computer solution of storm surges. *Proc. ASCE*, 88(WW3): 117–123.

[46] A.N. Sirkin, 1970. Hurricane surge computations by computer. *Proc. ASCE*, 96(WW2): 467–482.

[47] J. Isozaki, 1968–1970. An investigation on the variations in sea level due to meteorological disturbances on the coast of the Japanese Islands, I. On the accuracy of tide predictions. *Pap. Meteorol. Geophys.* 19: 401–426; II. Storm surges on the coast of the Japan Sea, *J. Ocean. Soc. Japan*, 24: 32–190; III. On the variation of daily mean sea level. *J. Ocean. Soc. Japan*, 25: 91–102; IV. Storm surges on the Pacific and Okhotsk Sea coasts of North Japan. *J. Ocean. Soc. Japan*, 25: 192–200; V. Storm surges on the coast of the Kanto and Tokai districts, *Pap. Meteorol. Geophys.*, 21: 1–32; VI. Storm surges on the coasts of the Inland Sea and Osaka Bay. *Pap. Meteorol. Geophys.*, 21: 291–321; VII. Storm surges on the coast of West Japan facing the Pacific and the East China Sea. *Pap. Meteorol. Geophys.*, 21: 421–448.

[48] P. Bruun, T.Y. Chiu, F. Gerritsen and W.H. Morgan, 1962. Storm tides in Florida as related to coastal topography. *Univ. Fla., Eng. Exp. Sta., Bull.*, 109.

[49] G. Marinos and J.W. Woodward, 1968. Estimation of hurricane surge hydrographs. *Proc. ASCE*, 94(WW2): 189–216.

[50] T. Kishi, M. Tominaga and I. Oeda, 1960. Studies on meteorological tides at the mouth of the Tone River. *Coastal Eng. Japan*, 3: 1–8.

[51] W.H. Munk, F. Snodgrass and M. Wimbush, 1970. Tides off-shore: transition from California coastal to deep-sea waters. *Geophys. Fluid Dyn.*, 1: 161–235.

[52] C.V. Gole and P.P. Vaidyaraman, 1966. Salinity distribution and effect of fresh-water flows in the Hooghly River. *Proc. 10th Conf. Coastal Eng.*, 2: 1412 – 1434.

[53] H.W. Partenscky and J.C. Warmoes, 1969. Damped co-oscillating tide with negative reflection at the end of the estuary. *Proc. Congr. Int. Assoc. Hydr. Res., 13th*, 3: 363–370.

Chapter 5

ESTUARINE PROBLEMS

An estuary is a zone, either in a waterway or part of the sea, where fresh and salt water are in the process of mixing. Two extremes can exist, firstly where the strata of the lighter fresh water is distinct from the heavier salt water underneath, and secondly where density changes exist throughout the depth and length of a river or embayment. The former condition occurs when tidal motion is minimal, so that mixing is effected only by the fresh-water outflow of the river. When tidal oscillations are large, compared with the river discharge, the turbulence present in the stratum close to the bed causes greater mixing at the interface of fresh and salt bodies.

The density difference between outflowing fresh water and the sea water itself causes the former to override the latter, which, in turn, penetrates up the embayment or river in the form of a saline wedge. The drag of the fresh water carries salt water from the interface or upper surface of the wedge seawards. This demand creates a landward current at the bed which increases as the toe of the wedge is approached. The static weight of the salt water in the wedge is balanced dynamically by the friction drag of the fresh water at the interface, the position of which can vary cyclically with the tide.

The stratification and concomitant currents create problems of sedimentation due to the combination of inward flow at the base of the wedge and outward flow at the river bed upstream of the wedge. It also causes difficulties in the discharge of pollutants to the sea, either from upriver or from within the estuary itself. These particular problems are difficult to quantify, and even when this is done for a particular estuary any relationship cannot be applied elsewhere without great caution being exercised.

SALT-WATER WEDGE

For a given river discharge a wedge will reach an equilibrium length as measured from the natural river mouth, or some suitable pseudo-mouth seawards of this, if the offshore zone is depressed across the continental shelf. To eliminate the effect of tidal currents such length measurements are generally made at, or referred to, the static condition on the turn of the tide, either at the end of the flood or the ebb. The excursion of water particles due to the tidal oscillation can be computed and this displaces the wedge upstream or downstream by a specific amount.

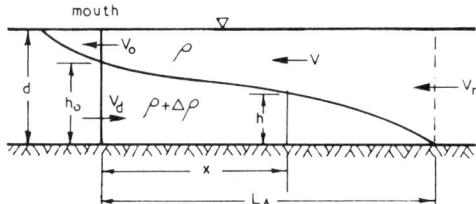

Fig. 5-1. Definition sketch of an arrested saline wedge.

The length as determined for static conditions refers to an arrested saline wedge, which is depicted as L_A in Fig. 5-1, where it is seen that: ρ is the density of the fresh water outflow; $\rho + \Delta\rho$ is density of the salt water of the adjacent sea; V_r is mean velocity of river flow just upstream of the toe of the wedge; V is the mean velocity of river flow at distance x from the mouth; V_o is the mean velocity of river flow at the mouth; d is the depth of the river (considered constant over the length of the wedge); h is the height from the river bed to the interface at distance x from the mouth; h_o is the height from the river bed to the interface at the mouth.

In Fig. 5-1 is also shown a vector V_d which would be the velocity of advance of the toe if a barrier between fresh and salt water were suddenly withdrawn. This has direct application in the penetration of salt water into locks, where run-off supplies fresh water in the landward side of the gates. But it also enters relationships in the kinematics of stratified fluids and is termed the densimetric velocity (V_d) defined as:

$$V_d = \sqrt{(\Delta\rho/\rho_m)gd} \qquad (5\text{-}1)$$

where ρ_m is a mean density $= (2\rho + \Delta\rho)/2$.

Since $\Delta\rho \doteq 0.03$ (or 1/32 by weight of salt, i.e., each cubic ft. of seawater contains 2 lb. salt) so $\rho_m = 1.015$ and $\Delta\rho/\rho_m \doteq 0.02$. The salinity of seawater near the coast can vary from place to place and from season to season and even with depth. Where the calculations are critical averages should be obtained from records over years, bur otherwise the figure of 0.02 can be accepted.

The densimetric velocity leads to two dimensionless parameters, namely the densimetric Froude number:

$$F_d = V_r/V_d \qquad (5\text{-}2)$$

and the densimetric Reynolds number:

$$R_d = V_d d/\nu \qquad (5\text{-}3)$$

where ν is the kinematic viscosity ($\doteq 11 \cdot 10^{-6}$ ft.^{2}/sec).

The stationary length of the wedge can be expressed in dimensionless terms [1] as:

$$L_A/d = f(F_d, R_d, d/b) \tag{5-4}$$

where b is the mean breadth of the water way.

Ozturk [2] has termed the cross-sectional proportions d/b as a "length correction factor", since any specific water-way has non-rectangular features as well as meanders which can cause mixing not normally introduced into model studies.

From extensive laboratory measurements on straight rectangular canals of varying cross-section, Keulegan [1] arrived at a relationship:

$$\frac{L_A}{d} = \left[\frac{0.88}{280(R_d)^{-1} + 0.148(R_d)^{-1/4}}\right](2F_d)^{-5/2} \tag{5-5}$$

for F_d from 0.15 to 0.6. Where $R_d > 10^7$ this reduces to:

$$L_A/d = 6.0(R_d)^{1/4}(2F_d)^{-5/2} \tag{5-6}$$

This was confirmed by measurements made on the Mississippi River where tidal action is negligible.

Ozturk has plotted the same data in such a way as to fix the powers of the Reynolds and Froude numbers so that for the Mississippi case:

$$L_A/d = 6.018(R_d)^{0.82}(F_d)^{-5/2} \tag{5-7}$$

The differences embodied in eq. 5-6 and 5-7 indicate the need for more research in this topic.

For two estuaries in Norway the constant in eq. 5-7 is replaced by 2.665 and 1.062, respectively, which would indicate the need to take measurements of the variables involved at some particular static stage in order to determine the "length" correction. Ozturk states: "The distortion in the velocity of flow caused by the configuration of the estuary bed should influence the friction existing between flow and the estuary bed, between brackish water and sea-water flowing opposite to each other."

Substitution of $\Delta\rho/\rho_m = 0.02$ into eq. 5-5 allows the variables L_A, d and V_r to be graphed as in Fig. 5-2, which should permit a rough calculation of L_A for a wide shallow estuary of essentially parallel sides and constant depth. It is clear from this figure that as V_r increases so L_A decreases.

Shi-igai [3] has derived an empirical relationship assuming the stress coefficient at the interface f_i is proportional to the n'th power of the parameter $1/R_d F_d^2$. Where $n = 0$, f_i is constant along the length of the wedge. When $L = 1$ the value of f_i is given by the variable shearing stress along an internal wave. By certain measurements it is possible to evaluate n for any given river and so determine L_A for this or any similar situation. Shi-igai also allows in his relationship for a sloping bottom to the waterway. This approach should be valuable when more data are available from the field for well stratified conditions.

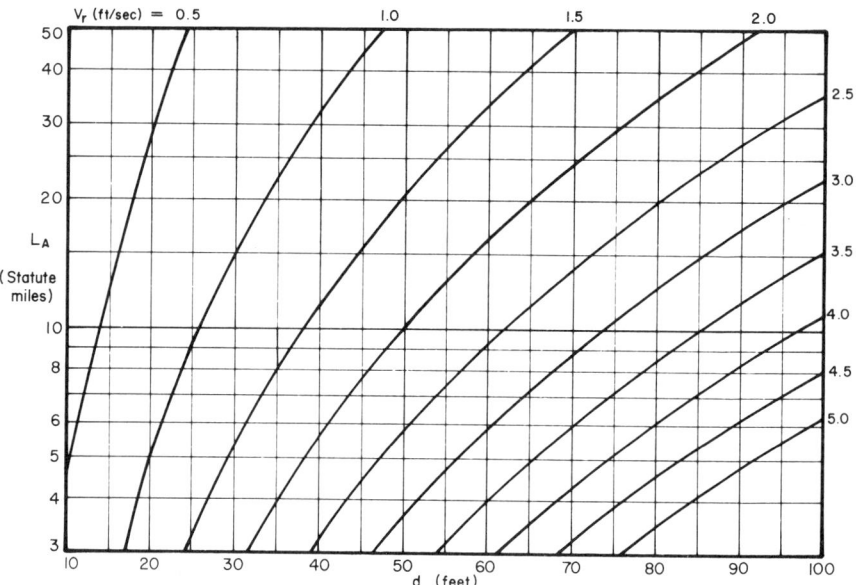

Fig. 5-2. Length of an arrested saline wedge from eq. 5-5.

The slope of the saline wedge is determined mainly by h_o and L_A (see Fig. 5-1), an empirical relationship of which is graphed in the upper section of Fig. 5-3. The height of the wedge at the mouth (h_o), for negligible salt-water mixing and critical conditions of the fresh-water flow at the entrance [4, 5] (i.e., $F_d = 1$ or $d-h = [V_r^2 d^2/(g \Delta \rho/\rho_m)]^{1/3}$) is given by:

$$\frac{h_o}{d} = 1 - \frac{V_r^{2/3}}{(gd\Delta\rho/\rho_m)^{1/3}} \tag{5-8}$$

which has been graphed in the lower portion of Fig. 5-3 for $\Delta\rho/\rho_m = 0.02$. It should be noted that h_o can be zero if V_r is large enough. From the inset of Fig. 5-3, it is seen that for a barrier to be installed some distance x upriver from the mouth it will have to be of height h to be effective. The benefit of river flow (V_r) is apparent in that the wedge may be minimised or actually excluded from the river. On the other hand, dredging deep channels for navigational purposes can produce greater penetration of salt water. With fluctuations in river discharge over the year, so the excursion of salt water upstream can vary throughout the seasons [6, 7].

Although the salt-water wedge may be static in its profile and location the water inside it is in constant motion. The fresh-water mean velocity increases up the wedge, from pure continuity considerations, and the drag on the salt water causes the interface to have a downstream velocity, as illustrated in Fig. 5-4A. This water demand results in an upstream flow near the bed, with an upward component for

SALT-WATER WEDGE

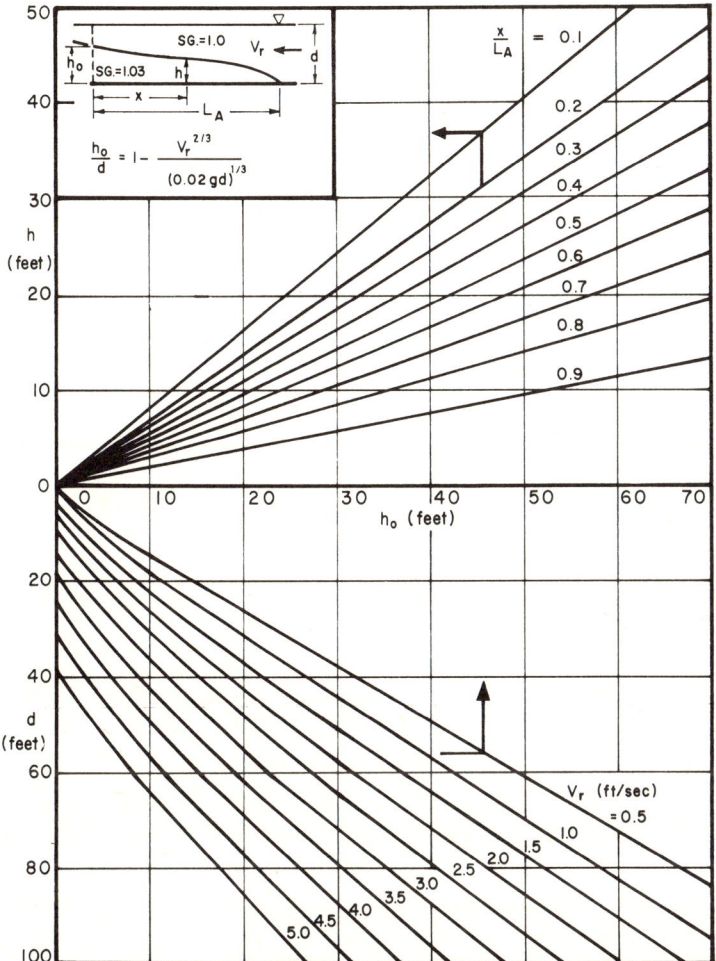

Fig. 5-3. Profile of an arrested saline wedge.

servicing the outflow near the interface. Except for near the toe of the wedge it could be accepted [1] that:

$$V_s/V = 1.06 \quad \text{and} \quad V_f/V = 0.63 \tag{5-9}$$

where V_s is the velocity at the water surface; V is the mean fresh water flow at some point along the wedge; and V_f is the velocity at the interface.

Mixing occurs at the interface when the fresh water velocity reaches a critical value (V_c), given by:

$$\frac{V_c}{(\nu_b g)^{1/3}} = C \left(\frac{\Delta \rho}{\rho}\right)^{1/3} \tag{5-10}$$

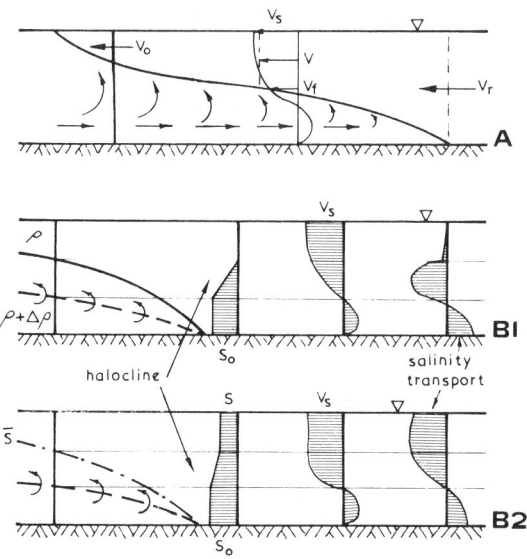

Fig. 5-4. A. Velocity vectors in the region of an arrested saline wedge; B. Velocity, salinity and salt transport distributions for: B1. highly stratified condition; B2. well-mixed conditions.

where ν_b is the kinematic viscosity for the salt-water near the bed; ρ is the density of the fresh-water near the surface; and C is a constant (= 7.3 for an arrested wedge).

Substituting $\nu_b = 11 \cdot 10^{-6}$ ft.^{2}/sec and $\Delta\rho/\rho = 0.03$, the value $V_c = 0.16$ ft./sec is obtained, which indicates that mixing is taking place almost continuously.

Carstens [8] observes that salt may be extracted from a wedge by diffusion and entrainment. The former mechanism involves a two-way exchange of water whilst the latter is one-way. The relative rate of each depends upon the turbulence differential in the fresh water and the lower salt layer. The more turbulent water mass will erode the less turbulent mass and entrain it. Thus, when the salt-water mass is stationary, due to little or no tidal oscillation, the outflowing fresh water takes more salt water with it. The greater the river discharge the greater the inward flow near the bed in order to supply this demand. Where turbulence is increased in the wedge, from tidal or wave action, so the transfer of salt water through the interface decreases and so the upstream velocity at the bed is reduced. It can be surmised from this that this velocity will vary throughout the tidal period as the turbulence differential at the interface changes.

Another interesting facet of this mixing process is the influence of different components of the turbulence spectrum on it. Minami [9] has made a theoretical estimation of salinity dispersion by analyzing the motion of "turbulons" or spheres of rotating liquid passing across the interface. The larger "turbulons" of liquid, with densities different from their surrounding medium, are more affected by gravity and are therefore pulled back into their original stratum. Minami concludes that the major dispersion is carried out by the smaller "turbulons" which are not so influenced by gravity. This is the opposite case to diffusion in homogeneous density flow.

Thus, the rate of salt-water transfer across the interface and the resultant upstream flow at the bed, depends upon the magnitude and nature of the turbulence in the upper and lower elements of the wedge. It will therefore vary greatly between shallow estuaries and deep ones such as fiords [8]. It will also vary with the capacity of the turbulence generators which are located at the surface (as breaking waves), at the interface (as breaking internal waves), at the bed (as roughness elements) at the sides of the bed (as wake producers). Thus, major differences exist between estuaries in their ability to promote mixing, so that great caution should be exercised in applying results from model to prototype and from one prototype to another.

When making salinity measurements in an estuary, in order to assess the degree of mixing, it would appear essential from the previous discussion to carry out such a survey over various types of tidal cycle. If the saline wedge extends a long distance upstream from the mouth, conditions can vary greatly at locations where the boundary conditions vary from the normal. Such features as sudden contractions or expansions of the waterway can generate turbulence in the lower stratum, which can influence the mixing process upstream or downstream of the zone depending upon the stage of the tide. Also, in long estuaries the phase difference of maximum current and maximum tidal height varies from the entrance to the head.

With the above complications in mind it is nevertheless instructive to express the degree of mixing by some parameter [40] such as:

$$K = \frac{\text{vol. fresh water per tidal cycle}}{\text{vol. salt water in flood tide}}$$

which results in the following classification:

$K = >0.7$ highly stratified
$K = 0.2-0.5$ partly mixed
$K = <0.1$ well mixed

Ippen [10] has illustrated depth distributions of salinity, velocity, and salinity transport which are reproduced in modified form in Fig. 5-4B, with *B1* representing highly stratified conditions and *B2* the well mixed condition. Tully [11] has termed the zone between the upper and lower regions of sensibly uniform salinity as the "halocline". In order to define this region he suggests the plotting of salinity to linear scale against depth to a logarithmic scale. At the lower limit of the halocline there is a continual upward transfer of salt water but no downward transfer of fresh water.

In either partly or well-mixed conditions the wedge is not stationary, it will oscillate backwards and forwards with the tide. Also the intrusion will be apparent mainly from the distributions in velocity and salinity across various planes along the length of the waterway. In well-mixed conditions the salinities at surface and bed may differ by only 15–25%. Such identification data will vary throughout the year as both tides and river discharge vary. But also long-term changes can be effected by decreasing fresh-water input by dam construction, or by deepening the waterway for navigational purposes.

The degree of vertical circulation can be determined by measuring simultaneous velocities at different locations, at various depths and over a tidal cycle. Each point

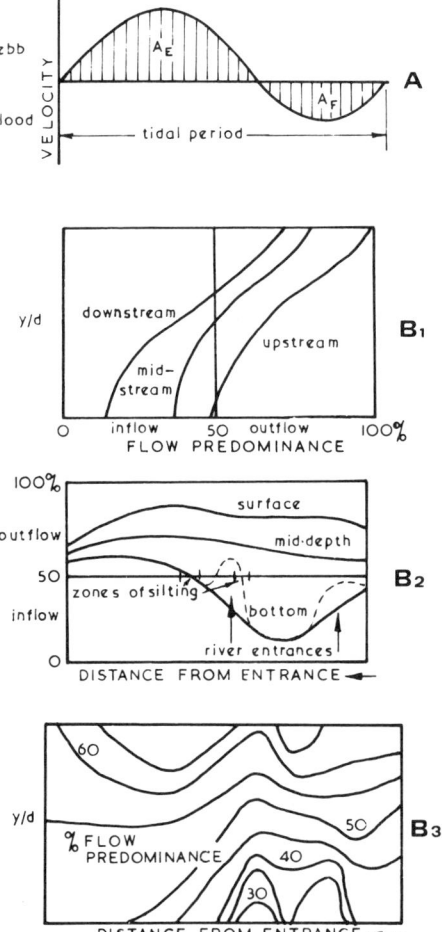

Fig. 5-5. A. Velocity balance at specific depth and specific distance from mouth; B. Flow predominance curves.

then provides a graph of velocity with time, either downstream or upstream, which is associated with the ebb and flood tide [12, 13]. Such a curve is illustrated in Fig. 5-5A, from which the areas A_F and A_E, for flood and ebb respectively, can provide a ratio $A_E/(A_E + A_F)$ which is the % ebb or downstream flow. These areas under the curve represent volumes of water per tidal cycle. The percentage so obtained from each point can be graphed in various ways, as illustrated in Fig. 5-5B. The % ebb flow can be graphed against either y/d or distance from mouth, with curves for position or depth depicted respectively (*B1* and *B2*). Isopleths of this % can be displayed in a longitudinal cross-section of the waterway as in Fig. 5-5B(3). From a graph such as *B2* zones of shoaling can be identified, since on the upstream side net flow near the bed is seawards and on the downstream side the net flow is landwards. Where tributaries feed into an estuarine reach of a river the net outflow is affected as shown by the dotted curves in Fig. 5-5B(2). The upstream flow is reduced by the extra mixing and may even be turned into a downstream velocity for some distance. Where the curve again passes through the 50% demarcation line it creates a site for siltation.

Even in a well-mixed estuary, where stratification is not likely to cause siltation and other problems, there is the mass transport effect from the tidal oscillation itself. Johns [14] has computed this net movement of water particles near the bed of a converging estuary and identified a zone of convergence for this mass transport. When applied to a specific case this zone was found to be the location of major shoals. Previously Abbott [15] had conducted a similar study on reaches of the Thames estuary where boundary layer effects were correlated with mud accretion.

MIXING PROCESSES

The quantification of dilution of either salt water or pollutants in an estuary is a very complex problem, the most frustrating aspect of which is the indeterminancy of a widely applicable diffusion coefficient over the length of a salt-water wedge. This coefficient appears in the relationship:

$$Q = DA \, \delta C/\delta x \qquad (5\text{-}11)$$

where Q is the rate of transfer of tracer; D is the diffusion coefficient; A is the area of fluid through which transfer is effected; $\delta C/\delta x$ is the rate of change of concentration through the surface in the direction normal to it (assumed as x in this case) Eq. 5–11 is known as the Fickian law.

Diffusion can occur even where the fluids with differing concentrations of tracer are static, since molecular interchange can take place. In this case the D used is the molecular diffusion coefficient. As soon as motion occurs, especially differential motion between the two fluid bodies, exchange of material is accelerated and a new value of D demanded.

It can be seen from the dimensions of the terms in eq.5–11 that D has the same dimension as kinematic viscosity (e.g., cm^2/sec), so that a dimensionless parameter D/ν is sometimes employed. It varies in turbulent flow conditions with the largest turbulons present and should probably be correlated with the area under the spectral density curve of turbulence (similar to $H_{1/3}$ for the wave spectrum). Such spectra have a directional quality so that the ability to diffuse will differ in the longitudinal, transverse and vertical planes. Besides this, the generation of macro-turbulence is determined by the shear stresses developed at the boundaries, of solid to liquid, liquid to liquid, and liquid to air. At the interface it is the difference in relative velocities of the fresh and saline media that determines the degree of mixing. Also, when the lower body of water is in motion, due to tidal oscillation, it will have its own turbulence structure, which will be transferred across the interface and so expedite mixing. The generation of waves on the upper layer of liquid can produce macro-turbulence which also aids the dispersion process.

Diffusion coefficients derived from pipe flow [16, 17] have been employed in channels of very shallow section (assumed as a two-dimensional problem) by means of a hydraulic radius (R) equivalence. This accounts for mixing in the longitudinal and vertical directions. For a logarithmic distribution of velocity in the vertical plane Elder [18] obtained the relationship:

$$D = 5.86 dU\sqrt{f} \qquad (5\text{-}12)$$

to which should be added his experimental evaluation of the lateral diffusion coefficient (constant = 0.23), so that:

$$D = 6.09 dU\sqrt{f} \qquad (5\text{-}13)$$

Elder's experiments gave a result in which the constant was 6.30. Sumer [19], using a velocity distribution suggested by Zagustin [20] obtained a longitudinal constant of 6.05 to which he added Elder's lateral value of 0.23 to obtain 6.28.

However, these discussions of decimal accuracy seem a little pointless when prototype measurements appear to give values some 10–40 times these theoretical and experimental values, as reported by Sooky [48], who gives three reasons for the discrepancies, namely:

(1) Velocity variations in the horizontal transverse direction were omitted. This may not be strictly correct since Elder [18] and Fischer [21] have derived such lateral diffusion coefficients. The anomaly may derive from the scale of turbulence in the horizontal plane with vortices the same size as the channel width. This cannot be reproduced to the same scale even in undistorted models of reasonable size.

(2) Coefficients derived from experiments are strictly applicable to straight channels of uniform rectangular cross-section. Natural streams, on the other hand, contain bends, islands, control structures, expanded zones, changes in depth, etc., which indicates that realistic coefficients can only emerge from field measurements.

(3) Transverse velocity components and secondary currents are omitted from the analysis. The latter can be induced by bends in a channel or differential velocities in non-circular sections of waterway.

To overcome partly the deficiencies outlined above, Sooky [48] has carried out a dispersion analysis of channels shaped as triangles of any symmetry or as segments of a circle. The resultant coefficient is then found by adding the longitudinal, vertical and transverse coefficients as obtained from an analysis of the velocity profiles. The results apply strictly to a homogeneous mixture, either upstream or downstream of a saline wedge, although they could be used with caution in a well-mixed estuary in which relatively uniform variations of salinity occur vertically and longitudinally. An additional factor might be added for the mixing capacity of the interface or density difference. However, the analysis highlights the influence of variable transverse depth and gives coefficients much closer to those found in nature.

Sooky assumed alternately a logarithmic velocity profile normal to the bed, and a power law profile in the form:

$$u = u_{max}(Z/d)^{1/n} \tag{5-14}$$

where u is the velocity at normal height Z from the bed; u_{max} is the maximum velocity in the same plane; d is the height to the surface in the same plane; n is the power which varies with Reynolds' number as defined by $4uR/\nu$.

The results so obtained are graphed in Fig. 5-6 for the logarithmic velocity variation. These approximated the curves of the power law profile when

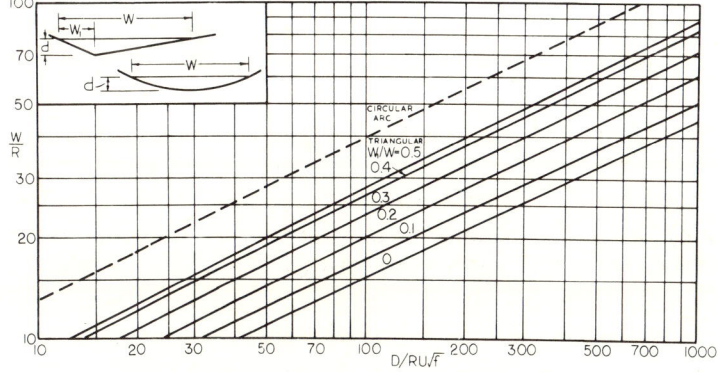

Fig. 5-6. Dispersion coefficient for logarithmic velocity profile normal to bed.

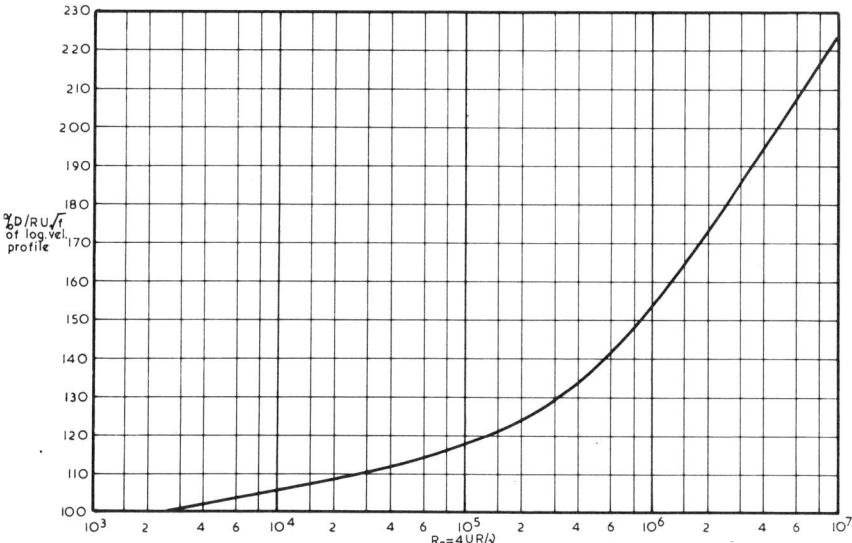

Fig. 5-7. Multiplication factor for dispersion coefficient for power-law/velocity profile.

$R_n = 3 \cdot 10^3$, so that for other flow velocities the curve of Fig. 5-7 can be utilised.

Comparison with laboratory and field measurements [29] showed similar trends regarding shapes and Reynolds number. The practical values exceeded the theoretical due to possible extra mixing by banks, changing cross-sections or stratification.

Harleman [22] has attempted to correlate the diffusion coefficient with energy levels of turbulence. For stratified conditions he found that the coefficient varied lengthwise along the channel. Ippen [10] has suggested even larger values due to the upward passage of water through the interface of the wedge. Using the largest possible value of D at the entrance the salinity distribution as at low water is given by:

$$\frac{\bar{S}}{S_o} = \exp(-V_r/2x'D)(x+x')^2 \tag{5-15}$$

where V_r is fresh-water river discharge; x is the distance upstream from river entrance where salinity is $\bar{S}$; x' is distance seaward of the river entrance where $\bar{S} = S_o$ (i.e., $x = -x'$). Values of x' can be derived from eq.5−20.

The effect of tidal motion in generating turbulence and in oscillating the saline wedge up and downstream can be studied by the reader through the references supplied [10, 22].

An alternative to using diffusion coefficients derived theoretically or experimentally, which are seen to be grossly in error, is to make measurements directly over a number of years in a prototype situation. Results may be applicable to similar sized and shaped rivers elsewhere, where river discharges and tidal range are commensurate. One such series of measurements is reported by Van der Burgh, [23] who has presented an empirical method for determining salt concentrations and intrusion length for a well-mixed saline intrusion. Because of its practical value it is outlined below.

Salinity measurements were made at points along the Rotterdam Waterway at the instant of low water slack (l.w.s.). This was when salinity was at its lowest, just before flood water starts to intrude the entrance. Measurements of surface, bottom and mean salinity were made on nine occasions over the years 1892 to 1967. During this period the mean depth increased due to dredging for navigational purposes.

From the longitudinal distributions under varying flow conditions the salinity at the entrance, or equivalent entrance, was extrapolated to give:

$$\bar{S}/S_o = \exp(-1.655 V_o) \qquad (5\text{-}16)$$

where S_o is the average salinity over the depth at the 10-m continental shelf contour expressed as a percentage of 19,000 p.p.m. or seawater concentration; and V_o = (volume of fresh water per tide)/(volume of flood water at mouth per tide).

For the purpose of calculating these volumes the width at the actual river mouth was assumed, with the 10 m depth, beyond which the shelf sloped away relatively steeply. Eq.5–16 is graphed in Fig. 5-8. It is noteworthy that the depth of channel did not enter the boundary condition, only the river flow and tidal prism.

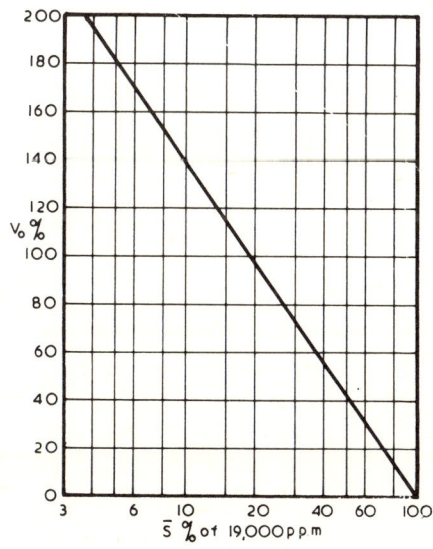

Fig. 5-8. Average salinity at entrance for given ratio V_o as in eq. 5-16. (After Ref. [23].)

The longitudinal distribution of mean salinity $\bar{S}$ was given by:

$$190 \Delta\bar{S}/\Delta x = -V_{ebb} f(\bar{S})/d^3 \tag{5-17}$$

where $\Delta\bar{S}/\Delta x$ is the change in $\bar{S}$ (expressed as % of 19,000 p.p.m.) with distance Δx (m) along the canal; and V_{ebb} (m/sec) = (volume of ebb flow per tide/area at mean tide stage × 44,700 sec (m/sec).

(Note that the volume of ebb flow equals the flood volume plus the fresh-water discharge. The term V_{ebb} is not strictly the ebb velocity since the ebb flow is completed in half the tidal cycle of 44,700 sec).

$f(\bar{S})$ is an empirical function of $\bar{S}$ such that $f(\bar{S}) = 0$ when $S = 0$ (fresh water) and 19,000 p.p.m. (salt water); d is the greatest depth (m) at the section distance x from the equivalent mouth.

Eq. 5-17 has been graphed in Fig. 5-9 as $\Delta\bar{S}/\Delta x\, V_{ebb}$ versus mean salinity $\bar{S}$ at distance x (m) from the entrance expressed as a percentage of 19,000 p.p.m., for various increments of depth d. From both the equation and the figure, the great influence of depth can be seen. The larger is d the slower is the salinity decrease and hence the greater the intrusion. Hinwood [24] has drawn attention to the adverse affects of channel dredging on salt-water intrusion. Not only does deepening decrease the mean velocity of fresh-water flow, but also a smaller fraction of the depth is influenced by this discharge.

The surface and bed records of salinity permitted Van der Burgh [23] to derive

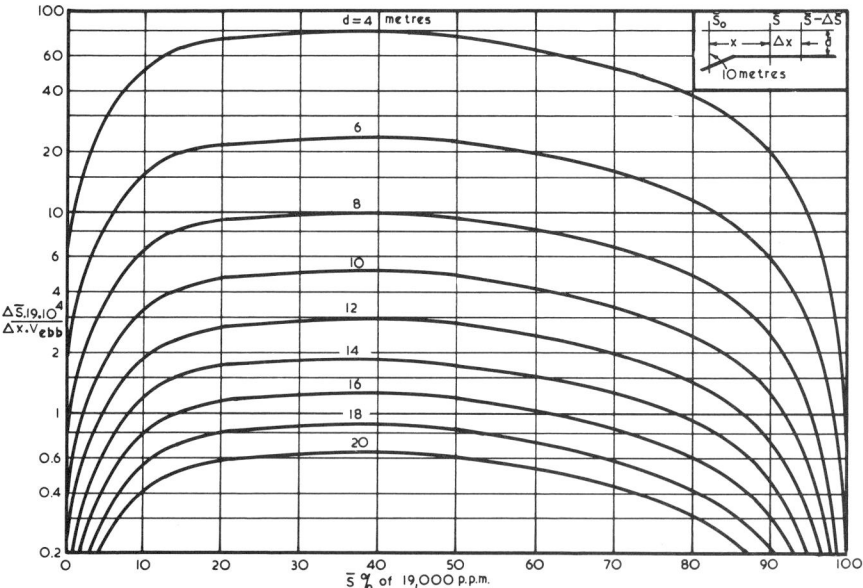

Fig. 5-9. Slope of salinity distribution curve. (After Ref. [23].)

MIXING PROCESSES 217

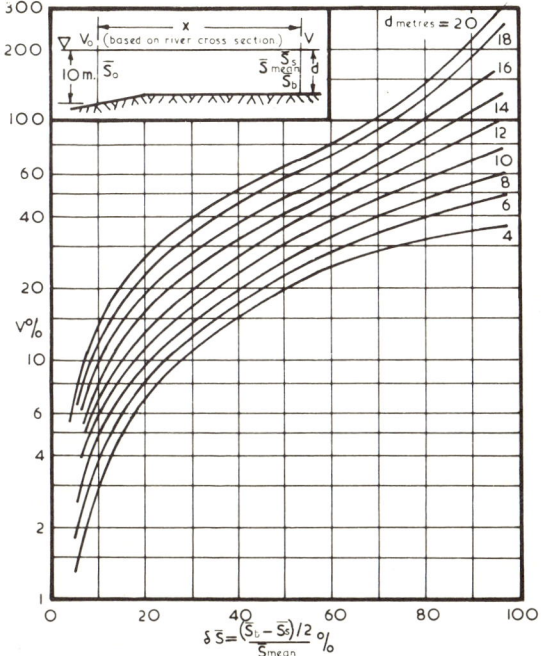

Fig. 5-10. Vertical salinity distribution along intrusion length. (After Ref. [23].)

an empirical relationship for the vertical distribution at any point along the salt-water intrusion length. This was supplied only by a graph whose values have been extrapolated to provide Fig. 5-10. The mean salinity $\bar{S}$ is considered to exist at mid-depth, and a uniform variation from the surface to the bed provides values of $\delta\bar{S}$. The ordinate ($V\%$) in Fig. 5-10 is the ratio of fresh water per tide to the tidal prism at the given local distance x from the equivalent mouth.

The locations of the salt concentrations as given above are for l.w.s. When the tidal flood occurs this bulk of water is forced upstream by the amount of tidal excursion, which is given by Ippen [10] as:

$$L_{hws} - L_{lws} = \left(\frac{T\sqrt{gd}}{2\pi} - L_{lws}\right)[1 - \exp(-H/d)] \tag{5-18}$$

where x is the distance from entrance to a specific salinity location (note that this is not the 10-m depth contour as in Van der Burgh's analysis, which by inference is x' seawards of entrance); T is the period of the tide being considered (normally T = 44,700 sec); d is the mean depth of waterway over which saline wedge occurs; and H is the tidal range at the entrance to the waterway; suffixes hws and lws refer to high and low water, respectively.

Eq. 5-18 assumes no reflection of the tide and no friction effect over the length

TABLE 5-I

Values of K in eq. 5-19

$\bar{S}/S_o$	0.01	0.02	0.03	0.04	0.05	0.1	0.2	0.3	0.4 -	0.5
K	3.03	2.80	2.65	2.53	2.45	2.15	1.80	1.55	1.35	1.18

of the salt-water intrusion, so that high water occurs simultaneously along this length of channel.

From eq. 5-15 Ippen has shown that the length of intrusion of water with a given mean salinity $(\bar{S})$ is given by:

$$L_{lws} = x'(K\sqrt{d/V_r x'} - 1) \qquad (5\text{-}19)$$

where K is a function of $\bar{S}/S_o$ as in Table 5-I, and V_r is the mean river velocity.

In order to apply eq. 5-19, both x' and D must be evaluated. At low-water-slack the mean salinity at the natural entrance ($x = 0$) will be below $\bar{S}_o$, with the hypothetical entrance pushed seawards an amount x' where $\bar{S} = S_o$. As the flood proceeds so in time will the entrance water have a mean salinity of S_o. If this occurs after time t from the commencement of flood, distance x' is given by [10]:

$$x' = \frac{T\sqrt{gd}}{2\pi}(1 - \cos 2\pi t/T) \qquad (5\text{-}20)$$

assuming $H/d \ll 1$ and no tidal reflection in the channel.

If the mean salinity at the entrance $(\bar{S})$ is measured at l.w.s. and distance x' is computed from eq. 5-20, the diffusion coefficient (D) in eq. 5-19 can be derived from eq. 5-15 which then becomes:

$$\frac{\bar{S}}{S_o} = \exp\left(\frac{V_r x'}{2D}\right) \qquad (5\text{-}21)$$

where V_r is expressed as a negative velocity.

For the more complex case of the co-oscillatory tide the reader is referred to the literature [10].

ESTUARY POLLUTION

The estuaries of the world are sites for some of the greatest industrial and commercial undertakings. Their proximity to the sea and relative shelter from it make them ideal for harbour location, with all the enterprises that benefit from easy access to maritime transport. Refineries, electric power plants, steel works and other facilities requiring cooling water are found along the banks of estuaries. Effluents from chemical engineering plants and from the high-density population also find their way into estuaries. With the rapid growth of industry the capacity of estuaries to discharge warm water and effluents to the open sea without undue concentration of obnoxious liquid is fast becoming overtaxed. Ecologists have become very vocal over preservation of natural balances in the environment, and rightly so, which should make the public conscious of the costs involved. This is having its effect on coastal engineering in that alternatives must be sought for marine outfalls. Ingenuity plus greater data collection must be used to minimise the increasing expenditure on clean housekeeping.

The problems of thermal pollution have been stressed [25], with suggestions for needed research. It is noted that generation capacity is doubling every decade and that nuclear plants produce 50% more waste heat than mineral-fired steam plants. Besides its direct impact on temperature of the estuary, heat accumulation can affect chemical and biochemical processes adversely and promote growth of flora, which in turn can influence these and other continuous mechanisms. Warm water can disrupt spawning cycles of fish and, in the extreme case, increase evaporation and so increase salinity in the estuary.

The research needs suggested [25] are: the development of comprehensive hydraulic analyses for application on a wide variety of estuaries; three-dimensional models covering the liquid and air phases over a wide water area; analysis and field measurements of initial dispersion from outfalls; determination of coefficients of dispersion in the liquid and air phases which are applicable over various seasons of the year.

The role of the engineer in the estuary pollution problem is to make sure that concentration of obnoxious liquids is below the hazard level either for humans or marine life in the area. The degree of treatment before discharge is dictated by the deleterious nature of an effluent initially, after mixing in the stream, and ultimately where it may accumulate as a sedimentary deposit. The initial mixing likely to take place on discharge from an outfall is treated in the next chapter, here the concern is with dispersion over the length, breadth and depth of the estuary during transit to the sea.

Collection of data for the prediction of dispersion in estuaries is more difficult than in rivers where uni-directional flow occurs. Besides the oscillatory flow due to tides (with differing ebb·and flood velocities, phase shifts between maximum velocities and maximum depths and changes in channel geometry with tidal stage) there are the complications of stratification and additional mixing due to vertical transport across the interface.

Outfalls may be located up-river or within the length of salt-water intrusion. They may be at the surface, mid-depth or near the bed. The effluent may have a positive, negative or neutral buoyancy. If deleterious its effects may be conservative or non-conservative [26]. The former may not be toxic in the concentrations as discharged or diluted in the estuary, but may become so when continually ingested by marine fauna which is later consumed by humans. Even chemicals rendered innocuous with time may, during their passage through an estuary, impede processes such as oxygen uptake of various substances, so establishing anaerobic conditions for marine life.

Dispersion model

Holley [27] has distinguished between diffusion and dispersion. The former he restricts to mixing from molecular and turbulence action, whereas the latter term he applies to spreading from velocity differentials across a channel cross-section. Both processes are related to other variables through a coefficient which is likely to vary greatly with scale and geometry of a waterway. The coefficient calculated or measured for dispersion would necessarily contain that for diffusion also [28].

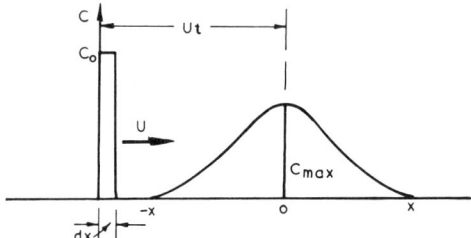

Fig. 5-11. Definition sketch of batch discharge of pollutant.

It is instructional at first to consider a batch source of pollutant in an uni-directional flow without any shear present. As depicted in Fig. 5-11 a mass M of tracer is introduced over a cross-section A of stream at time $t = 0$. The initial concentration C_o is given by:

$$C_o = M/\rho A dx \qquad (5\text{-}22)$$

where M is the mass of tracer in batch; ρ is the density of the ambient liquid; dx is the longitudinal thickness of liquid containing M.

This batch of partly diluted pollutant is diffused as it is carried downstream with a velocity U, until, at distance Ut from the source, it is spread over a length $\pm x$ from the point of maximum concentration C_m (see Fig. 5-11). Assuming the total mass of pollutant M is conserved, the concentration (C) varies in a Gaussian curve given by [17]:

$$C = \frac{M}{\rho A (4\pi Dt)^{1/2}} \exp \frac{-(x - Ut)^2}{4Dt} \qquad (5\text{-}23)$$

where D is the diffusion coefficient for this particular kinematic condition. For the case of continuous injection numerical methods are suggested [30].

Now consider tidal oscillation in a rectangular channel where the density is uniform. This would apply to those sections of an estuary upstream or downstream of the brackish water zone or even in this zone if the fresh water flow was negligible.

Ippen has applied Taylor's pipe solution to rectangular channels and using $u = 2u_{max}/\pi$ derived:

$$D = 4.55 f^{1/2} u_{max} d \qquad (5\text{-}24)$$

This assumes that the bulk of the mixing is due to shear velocity in the vertical plane. Holley et al. [28] have derived an expression for the transverse coefficient which involves a knowledge of the velocity distribution across the waterway. But for continuous releases of pollutant (the normal situation) the concentration distribution after many tidal cycles is not sensibly different if the transverse coefficient is 10 times greater than the vertical. So that eq. 5-24 is applicable.

For the case of triangular and circular arc channels, as depicted in Fig. 5-6, replacement of u by $2u_{max}/\pi$ should give a comparative value of D. To apply the Reynolds number amplification factor of Fig. 5-7 a similar substitution in $R_n = 4 uR/\nu$ would be required.

Within a region of salinity intrusion where the density is not uniform, either depth-wise or lengthwise, mixing is aided by vertical currents. Harleman [29] has shown that the dispersion coefficient is highly variable, rising swiftly at the toe of the intrusion and then slightly from there to the mouth. Theoretical approaches to D are impossible, so that field measurements must be made over a range of river flow conditions. This is accomplished by measuring mean concentrations at three locations equal distances (Δx) apart along the river. The dispersion coefficient for the central location is then given by [29]:

$$D = \frac{V_r C_x 2\Delta x}{(C_{x+\Delta x} - C_{x-\Delta x})} \qquad (5\text{-}25)$$

where V_r is the mean velocity of fresh water flow just upstream of the intrusion; C_x is the mean concentration at point distance x from the toe of the intrusion; Δ_x is longitudinal distance to the upstream and downstream points; and $C_{x + \Delta x}$ and $C_{x - \Delta x}$ are the mean concentrations at the downstream and upstream points, respectively.

The concentrations should be measured for both hws and lws and an average taken. Then the D obtained is used as for the uniform density case over a limited length of channel.

With all the many complications and profuseness of data to be collected, numerical methods and computer application have become necessary [30–34].

Tidal prism concept

To obviate the difficulties of assessing dispersion coefficients over the length of a salt-water intrusion, an approach has been used which compartmentalises the waterway and considers mixing of batches of salt and fresh water in each during a tidal cycle. Ketchum [35] proposed such a model by dividing the estuary into segments with lengths equal to the average excursion of water particles in a flood tide. At high-tide level the water in each segment is considered as fully mixed. During the ebb an exchange takes place between adjacent segments. The requirement of complete mixing during each tide infers large tidal oscillations or relatively small river discharge.

Any segment at high tide contains the same volume as an adjacent downstream segment at low tide. The length of the innermost section is defined by the tidal range containing the volume of fresh water (V_r) contributed during a tidal cycle. Thus, in Fig. 5-12 the length L_o is given by:

$$\overline{V}_r = L_o H = V_r Td \tag{5-26}$$

and subsequent volumes of segments by:

$$V_1 = V_o + \overline{V}_r \tag{5-27}$$

$$V_2 = V_1 + P_1 = V_o + \overline{V}_r + P_1 \tag{5-28}$$

$$V_3 = V_2 + P_2 = V_o + \overline{V}_r + P_1 + P_2 \tag{5-29}$$

so that in general:

$$V_n = V_o + \overline{V}_r + \sum_1^{n-1} P \tag{5-30}$$

In eq. 5-27 to 5-29 the P terms are local tidal prisms or volumes of water in segments of length L and tidal range H.

The proportion of polluted (fresh) water removed from the n-th segment each ebb tide is given by:

$$\Delta_n = P_n/(P_n + V_n) \tag{5-31}$$

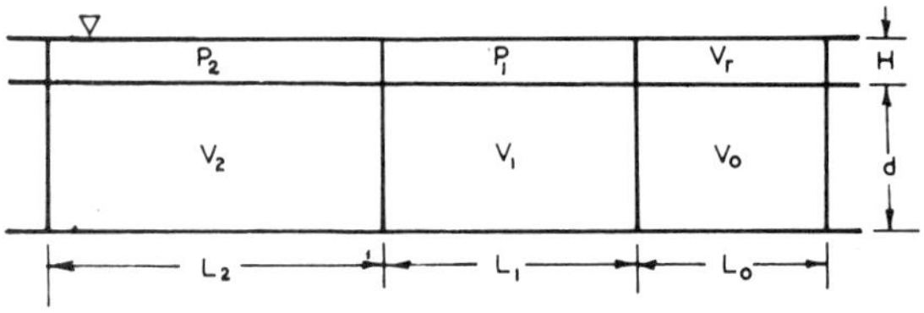

Fig. 5-12. Definition sketch of tidal prism concept.

For a constant river flow the total volume of river water (Q_n) retained in the n-th segment at high tide after many cycles is:

$$Q_n = \overline{V}_r / \Delta_n \qquad (5\text{-}32)$$

The above equations assume complete mixing in each segment during a tidal cycle. Should this not be effected it implies that the lower levels of water have their original salinity. In this case it is convenient to consider a false bottom to the river at a depth h from the surface so that the exchange ratio Δ_n is given by:

$$\Delta_n = \frac{P_n}{P_n + V_n} \frac{d}{h} \qquad (5\text{-}33)$$

If the river water contains a concentration C_o of pollutant, and this is conserved during the passage to the sea, the concentration at subsequent segments can be obtained from:

$$C_n = C_o \Delta_n \qquad (5\text{-}34)$$

The time for the river flow volume (V_r) to flush through the n-th segment in terms of tidal cycles is given by:

$$N_n = \frac{1}{\Delta_n} = \frac{Q_n}{\overline{V}_r} \qquad (5\text{-}35)$$

so that the flushing time for the whole estuary is:

$$N_T = \sum_1^n N_n = \sum_1^n 1/\Delta_n \qquad (5\text{-}36)$$

Ketchum's method is satisfactory only where $n > 10$ and the tidal prism is some scores of times greater than the river flow.

Preddy [36] proposed another segmental analysis which involves two integral equations that have detracted from its widespread use [37]. Pritchard [38] has presented what he terms is a "two-dimensional box model", in that his segments are divided into an upper and lower layer. Material is exchanged between bottom and end boundaries of each half box. If m segments are chosen then 2-m equations are available to solve an equal number of unknowns. Di Toro [37] has applied the segmental approach using the theory of Markov chains, basing his choice of matrix on the maximum entropy principle of information theory and statistical mechanics. For m segments the 3-m simultaneous equations were found soluble by computer in less than 1 min.

ESTUARY SEDIMENTATION

Because of the stratification or differential densities throughout an estuary the currents averaged over a tidal cycle at the surface and the bed will vary. This distribution of velocity alters along the length of the estuary [39], as depicted in Fig. 5-13, where it is seen that the mean velocities near the bed change from upstream to downstream at the mouth and beyond the toe of the intrusion, respectively. Whatever type of sediment is being transported, bed load or wash load, any that is in the vicinity of the bed will be carried towards the null-point, from both upstream and downstream. The correlation of siltation and mean zero flow near the bottom has been discussed already and illustrated in Fig. 5-5B2.

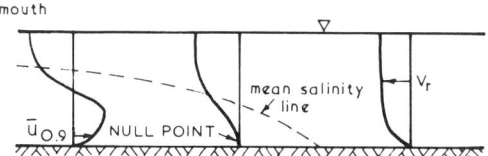

Fig. 5-13. Vertical distributions of velocities averaged over a tidal cycle.

In order to predict changes in the null-point with variations in fresh-water flow and compare data from different estuaries, it is necessary to correlate some parameter with the distance of this zone from the mouth. Such an analysis has been presented by Harleman and Ippen [39], the conclusions of which are outlined below.

Velocities ($\bar{u}_{0.9}$), both positive and negative, are recorded at about 0.9 depth from the surface during a complete tidal cycle at a number of points (x) along a waterway. These are averaged and are likely to have distributions as shown in Fig. 5-13. During the same period of time the fresh-water flow V_r is recorded. At each station x the mean salinity S_x, throughout the depth and the tidal cycle, is also determined. A plot of S_x/S_o (where S_o is the ocean basin salinity) versus distance x from the natural mouth of the rectangular waterway of mean depth d gives L_i the length of salt-water intrusion or distance to the station where $S_x/S_o = 1\%$ (see Fig. 5-14A).

The ratios U_{09}/V_r and x/L_i are then plotted on what is termed a "velocity predominance curve", as illustrated in Fig. 5-14B. Similar curves were given previously for various depths, but containing actual distances x, instead of the dimensionless ratio x/L_i (see Fig. 5-5B, 2). Where the $U_{0.9}/V_r$ line intersects the zero line gives the null-point at which siltation has been shown to occur [40]. Such a curve is useful in predicting where present shoaling is taking place, but does not serve for such prediction when changing river flow (V_r) or depth (d) occur.

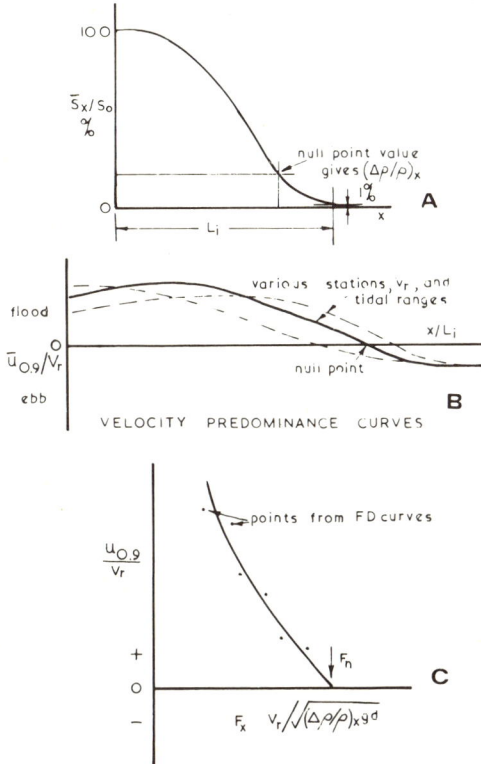

Fig. 5-14. A. Typical mean salinity distribution curve along waterway. B. Velocity predominance curve with dimensionless length. C. Bottom velocity correlation with densimetric Froude number.

For the determination of the null-point under these different circumstances the densimetric Froude number at the nullpoint must be assessed. This then has a unique value for a waterway, which can be used to find the influence of changing (V_r) and (d). This Froude number is given by:

$$F_x = V_r / \sqrt{(\Delta \rho / \rho)_x g d} \qquad (5\text{-}37)$$

where $(\Delta \rho / \rho)_x$ is the density difference between the saline water at the x station and fresh water, and is given by [39]:

$$\left(\frac{\Delta \rho}{\rho}\right) x = \frac{0.7}{1{,}000{,}000} \bar{S}_x \qquad (5\text{-}38)$$

where $\bar{S}_x$ is expressed in p.p.m. Eq. 5-38 is a general relationship for dilute salt solutions.

For the given V_r, d, and $(\Delta \rho / \rho)_x$, values of F_x can be computed and plotted against $U_{0.9}/V_r$, as illustrated in Fig. 5-14C. Where this is extrapolated down to the value $U_{0.9}/V_r = 0$ the null-point Froude number (F_n) is determined, which is unique to the waterway. The density ratio for this condition can be expressed as:

$$\left(\frac{\Delta \rho}{\rho}\right)n = \frac{1}{(F_n)^2} \frac{V_r^2}{gd} \tag{5-39}$$

and since from eq. 5-38 it can also be expressed as:

$$\left(\frac{\Delta \rho}{\rho}\right)n = \left(\frac{\bar{S}_n}{S_o}\right)\frac{0.7 S_o}{1,000,000} \tag{5-40}$$

The salinity at the null-point under new conditions can be determined by:

$$\frac{\bar{S}_n}{S_o} = \frac{1,000,000}{0.7 S_o F_n^2} \frac{V_r^2}{gd} \tag{5-41}$$

In eq. 5-41 values of S_o and F_n can be accepted as fixed by the nature of the waterway and its geographic location. Changes in V_r and d can provide a new S_n/S_o value in a new salinity distribution curve, which is dictated by the changed values of V_r and d. Where this occurs in terms of x/L_i can provide the location of the new

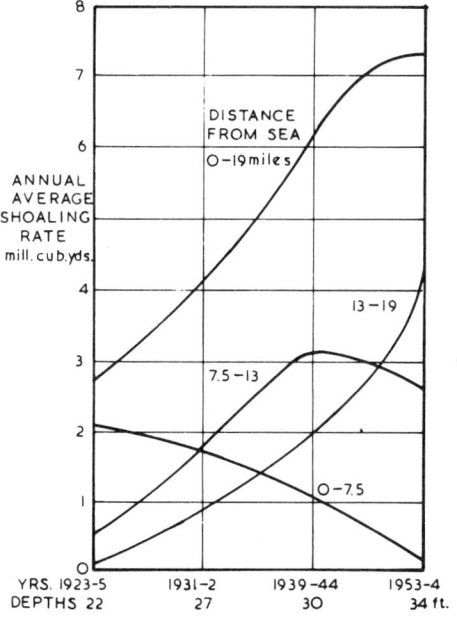

Fig. 5-15. Shoaling rates in Savannah estuary (After Ref. [40].)

null-point and therefore the likely location of sedimentation. Salinity distribution calculations have been treated in a previous section of this chapter.

The consequences of not carrying out such an analysis can be gauged from the experience of Savannah Harbour [40], sited some 17 miles upriver from the ocean. Although the fresh-water flow has remained sensibly static over the years, the river has been dredged from 22 to 34 ft. depth to cope with maritime traffic. This has forced the null-point from the entrance area up to the region of the harbour itself. The progressive deterioration in the condition is presented in Fig. 5-15, where it is seen that the total siltation has more than doubled over the 30 years that measurements have been made.

Sediment entering an estuarine situation, from either upstream or from the sea, can be either bed load or wash load. The proportion of each depends upon the sediment and flow characteristics. Bedload requires some minimum shear flow before it becomes significant and is likely to be spasmodic in character. Meade [41] mentions cases of rivers discharging the bulk of their annual sediment load in a few days of flood. Although this tends to accumulate at the toe of the saline wedge, likely to form at such time, the wedge may in fact be swept from the waterway temporarily. However, the sediment, particularly the coarser material, will be deposited in the vicinity of the mouth and be swept back by subsequent wedge action in conjunction with mass transport of tides and waves. The wash load, on the other hand, will be spread far out to sea during such floods. At more normal periods these finer particles can be affected by saline conditions along the wedge and be caught in the upstream flow near the bed. Under strong stratification coarse particles accrete near the toe of the wedge and the fine particles are swept to the downstream edge of this mound [40].

The ability of estuaries to silt up through the induction of sediment from the sea cannot be questioned. Evidence is growing continually on the source of "recent" marine deposits found well upstream of estuary mouths. Byrne and Kulm [42] report that material from beaches outside the entrance to Yaquina Bay, Oregon, more than six miles upstream. Such deposition is expedited during the winter flood season when littoral drift and onshore winds are greatest. Any plans to increase fresh-water flow through a waterway in order to improve water supply, disperse pollutants, or to sweep out sediment, should be investigated thoroughly, to see that no new and perhaps more grave siltation problem is not generated. Simmons [12] has shown that inward flow near the bed maximizes before it is reduced to zero by the overwhelming force of flooding water. Meade [41] cites several cases of marine material working its way upstream, including that of Savannah estuary where, of the annual deposition of 5.5 m cubic metres only 1.5 m could be accounted for by the river itself. Of the remaining 4 m some was material previously dredged and dumped near the mouth and some from offshore deposits. Meade also poses the major questions: "Of the river sediment that is brought in, how much is trapped permanently in the estuary, and how much is carried out to sea? Over the long term, is more sediment carried out the mouth of the estuary than into it, or vice versa?"

Ippen [43] has stated that major problems are caused by colloids which are silts and clays with particle sizes of 1 micron (μ) (i.e., 1/1000 mm). Of these about a quarter come from upstream, the remainder from within or seawards of the estuary. It has previously been considered that flocculation of fine material from the river was the prime cause of estuary siltation, but now it is realised that precipitation of salts from sea water is a major source of material. Such deposits will be found to have a high percentage of calcium carbonate. This may be pure precipitate or shell debris of marine animals, which can readily use $CaCO_3$ in warm shallow water.

Where the geology of the river catchment provides ample clay products, the electro-chemical process known as "flocculation" takes place in the brackish water of the salt-water intrusion. In such an electrolyte the ions neutralise the normal repelling charge and so particles coalesce and form flocs which settle to the bed. The rate of such formation depends upon many factors, including hydraulic, chemical and sedimentary [43].

Abecasis [44] has emphasized the change in sedimentation along the length of an estuary. Due to the maximum current from the tidal flood occurring between half and high tide levels it acts on the banks. On the other hand, since the ebb flow is optimized at the half to low tide level the ebb acts more on the bed. Near the mouth the ebb and flood are more-or-less sinusoidal, whereas upstream the flood period is shorter and the ebb longer. Thus the time to act on bank or bed varies along the estuary. At the mouth the inward flood flow is radial and therefore can influence material coming from along the coast. The ebb discharge is more jet-like and hence deposits material along the centre-line of the outlet. However, with the longer waves arriving normal to the coast this sand can readily be brought back into the estuary, as discussed in other chapters.

ESTUARY CONTROL

The physiographic features termed estuaries are large by any engineering standards. They are zones of great commercial interest, with many conflicting demands being made upon them. Whilst some need deeper channels to transport goods, others need high velocities to disperse effluents. The coastal margins must serve industry and recreation alike. Such a situation demands a continuous overall supervisory body which can marshal all the scientific and engineering knowledge required in the solution of problems, or the avoidance of some before they arise. For a thorough discussion of estuary management the reader is referred to the proceedings edited by Lauff [45].

In respect to sedimentation, comprehensive surveys of an estuary are required before detailed studies are commissioned. Such an overview has been provided by Fleming [46] of the Clyde estuary, where sediment supply from the rivers draining into the estuary has been assessed and compared to present rates of siltation and dredging. The author concludes: "The development and operation of major harbours and navigational waterways should be carried out with a complete understanding of the inherent sediment transport problems and sedimentation patterns. The sediment problem should never be under estimated."

One of the major aspects of dredging in estuaries is the availability of dumping grounds. Whilst land reclamation is ideal from the point of view of preventing recirculation of material, the mud so available in estuaries is unsuitable for foundations of major structures. Cable [47] has discussed disposal practices in estuaries and concludes as follows: "The disposal of dredged material is an increasingly complex problem. The desire to reduce repetitive redredging of materials in the channels has intensified efforts to provide positive retention of dredged materials where they are deposited... The emphasis on clean waters has redirected planning towards alternative disposal techniques which minimize pollution aspects of some methods of dredgings. These alternatives probably will result in more costly disposal methods to provide accepted means of accomplishing required dredging work."

Sediment transported by a river as a natural mechanism cannot be considered as a pollutant, it is just a fact of life to be measured and dealt with by engineers. However, when additional material is thrown into suspension by construction works, dredging, or mining operations, it may throw the situation out of equilibrium and cause an ecological imbalance [25]. Sediment may also be a pollutant if it serves as a vehicle for contaminants such as pesticides or organic material [26].

There is still much to be learned about the mechanism of density currents, flocculation, dispersion, pollutants, velocity distributions and levels of turbulence [25]. Some of this search for knowledge can be gained in the laboratory, but due to the interaction of so many factors field measurements are required in the great variety of estuary shapes and marine conditions.

Comprehensive research teams are paramount as indicated in the following quotation [25] from a committee on Tidal Hydraulics of the American Society of Civil Engineers: "The problems associated with thermal and sedimentary pollution in tidal waters must necessarily be considered from a broad viewpoint. Each problem has characteristics that are of engineering, physical, chemical and biological importance. In an estuary, one clearly cannot be concerned only with the tidal hydraulics, but must also know the nutrient ranges and cycles, the flora and fauna that are supported within the existing ranges, and how much change can be tolerated if a desired condition is to be maintained. When predicted changes reach the point that one or more species of plant or animal are eliminated, then an understanding of the reliance of remaining biota on the vanishing species is desirable."

EXAMPLES

1

Compute the extent of salt-water intrusion on a river debouching to the Mediterranean Sea, when its discharge is 150,000 cfs, average width 1/2 mile, average depth 42 ft. If a fresh-water intake for irrigation were to be installed 5 miles upstream from the mouth, what is the maximum depth to which it should be taken so as to preclude induction of brackish-water (assume $\Delta\rho/\rho_m = 0.02$).

Because the tidal range can be accepted as negligible in the Mediterranean Sea, the intrusion can be considered as an arrested saline wedge.

From Fig. 5-2, for $V_r = 150{,}000/(5280/2)42 = 1.35$ ft./sec, and $d = 42$ ft, $L_A = 17$ miles.

From Fig. 5-3, $h_o = 25$ ft., and for $x/L_A = 5/17 = 0.294$, $h = 15$ ft., so that the suction of the intake can be located at $42 - 15 = 27$ ft. depth.

However, if the river discharge drops below 150,000, to say 100,000 cfs, $V_r = 1.0$, $L_A = 40$ miles, $x/L_A = 0.125$, $h = 22$ ft., so that the desired depth is only 20 ft. The minimum flow during the irrigation period should be used for such planning.

Besides this caution, the intake should be placed high enough above the interface for the suction not to draw up the salt water. This aspect is treated in Chapter 6.

2

For the estuary in Example 1, find the tidal range necessary to turn it from a well mixed case to a partially mixed one, and then to a highly stratified estuary.

The criterion for a partially mixed estuary is:

K = vol. fresh-water per tide/vol. flood-water per tide ≥ 0.2.

Fresh-water discharge per unit width = $150{,}000/2{,}640 = 57$ cfs, vol. per tide = 57×44700 cub. ft.

Vol. flood-water = $u_{ave}(d + H/4)T/2$ (assuming average depth over flood tide = $d + H/4$)

$= u_{max} \dfrac{2}{\pi}(d + H/4)\dfrac{T}{2} = \dfrac{Hg^{1/2}}{2d^{1/2}}\dfrac{2}{\pi}(d + H/4)\dfrac{T}{2}$ from eq. 4-22 of *Coastal Engineering*, I) = $H(32.2)^{1/2}(42 + H/4)\,44700/2\pi(42)^{1/2}$.

$K = 0.2 = \dfrac{57 \times 44{,}700 \times 2\pi(42)^{1/2}}{H(32.2)^{1/2}(42 + H/4)\,44700}$, $H = 3.7$ ft.

For a highly stratified estuary $K \geq 0.7$, in which case $H = 13$ ft.

3

Compare the dispersion coefficients suggested for a waterway 2000 ft. wide when considered as rectangular, triangular, or the segment of a circle, with the same cross-sectional area of 80,000 ft^2 and the same mean flow velocity of 1 ft./sec.

(1) Rectangular. $d = 80{,}000/2000 = 40$ ft. From eq. 5-13, $D = 6.30\,du\sqrt{f} = 6.30 \times 40 \times 1\sqrt{f} = 252\sqrt{f}$.

(2) Triangular. Depth at apex $d = 80{,}000/(2000/2) = 80$ ft. $W/R \doteq 2000/(80{,}000/2000) = 50$. From Fig. 5-6, for $W_1/W = 0.5$, $D/Ru\sqrt{f} = 315$, $D = 315 \times 40 \times 1\sqrt{f} = 12{,}600\sqrt{f}$; for $W_1/W = 0$. $D/Ru\sqrt{f} = 1300$, $D = 1300 \times 40 \times 1\sqrt{f} = 52{,}000\sqrt{f}$.

(3) Segment of circle. From eq. 3-4, $W^2 = (2.44\,A^2/d^2) - 4d^2$, $2000^2 = (2.44 \times 80{,}000^2/d^2) - 4d^2$, $d = 62.3$ ft. $W/R \doteq 2000/(80{,}000/2000) = 50$. From Fig. 5-6, $D/Ru\sqrt{f} = 160$, $D = 160 \times 39.5 \times 1\sqrt{f} = 6300\sqrt{f}$.

Based upon these theoretical and experimental analyses it would appear preferable to discharge pollutants into non-rectangular sections of waterway for greater dispersion.

EXAMPLES   231

For a power-law velocity distribution an amplification is needed which varies with $R_n = 4uR/\nu = 4 \times 1 \times 40/11 \cdot 10^{-6} = 1.45 \cdot 10^7$.

From Fig. 5-7, factor is 2.3, so that the triangular and circular arc coefficients are more than doubled.

The difference between these shapes and the rectangular channel, as far as dispersion is concerned, requires prototype verification.

4

At low-water/slack (lws) in the entrance to a waterway 10 m deep the mean salinity over the whole depth is 11,000 p.p.m. when the average river flow velocity upstream is 0.3 m/sec. Find the mean simultaneous salinity at a distance of 10 km upstream from the mouth. It is known that 1 hour after lws the mean salinity at the entrance is the sea value of 19,000 p.p.m. If this semi-diurnal tide has a range of 1 m, find the location of the zone where salinity is similar to that at the 10 km station at lws. Find the location of the zone where mean salinity is 1% of the sea value at lws and hws.

Eq. 5-20 gives $x' = \dfrac{T\sqrt{gd}}{2\pi}(1-\cos 2\pi t/T)$, $x' = \dfrac{44,700\sqrt{9.81 \times 10}}{2\pi}(1-\cos 360 \times 1/12.42) = 8.9$ km.

Eq. 5-21 gives $\bar{S}/S_o = \exp -\dfrac{V_r x'}{2D}$, so that $\dfrac{11,000}{19,000} = \exp -\dfrac{0.3 \times 8,900}{2D}$, $D = 2450$ m²/sec, c.f. eq. 5-13, $D = 6.3\, du\sqrt{f} = 6.3 \times 10 \times 0.3\sqrt{0.03} = 3.27$ m²/sec.

Eq. 5-15 gives $\bar{S}/S_o = \exp -\dfrac{V_r}{2x'D}(x+x')^2$, so that $\bar{S}/S_o = \exp -\dfrac{0.3(8.900+10,000)^2}{2 \times 8,900 \times 2450} = e^{-2.46} = 0.0854$, $S = 1620$ p.p.m.

Eq. 5-18 gives $L_{hws} - L_{lws} = \left(\dfrac{T\sqrt{gd}}{2\pi}\right) - L_{lws}(1-e^{-H/d})$, so that $L_{hws} - 10,000 = \left(\dfrac{44,700\sqrt{9.8 \times 10}}{2\pi} - 10,000\right)(1-e^{-1/10}) = 62,000 \times 0.1$. $L_{hws} = 10 + 6.2 = 16.2$ km (distance to $\bar{S} = 1620$ p.p.m.).

Eq. 5-19 gives $L_{lws} = x'\left(k\sqrt{\dfrac{D}{V_r x'}} - 1\right)$, for $\bar{S}/S_o = 1\%$, Table 5-I gives $K = 3.03$, $L_{lws} = 8,900\left(3.03\sqrt{\dfrac{2450}{0.3 \times 8900}} - 1\right) = 16.9$ km (distance to $\bar{S} = 1\% S_o$).

From eq. 5-18, $L_{hws} - 16,900 = (72,000 - 16,900)\,0.1$, $L_{hws} = 22.4$ km (distance to $\bar{S} = 1\% S_o$).

5

A river can be considered to be of uniform cross-section and average depth at the deepest points of 12 m. The fresh water flow can be equated to a mean velocity of 9.4 m/sec, the average ebb flow is 1.4 m/sec and the flood flow 1.0 m/sec. Determine the intrusion length from the entrance to the stations where mean salinity is 5% and 1% of the full marine salinity. Also find the surface and bottom salinities 5 km upstream from the entrance.

$V_o = \dfrac{\text{fresh-water/tide}}{\text{tidal prism at mouth}} = \dfrac{V_r dT}{V_f dT/2} = \dfrac{0.4 \times 2}{1.0} = 0.8$.

From eq. 5-16, $\bar{S}/S_o = \exp(-1.655 \times 0.8) = 0.265$ (or from Fig. 5-8). From Fig. 5-9, determine $\bar{S}/S_o\%$ and ΔS progressively with $V_{ebb} = 1.4$ m/sec as in Table 5-II. Mean salinity at 5 km station = .16.3% S_o.

From Fig. 5-10, for $V = V_o = 0.8$ along intrusion length and $d = 12$ m, $d\bar{S} = 86\%$, i.e., $\bar{S}_b\% = 16.3 + 0.86 \times 16.3 = 30.3\%$, $\bar{S}_b = 5,750$ p.p.m. and $\bar{S}_s\% = 16.3 - 0.86 \times 16.3 = 2.3\%$, $\bar{S}_s = 440$ p.p.m.

TABLE V-2

Computations of salinity for Example 5

x	$\bar{S}/S_o\%$	$\dfrac{\Delta \bar{S} \, 19 \cdot 10^4}{\Delta \times V_{ebb}}$	$\Delta \bar{S}\%$	x	$\bar{S}/S\%$	$\dfrac{\Delta \bar{S} \, 19 \cdot 10^4}{\Delta \times V_{ebb}}$	$\Delta \bar{S}\%$
0	26.5	2.9	2.1	11000	7.1	1.4	1.0
1000	24.4	2.8	2.1	12000	6.1	1.2	0.9
2000	22.3	2.8	2.1	13000	5.2	1.1	0.8
3000	20.2	2.7	2.0	14000	4.4	1.0	0.7
4000	18.2	2.6	1.9	15000	3.7	0.9	0.6
5000	16.3	2.5	1.8	16000	3.1	0.8	0.6
6000	14.5	2.4	1.8	17000	2.5	0.7	0.5
7000	12.7	2.2	1.6	18000	2.0	0.6	0.4
8000	11.1	2.0	1.5	19000	1.6	0.5	0.4
9000	9.6	1.8	1.3	20000	1.2	0.4	0.3
10000	8.3	1.6	1.2	21000	0.9		

5% S_o of station between 13 and 14 km from entrance; 1% S_o at station between 20 and 21 km from entrance.

6
For the river in Example 5 determine the dispersion coefficient at 1 km and 13 km from the entrance.

Eq. 5-25 gives $D = \dfrac{V_r C_x \, 2 \Delta x}{(C_{x+\Delta x} - C_{x-\Delta x})}$, so that $D_1 = \dfrac{0.4 \times 24.4 \times 2000}{(26.5 - 22.3)} = 4555$ m^2/sec from values in Table 5-II: $D_{13} = \dfrac{0.4 \times 5.2 \times 2000}{(6.1 - 4.4)} = 2455$ m^2/sec; values of D decrease upstream.

7
A long estuary of uniform rectangular section, with a depth of 25 ft. and tidal range 3.5 ft., has a dispersion coefficient (D) of 10,000 ft.^{2}/sec. A batch of pollutant is pumped into the estuary, which is seen to be well mixed transversally after three tidal cycles. At this time the concentration is recorded. What percentage of this concentration could be expected another four tidal cycles later at this point, assuming the river flow to be negligible compared to the tidal currents.

Mean tidal current $u = u_{max} 2/\pi = \left(\dfrac{Hg^{1/2}}{2 \, d^{1/2}}\right) \dfrac{2}{\pi} = \dfrac{3.5(32.2)^{1/2}}{\pi \, 25^{1/2}} = 1.26$ ft./sec, from eq. 4-22 of *Coastal Engineering, I*.

Eq. 5-23 gives: $C = \dfrac{M}{\rho A (4 \pi D t)^{1/2}} \exp - \dfrac{(x - ut)^2}{4 D t}$.

After time $3T$ the initial concentration is recorded, so that:

$C_1 = \dfrac{M}{\rho A (4 \pi D \, 3T)^{1/2}} \exp - \dfrac{(0 - 1.26 \times 3T)^2}{4 D \times 3 T}$

After another time of $4T$ (total $7T$), the concentration is given by: $C_2 = \dfrac{M}{\rho A (4 \pi D 7 T)^{1/2}} \times \exp - \dfrac{(0 - 1.26 \times 7T)^2}{4 D 7 T}$, so that: $\dfrac{C_2}{C_1} = \dfrac{1}{(7/3)^{1/2}} \exp \left[\dfrac{(1.26)^2}{4D} T(-7 + 3)\right] = \dfrac{1}{(7/3)^{1/2}} \exp(-7.08) = 0.655(0.0012) = 0.000786$ or 0.08%.

PROBLEMS

8

An estuary, for which the mean salinity averaged over depth and tidal cycle is as given in Table 5-III, has been analyzed to give a null-point Froude number (F_n) of 0.25. Where would you expect siltation to occur if the mean depth were 25 ft. and the average river flow were 0.5 ft./sec? To what location would this accretion shift if a 35 ft. deep channel were dredged along the length of the salt-water intrusion? During the winter time the river discharge quadruples the value given. On these occasions, where would the accretion be expected for either channel depth? Assume the longitudinal salinity distribution does not change with river discharge greatly.

TABLE 5-III

Data for Example 8

Distance from entrance (1000's ft.)	10	20	30	40	50	60	70	80
$\bar{S}$ (p.p.t.)		23	20	16	12.5	9	6	3	1

From eq. 8-41, $\bar{S}_n = \dfrac{1000}{0.7(F_n)^2} \dfrac{V_r^2}{gd}$ for $\bar{S}_n$ in parts per thousand. $\bar{S}_n = \dfrac{1000(0.5)^2}{0.7(0.25)^2\, 32.2 \times 25} =$
7.04 p.p.t., which indicates silting about 57,000 ft. from entrance (10.7 miles).

For a 35 ft. channel $\bar{S}_n = 7040 \times 25/35 = 5.03$ p.p.t., which occurs at 63,000 ft. (12 miles).

For $V_r = 2$ ft./sec, $\bar{S}_n = 28.16$ and 20.12 p.p.t., respectively which cause silting at the entrance and 20,000 ft. from it.

The values obtained should be compared with the curves in Fig. 5-15.

PROBLEMS

1

A river has a discharge of 30,000 cfs into an almost tideless sea. It can be considered uniform in its rectangular cross section of 1000 ft. width and 30 ft. depth. Assuming the seawater to have a specific gravity of 1.03 and a kinematic viscosity of $11 \cdot 10^{-6}$ ft.^{2}/sec find the distance of penetration of the salt-water wedge. If a barrier were to be constructed across the river to prevent salt contamination beyond 10 miles from the mouth, to what height above the bed should it be constructed? What is the maximum surface velocity experienced in the river?

2

A long 40 ft. deep river flowing into the Black Sea has a mean river velocity variation from 0.75 to 2.5 ft./sec throughout the year. Using the normal fluid characteristics for the waters concerned, determine the intrusion length of the saline water for these two extreme conditions. Compute also the shift in position of the toe of the wedge when the river is dredged to an equivalent depth of 50 ft. What is the depth of fresh water at the mouth for the above conditions? At what period would you expect the greatest siltation and where will this occur?

3

A river channel has a constant cross-section and a uniform depth of 10 m over the length to be considered. For tidal current calculations it can be considered of infinite length. The tidal range at the mouth is 2 m and the velocity of the fresh-water river flow is 0.315 m/sec. Compute the average salinity as % of 19000 p.pm. at the entrance and at a point 14 km upstream of the mouth. If the bed roughness has a dimension of 1 m, determine the average flood velocity at this upstream location and so find the salinity at the surface.

4

Find the tidal range necessary to turn the estuary in Problem 1 into a partially mixed case.

5

It has been mentioned previously that channels in the mouths of rivers might be dredged to triangular form in order to refract waves to the banks and so reduce wave action in a harbour upstream. What other useful purpose can it serve when salt-water intrusion is present? Which is the preferred shape, the apex to one side or in the centre of the channel?

6

Why are dispersion coefficients obtained by analysis or hydraulic models so much less than those measured in prototype streams. Discuss methods available for determining such coefficients.

7

A river half a mile wide and 30 ft. deep experiences semidiurnal tides, besides its fresh-water flow of 2 ft./sec. A dye is introduced into the river at a certain point and can be considered to be uniformly spread across the river section after three tidal cycles. At this time of high-water slack the concentration is measured at the point of insertion. It is again measured another three complete tidal cycles afterwards at the same location. If the second concentration is 1% that of the first, determine the dispersion coefficient of the river for such mixing conditions. Assume that the losses due to chemical action, boundary deposition etc. are negligible.

8

There are two locations for a sewage outfall which are of similar cost. One is within the salt-water wedge of an estuary, and the other is upstream of its toe. The outfall is to lie on the bed of the river in which the tidal oscillation is negligible. Which solution would you choose, and why?

9

To determine the length of salt-water intrusion in an estuary with tidal action, what information would you require?

10

Mean salinity is determined over a tidal cycle at the entrance to a waterway at low-water slack, which was found to be 15,000 p.p.m. Fifty minutes later as the flood tide came in the mean salinity reached the sea value of 19,000 p.p.m. The mean river velocity at this time in the 32 ft. depth channel was 1.2 ft./sec. Find the mean salinity at a station 5 miles upstream from the mouth both at lws and hws, if the semi-diurnal tide has a range of 4.5 ft. Locate the zone where the salinity is 5% of the seawater value. Compare the dispersion coefficient calculated with that of a triangular cross section, using a Darcy-Weisbach friction factor $f = 0.03$. Discuss any discrepancy. Discuss the effect on the salinity at the 5-mile station if a barrage were constructed across the waterway, say, 10 miles from the mouth.

11

A canal of rectangular cross-section has a mean depth of 8 m. The fresh-water flow is 0.3 m/sec and the tidal range of the semi-diurnal tide is 1 m. Find the intrusion length for 2% of the seawater salinity of 19,000 p.p.m. What are the surface and bottom salinities at this point? Plot also the dispersion coefficient against length of intrusion.

12

The landward velocity averaged over a tidal cycle at a height of 1/10 depth from the bed has been recorded at a number of stations and at various tidal and river flow conditions in a waterway. Values of $U_{0.9}/V_r$ are listed in Table 5-IV for concurrent values of densimetric Froude number at the same stations (F_x). The mean salinity distributions along the river have been averaged over all these tidal and river fluctuations, for which the results are also listed in terms of % seawater salinity (19,000 p.p.m.). For a river flow mean velocity of 0.2 ft./sec and a mean depth of 25 ft., find the site of likely siltation if the waterway is of reasonably constant width. What is the possible place of deposition if a 30-ft. channel is dredged along the channel? If the river flow is tripled during flood time, where is the optimum accretion likely to occur?

TABLE 5-IV

Data for Problem 12

$U_{0.9}/V_r$	3.6	3.1	4.3	3.5	2.7	3.3	1.9	1.4	0.81	0.6	0.3	0.3
F_x	0.045	0.049	0.051	0.060	0.076	0.095	0.11	0.13	0.18	0.24	0.25	0.29

x (1000's ft.)	0	2	4	6	8	10	12	14	16	18
$\bar{S}/S_0$	0.95	0.87	0.75	0.62	0.45	0.30	0.19	0.10	0.05	0.02

13

Surface water is used from a river for treatment and final water supply to a town nearby. It is proposed to dam the river for power generation some distance upstream, but still within the influence of the tide. What effect will this have on the quality of the water for the town? Give reasons for your answer.

14

In an estuary with tidal flow and variable density along its length, indicate how you would assess the diffusion coefficient at any place. How does this coefficient vary along the length of the salt-water intrusion? What will influence the location of this distribution in the longer term?

15

As a consultant you are asked to locate zones of possible accretion in an estuary. Describe the measurements you would have made and the use you would make of these data.

16

What benefits would result from the formation of a regional body to study the problems of an estuary? What should be the upstream and coastal limits of its jurisdiction for such a body to be effective?

17

In a triangular-shaped channel it is possible to dispose of an effluent at the surface, either at the alignment of the deepest point or near the bank. Which is to be preferred for the swiftest dispersion downstream? Give reasons for your choice.

18

Discuss briefly how the factors listed below affect the salinity distribution along a river and estuary of medium size:
(*1*) fresh-water discharge; (*2*) water traffic; (*3*) depth at the river mouth; (*4*) salinity of seawater; (*5*) wind; (*6*) tides; (*7*) temperature.

19
Write a report on the need for the engineer to cooperate with the scientist in order to produce the most desirable changes in an estuary. Suggested headings are:
(*1*) The nature of engineering problems in estuaries; (*2*) The manner in which solutions of *1* can upset the ecological balance; (*3*) The types of solutions which can improve natural conditions; (*4*) The types of solutions which can cause deterioration; (*5*) The organisations and bodies that should be contacted to discuss the merits of various proposals; (*6*) The distribution of financial responsibility for the benefits to be gained.
Reading for this project should include parts of Ref. 45. All topic headings should include salinity, pollution and sedimentation.

REFERENCES

[1] G.H. Keulegan, 1966. The mechanism of an arrested saline wedge. In: A.T. Ippen (Editor), *Estuary and Coastline Hydrodynamics*. McGraw Hill, New York, N.Y., pp. 546–574.
[2] Y.F. Ozturk, 1970. Seawater intrusion length in stratified estuaries. *Water Res.*, 4: 477–484.
[3] H. Shi-Igai, 1966. Experimental and theoretical modelling of salines wedges. *Proc. Congr. IAHR, 13th*, 3: 29–36.
[4] H. Stommel and H.G. Farmer, 1952. Abrupt change in width in two-layer open-channel flow. *J. Mar. Res.*, 11: 205–214.
[5] J.B. Schijf and J.C. Schonfeld, 1953. Theoretical considerations on the motion of salt and fresh water. *Proc. Minn. Int. Hydr. Conv., 1953*, p. 321.
[6] C.V. Gole and P.P. Vaidyaraman, 1966. Salinity distribution and effect of fresh-water flows in the Hoogly River. *Proc. 10th Conf. Coastal Eng.*, 2: 1412–1434.
[7] C.V. Gole and V.S. Thakar, 1969. Progressive salinity intrusion during the dry season in the Hoogly estuary. *Proc. Congr. IAHR, 13th*, 3: 275–282.
[8] T. Carstens, 1970. Turbulent diffusion and entrainment in two-layer flow. *Proc. ASCE*, 96 (WW1): 97–104.
[9] I. Minami, 1969. On vertical salinity dispersion into fresh water over salt water due to turbulence. *Proc. Congr. IAHR, 13th*, 3: 381–387.
[10] A.T. Ippen, 1966. Salinity intrusion in estuaries. In: A.T. Ippen (Editor), *Estuary and Coastline Hydrodynamics*. McGraw Hill, New York, N.Y., pp. 598–629.
[11] J.P. Tully, 1953. On structure, entrainment and transport in estuary embayments. *J. Mar. Res.*, 17: 523–535.
[12] H.B. Simmons, 1955. Some effects of upland discharge on estuarine hydraulics. *Proc. ASCE*, 81: Sep. No. 792.
[13] E.A. Schultz and H.B. Simmons, 1957. Fresh-water–salt-water density currents, a major cause of siltation in estuaries. *PIANC 19th Int. Nav. Congr.*, Sect. II Comm. 3: 43–64.
[14] B. Johns, 1967. Tidal flow and mass transport in a slowly converging estuary. *Geophys. J.R. Astr. Soc.*, 13: 377–386.
[15] M.R. Abbott, 1960. Boundary layer effects in estuaries. *J. Mar. Res.*, 18: 83–100.
[16] G.I. Taylor, 1953. Dispersion of soluble matter in solvent flowing slowly through a tube. *Proc. R. Soc.*, A219: 186–203.
[17] G.I. Taylor, 1954. The dispersion of matter in turbulent flow through a pipe. *Proc. R. Soc.*, A223: 446–468.
[18] J.W. Elder, 1959. The dispersion of marked fluid in turbulent shear flow. *J. Fluid Mech.*, 5: 544–560.
[19] M. Sumer, 1969. On longitudinal dispersion coefficient for a broad open channel. *J. Hydr. Res.*, 7: 129–135.

REFERENCES

[20] K. Zagustin, 1968. Sediment distribution in turbulent flow. *J. Hydr. Res.*, 6: 163–172.
[21] H.B. Fischer, 1967. The mechanics of dispersion in natural streams. *Proc. ASCE*, 93(HY6): 187–216.
[22] D.R.F. Harleman, 1966. Diffusion processes in stratified flow. In: A.T. Ippen (Editor), *Estuary and Coastline Hydrodynamics*. McGraw Hill, New York, N.Y., pp. 575–597.
[23] P. van der Burgh, 1968. Prediction of the extent of saltwater intrusion into estuaries and seas. *J. Hydr. Res.*, 6: 267–288.
[24] J. Hinwood, 1964. Estuarine salt wedges. *Dock Harbour Auth.*, 45: 79–83.
[25] Comm. Tidal Hydr., 1970. Research needs on thermal and sedimentary pollution in tidal waters. *Proc. ASCE*, 96(HY7): 1539–1548.
[26] J.H. Allen, D.I.H. Barr, W. Frazer and A.A. Smith, 1966. Effect of local conditions on effluent disposal in coastal waters. *Proc. 10th Conf. Coastal Eng.*, 2: 1375–1385.
[27] E.R. Holley, 1969. Unified view of diffusion and dispersion. *Proc. ASCE*, 95(HY2): 621–631.
[28] E.R. Holley, D.R.F. Harleman and H.B. Fischer, 1970. Dispersion in homogeneous estuary flow. *Proc. ASCE*, 96(HY8): 1691–1709.
[29] D.R.F. Harleman, 1966. Pollution in estuaries. In: A.T. Ippen (Editor), *Estuarine and Coastal Hydrodynamics*. McGraw Hill, New York, N.Y., pp. 630–647.
[30] D.R.F. Harleman, C.H. Lee and L.C. Hall, 1968. Numerical studies of unsteady dispersion in estuaries. *Proc. ASCE*, 94(SA5): 897–911.
[31] F.D. Masch and H.J. Shankar, 1969. Mathematical simulation of two-dimensional horizontal convection-dispersion in well-mixed estuaries. *Proc. IAHR Congr. 13th*, 3: 293–301.
[32] A.M. Kamel, 1970. Flushing pattern of non-reactive effluents. *Proc. 12th Conf. Coastal Eng.*, 3: 1867–1885.
[33] J.J. Dronkers, 1969. Tidal computations for rivers, coastal areas, and seas. *Proc. ASCE*, 95(HY1): 29–77.
[34] H.B. Fischer, 1970. A method for predicting pollutant transport in tidal waters. *Univ. Calif., Water Res. Centre, Contrib.*, 132.
[35] B.H. Ketchum, 1951. The exchange of fresh and salt waters in tidal estuaries. *J. Mar. Res.*, 10: 18–38.
[36] W.S. Preddy, 1954. The mixing and movement of water in the estuary of the Thames. *J. Mar. Biol. U.K.*, 33: 645–662.
[37] D.M. di Toro, 1969. Maximum entropy mixing in estuaries. *Proc. ASCE*, 95(HY4): 1247–1271.
[38] D.W. Pritchard, 1969. Dispersion and flushing of pollutants in estuaries. *Proc. ASCE*, 95(HY1): 115–124.
[39] D.R.F. Harleman and A.T. Ippen, 1969. Salinity intrusion effects in estuary shoaling. *Proc. ASCE*, 95(HY1): 9–27.
[40] H.B. Simmons, 1969. Salinity effects on estuarine hydraulics and sedimentation. *Proc. IAHR Congr., 13th*, 3: 311–325.
[41] R.H. Meade, 1969. Landward transport of bottom sediments in estuaries of the Atlantic coastal plain. *J. Sed. Petrol.*, 39: 222–234.
[42] J.V. Byrne and L.D. Kulm, 1967. Natural indicators of estuarine sediment movement. *Proc. ASCE*, 93(WW2): 181–194.
[43] A.T. Ippen, 1966. Sedimentation in estuaries. In: A.T. Ippen (Editor), *Estuarine and Coastline Hydrodynamics*. McGraw Hill, New York, N.Y., pp. 648–672.
[44] F.M.M. Abecasis, 1969. Hydraulic and sedimentary problems in tidal estuaries. *Proc. IAHR Congr., 13th*, 3: 249–256.
[45] G.H. Lauff (Editor), 1970. Estuaries. *Am. Assoc. Advan. Sci. Publ.*, 83.
[46] G. Fleming, 1970. Sediment balance of Clyde estuary. *Proc. ASCE*, 96(HY11): 2219–2230.
[47] C.C. Cable, 1969. Optimum dredging and disposal practices in estuaries. *Proc. ASCE*, 95(HY1): 103–114.
[48] A.A. Sooky, 1969. Longitudinal dispersion in open channels. *Proc. ASCE*, 95(HY4): 1327–1346.

Chapter 6

MARINE HYDRAULIC WORKS

Works implied in this title consist of marine outfalls for sewage and other effluents plus outlets and intakes for cooling water for power and industrial plants. The aim of the coastal engineer in any discharge to the sea or waterway is to promote the maximum of mixing of the noxious liquid with the receiving body of water. Certain limits of concentration may be established by health authorities, which will serve as criteria in design of any outfall. In respect to cooling water intakes the aim is to induct water at the lowest temperature, but with the least amount of suspended matter in the form of sand or seaweed.

With the increased concentration of industry and population on coasts and waterways, for the sake of cooling water and sea transport, greater demands are being made on the coastal waters for the disposal of liquid effluent. At the same time these beach areas are required for recreation of the populace, so that higher standards of purity are being demanded for river, estuarine and coastal waters. This situation has promoted the use of multiple ports in outfalls [1], so that warm water or effluent alike can be well mixed with the receiving liquid. Because such diffusion takes some time and distance to be effective the hydraulics of mixing tubes will be outlined.

Since most effluents are lighter than the salt or brackish water into which they are discharged, being fresh-water sewage or warmed salt-water, they will rise during the mixing process. If the receiving water is of uniform density the plume so formed will extend to the surface. However, if the density decreases from the bed to the surface the mixing effluent may reach the same value as its surroundings at some depth below the surface. This situation is desirable because polluted water at the surface is readily transferred to banks and beaches by wind stress and mass-transport due to wave action. On the coast a mid-depth accumulation of effluent might well be carried seawards by the mass-transport generated by swell which continually rolls into shore.

Power plants, for the sake of economy, must have their outfalls and intakes for cooling water placed in close proximity. It is necessary, therefore, to check on the horizontal and vertical extent of the warm water outflow to see that it is not recirculated into the system, otherwise the body of water involved will become hotter and hotter and so create an ecological nuisance. Longshore currents under all possible wave and tide conditions should be examined before the final decision on location are taken. Fortunately, the warm water if discharged at the surface will spread out there and not mix readily with the lower cooler water at depth, which can then be used, by the construction of a suitable mid-depth intake.

For convenience of discussion the chapter is divided into "sewage disposal" and "cooling water structures" because the former is concerned with discharge at depth and the latter with discharge at the surface and intake at depth. Only the hydraulic problems are treated since other aspects of design, such as forces and sedimentation, can be gleaned from other chapters in this treatise.

SEWAGE DISPOSAL

Although the waters of coastal zones and estuaries are in continual motion, due to tidal currents, the analysis to follow will assume stagnant ambient fluid, except in the specific discussion on the influence of currents. A design based upon this adverse condition is reasonable since at high- or lower-water slack the receiving liquid is stagnant for a period of 2 or 3 hours, in which time a sizable volume of effluent could accumulate.

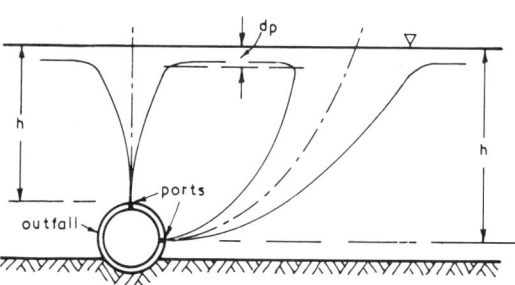

Fig. 6-1. Definition sketch of sewage outfall.

An outfall with discharging ports is depicted in Fig. 6-1, where the two limiting conditions of a vertical and horizontal jet are displayed. At the outlet the jet contains a certain kinetic energy in the form of momentum, either vertical or horizontal. Due to the density difference a certain potential energy is exerted in the jet in the form of a buoyant uplift. As the jet travels through the stagnant liquid it draws some of it in, so increasing the area through the jet and diluting its contents. The further from the port the greater is the proportion of buoyant force to vertical momentum, any horizontal momentum remaining steady, although the velocity associated with this approaches zero as the area of the jet increases.

Many workers have supplied analyses and experimental data on this topic, but those who have provided comprehensive reports suitable for ready application are Abraham [2] plus Fan and Brooks [3]. Both of these are mathematical treatments which include constants derived from experimental work of their own or of other workers. Abraham treats the vertical and horizontal jets only in ambient fluids of constant or variable density, either greater or less than the effluent. The experimental coefficients, which differ near to and far from the port, are assumed to change suddenly at mid-path. Fan and Brooks, on the other hand, have included in their computerised equations suitable variations of these constants. They also generalised the analysis to cover discharge at any angle to the horizontal and a linear density difference from bed to surface. Both reports contain graphs for circular orifices and continuous slots. The following discussion uses liberally the end results of these authors, for the detailed analyses the reader is referred to the original reports [2,3].

As indicated in Fig. 6-1, the jets arriving at the surface in stagnant water will be deflected through 90° and spread laterally. This spreading process will be discussed more fully further on, suffice it to say that there will be some thickness (d_p) of diluted effluent at the edge of the rising plume as it enters this layer. The depth of water through which mixing may take place is therefore restricted to $h-d_p$ where h is the distance from the port outlet to the surface. For circular jets Rawn and Palmer [4] found $d_p/h = 1/12$, whereas Hart [5] concluded this ratio was 1/6. For a slotted exit tests have shown $d_p/h = 1/4$. Since in many cases the jets of multiport outfalls will combine to form the equivalent of a slotted exit it is suggested that a mean figure of $d_p/h = 1/5$ serve for the determination of diffusion distance.

Tidal currents will oscillate this surface mass and the macro-turbulence so generated will aid in its mixing with water beneath. Waves, particularly breaking ones, will also expedite the dispersion of this surface layer in the general body of water. Wind stress can spread the layer down wind and in this sense can cause pollution in zones prior to its adequate mixing by the previously mentioned processes. Horizontal currents will deflect the rising jet and so influence the diluting mechanism, even a very small current can have a significant effect [8]. At present it will be considered only that tidal action removes the mass accumulated during a high- or low-water-slack period, so that the whole volume of ambient fluid does not become saturated with effluent.

The general assumptions made in the analysis are as follows [3]:

(*1*) The fluids are incompressible.

(*2*) Density variations throughout the plume are small compared to ambient density.

(*3*) The density is a linear function of either salt concentration or heat content above the reference level.

(*4*) Flow is fully turbulent (i.e., no Reynolds number influence).

(*5*) Longitudinal turbulent transport is small compared with convective transport.

(*6*) Pressure inside and outside the plume is hydrostatic.

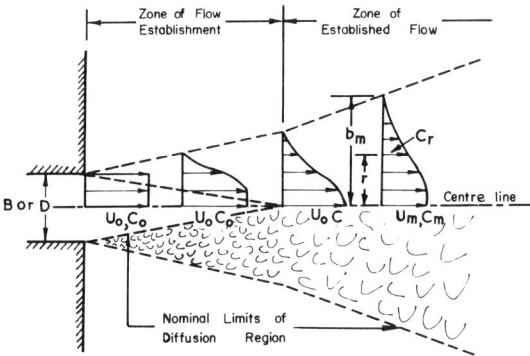

Fig. 6-2. Definition sketch of flow establishment.

(7) Effect of curvature of jet trajectory or of outside plume boundaries is negligible.

(8) The analyses strictly to zones of established flow in the jet, where velocity and concentration profiles transverse to the jet trajectory are similar and of Gaussian distribution. Allowance is to be made for flow establishment.

The zone of flow establishment is depicted in Fig. 6-2, where it is seen that the constant-velocity core is reduced in cross-section by shear with the stagnant ambient fluid. The distance to the point where the central velocity is that of the jet (U_o), but where the velocity distribution is similar to that of a Gaussian error curve, is the length for flow establishment. It is from this point onwards that the analysis to follow applies. A similar kind of distribution applies to concentration so that a length of flow establishment for this variable can also be found. The lengths for either variable differ by only one or two orifice diameters (D), or slot widths (B), so that expressions for the velocity distribution only will be used. This is given by Albertson et al. [9] as:

$$L_n = 6.2D \tag{6-1}$$

where L_n is measured along the axis of the jet no matter what angle it is to the horizontal, or whether the ambient fluid is of uniform or variable density. In the case of uniform density the dilution (S) taking place from the nozzle to the end of flow establishment is given in terms of concentration (C) by:

$$\frac{S_o}{S} = \frac{C}{C_o} = \frac{1+\lambda^2}{2\lambda^2} \tag{6-2}$$

where S_o and C_o are the dilution and concentration respectively of the effluent at the nozzle exit and λ is the spreading ratio between buoyancy and velocity profiles; or λ^2 is the turbulent Schmidt number. Rouse et al. [10] have suggested $\lambda = 1.16$ for circular jets, so that:

$$C_o/C = 1.15 \qquad (6\text{-}3)$$

This ratio of 1.15 must be used for dilutions given by the analysis centred on the end of flow establishment in order to refer them to conditions at the nozzle exit.

Similar equations have been derived for a slot, namely:

$$L_s = 5.2B \qquad (6\text{-}4)$$

$$\frac{S_o}{S} = \frac{C}{C_o} = \sqrt{\frac{1+\lambda^2}{2\lambda^2}} \qquad (6\text{-}5)$$

which for $\lambda = 0.89$ [10] and a modification for volume flux [3] approximates unity for a wide range of conditions, so that:

$$C_o/C = 1 \qquad (6\text{-}6)$$

The above equations apply for jets discharging at any angle into either uniform or non-uniform density media. They have been incorporated into the graphs to be presented, even though the corrections for this zone of flow establishment are slight compared to the large dilutions effected in the main trajectory to the surface.

Circular jet in uniform density medium

A jet of diameter D is depicted in Fig. 6-3 issuing with a velocity U_o at an angle θ_o to the horizontal into an ambient liquid of uniform density ρ_s. At the nozzle the density and concentration of the effluent is ρ_o and C_o respectively. At the end of flow establishment ($6.2\,D$ from the nozzle) the centre-line concentration C is given by eq. 6-3, whilst at a height y above this point the centre-line concentration C_m is given by the curves in Fig. 6-3. These vary with y/D, θ_o and Froude number F defined by:

$$F = U_o / \sqrt{\frac{\rho_s - \rho_o}{\rho_s} gD} \qquad (6\text{-}7)$$

The graphs of Fan and Brooks [3] from which Fig. 6-3 (and following figures) has been prepared, were presented with an abscissa of dimensionless momentum flux which at the nozzle (M_o) is related to F by:

$$F = \frac{\lambda}{2^{1/4}\alpha^{1/2}} M_o^{5/4} \qquad (6\text{-}8)$$

where α is the coefficient of entrainment, assumed from Rouse et al. [10] to be 0.082.

This has been noted since Fan and Brooks state that if new values of α or λ

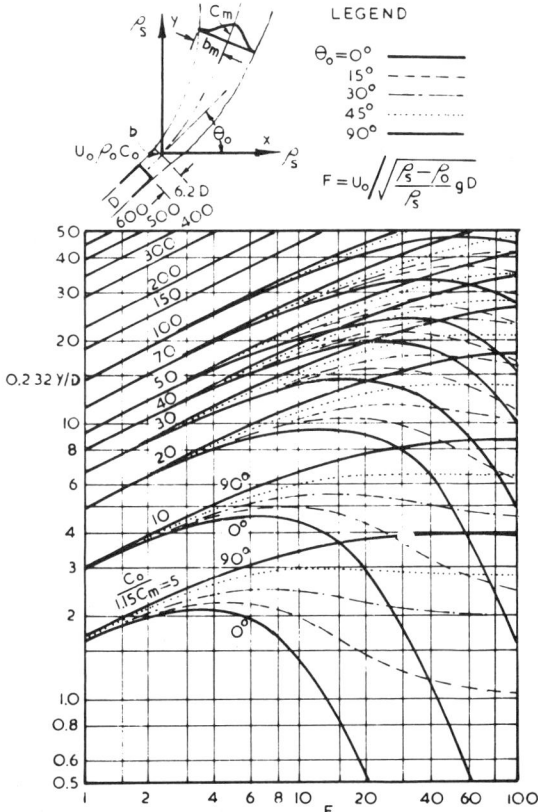

Fig. 6-3. Dilution of lighter circular jet in ambient fluid of uniform density.

emerge from future tests then a new multiplying factor can be derived to be used with the same graphs. For the sake of simplicity the values of $\alpha = 0.082$ and $\lambda = 1.16$ have been employed, resulting in the abscissa of F, which, in fact, Fan and Brooks inserted as an alternative scale. For $F = \infty$ a value of $\alpha = 0.057$ is suggested.

To obtain the vertical height from the nozzle to the trajectory point where the concentration is C_m the component of distance for flow establishment must be added, that is:

$$y_n/D = y/D + 6.2 \sin \theta_0 \qquad (6\text{-}9)$$

The dilution curves (C_o/C_m) in Fig. 6-3 for $\theta_0 = 90°$ rise consistently with F, whereas those for $\theta_0 \leqslant 45°$ reach a peak y/D and then fall, thus giving two locations at which the same dilution or concentration occurs. At small F the jet rises almost vertically from the nozzle and reaches a given dilution at a specific height (e.g., $F = 1$, $C_o/1.15\, C_m = 10$ at $2.32\, y/D = 3$). As the jet discharge is increased, and so

TABLE 6-I

Coordinates of trajectory and jet width for circular jet in uniform density medium

θ_o	$0.232 y/d$	Values of $0.232 x/D$ (upper) amd b_m/b (lower)							
		F 1	2	4	8	16	32	64	128
0	5	0.5	1.5	2.5	4.5	7.5	12.5	20.5	33.5
		3.5	4.0	4.5	6.0	8.5	13.0	20.5	33.0
	10	0.7	1.6	3.0	5.5	9.7	16.2	26.5	42.0
		6.5	7.0	7.5	8.5	11.0	17.0	26.5	41.5
	20	1.0	2.0	3.5	6.5	11.9	20.0	33.5	53.7
		12.0	12.5	13.0	14.5	17.0	22.5	33.5	52.0
	30	1.0	2.0	3.5	7.5	13.0	22.5	37.6	61.2
		19.0	19.5	20.0	21.0	23.0	29.0	39.0	60.0
	40	1.0	2.0	3.5	7.6	14.0	24.5	41.0	67.0
		25.0	25.5	26.0	27.0	29.5	34.0	45.0	64.0
	50	1.0	2.0	3.5	7.9	14.5	26.4	44.0	72.0
		30.5	31.0	31.5	33.0	35.0	40.5	51.0	68.0
15	5	0.5	1.4	2.4	4.0	6.6	10.0	12.6	16.5
		3.5	4.0	4.5	5.5	8.0	11.0	14.0	18.0
	10	1.0	1.5	3.0	5.0	8.5	13.4	19.2	27.0
		6.5	7.0	7.5	8.5	11.0	15.0	20.0	29.0
	20	1.0	1.8	3.5	6.0	10.5	17.5	27.2	40.0
		12.0	13.0	13.5	14.5	17.0	21.0	29.5	41.5
	30	1.0	2.0	3.6	6.9	11.8	20.0	32.0	48.5
		18.0	18.5	19.0	20.5	22.0	27.0	36.0	50.0
	40	1.0	2.0	3.8	7.1	13.7	21.9	35.6	55.2
		25.0	25.5	26.0	27.0	29.0	33.0	41.5	58.0
	50	1.0	2.0	3.9	7.4	13.5	23.2	37.5	60.5
		30.5	31.0	31.5	33.0	34.0	38.0	46.0	63.0
30	5	0.5	1.0	1.9	3.1	4.5	6.2	7.5	8.5
		3.5	4.0	4.5	5.5	6.5	8.5	9.5	10.5
	10	0.6	1.1	2.3	4.1	6.6	10.0	13.4	16.0
		6.5	7.0	7.5	8.5	10.0	13.0	17.0	20.0
	20	0.6	1.2	2.7	5.3	8.9	14.0	20.8	27.2
		12.5	13.0	13.5	14.5	16.0	20.0	26.0	32.0
	30	0.7	1.3	3.0	5.7	10.1	16.5	25.5	35.7
		18.0	18.5	19.0	20.5	22.0	26.0	32.0	42.0
	40	0.8	1.4	3.1	6.0	11.0	18.4	29.0	42.2
		25.0	25.5	26.0	27.0	28.5	32.0	39.0	50.5
	50	0.9	1.5	3.2	6.4	11.5	19.5	31.5	47.0
		30.5	31.0	31.5	32.0	34.0	38.0	44.0	55.0

TABLE 6-I (continued)

θ_o	$0.232 y/d$	Values of $0.232 x/D$ (upper) and b_m/b (lower)							
		F 1	2	4	8	16	32	64	128
45	5	0.3	0.9	1.5	2.4	3.2	4.2	4.8	5.2
		3.5	4.0	4.5	5.0	6.0	7.0	8.0	8.5
	10	0.4	1.0	1.9	3.2	5.1	7.0	8.8	9.6
		6.5	7.0	7.5	8.0	9.5	12.0	14.0	15.0
	20	0.5	1.1	2.3	4.0	6.9	10.6	14.7	17.5
		12.5	13.0	13.5	14.0	15.5	18.0	22.0	26.0
	30	0.6	1.2	2.4	4.7	7.9	12.7	18.6	24.5
		18.0	18.5	19.0	20.0	21.5	25.0	30.0	35.0
	40	0.7	1.3	2.5	5.0	8.5	14.2	21.5	29.7
		25.0	25.5	26.0	26.5	27.5	30.5	36.0	43.0
	50	0.7	1.4	2.6	5.1	9.0	15.2	24.0	34.1
		30.5	31.0	31.5	32.0	33.0	36.0	42.0	51.0

F, so is the height y/D for the same dilution until F is large enough for the trajectory length to permit enough mixing for the dilution to become similar at the same height (e.g., $F = 20$ gives same conditions as before). Sharp [11] has described the mechanism for small F as that of "starving" the jet. In this situation the buoyant force predominates and stretches the system, so causing gusts which expedite mixing. The condition for minimal dilution is therefore the Froude number at which the jet reaches the surface on the point of being starved. Operation at these maxima on the curves in Fig. 6-3 should be avoided. Conditions should be chosen either side of these F values, preferably smaller ones since these demand less power. It is seen that for any given Froude number the greatest dilution is achieved by the horizontal jet, but other angles must be considered so as to prevent interaction of jets, which accumulates concentration. This will be discussed later.

The coordinates of the trajectory with respect to the origin of established flow are given in Table 6-I in terms of x/D and y/D. The width ratio b_m/b (see Fig. 6-3), where b is the half jet width at the end of flow establishment ($= D/\sqrt{2}$), and b_m is half width where the control concentration is C_m, is also listed. The concentration at any radius r from the centre-line (see Fig. 6-2) is given by the Gaussian distribution:

$$C_r/C_m = \exp[-r^2/(\lambda b_m)^2] = \exp[-0.74(r/b_m)^2] \qquad (6\text{-}10)$$

Circular jet in non-uniform density medium

A vertical jet of diameter D is illustrated in Fig. 6-4 issuing with conditions U_o,

SEWAGE DISPOSAL 247

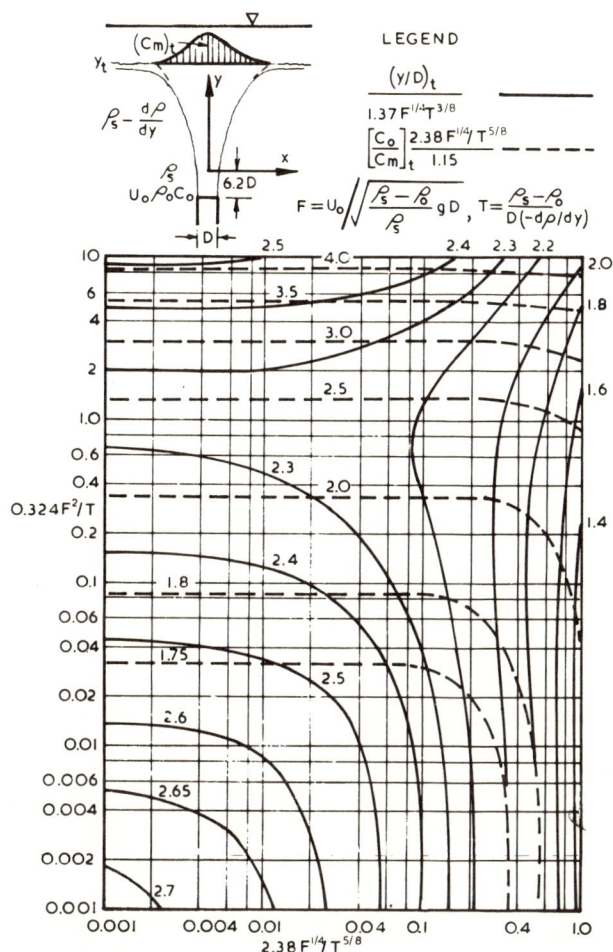

Fig. 6-4. Dilution and height at limit level of vertical circular jet in ambient fluid of non-uniform density.

ρ_o and C_o into a liquid with density ρ_s at the origin of established flow and decreasing with height at the rate of $-d\rho/dy$. The decrease in density of the effluent as it draws in the denser liquid at the lower levels may cause its centre-line value to equal that of the surroundings before the surface is reached (i.e., at a terminal distance y_t +6.2 D above the nozzle where the centre-line concentration is $(C_m)_t$. Besides the Froude number as defined in eq. 6-7 a new dimensionless stratification parameter is introduced.

$$T = \frac{\rho_s - \rho_o}{D(-d\rho/dy)} \tag{6-11}$$

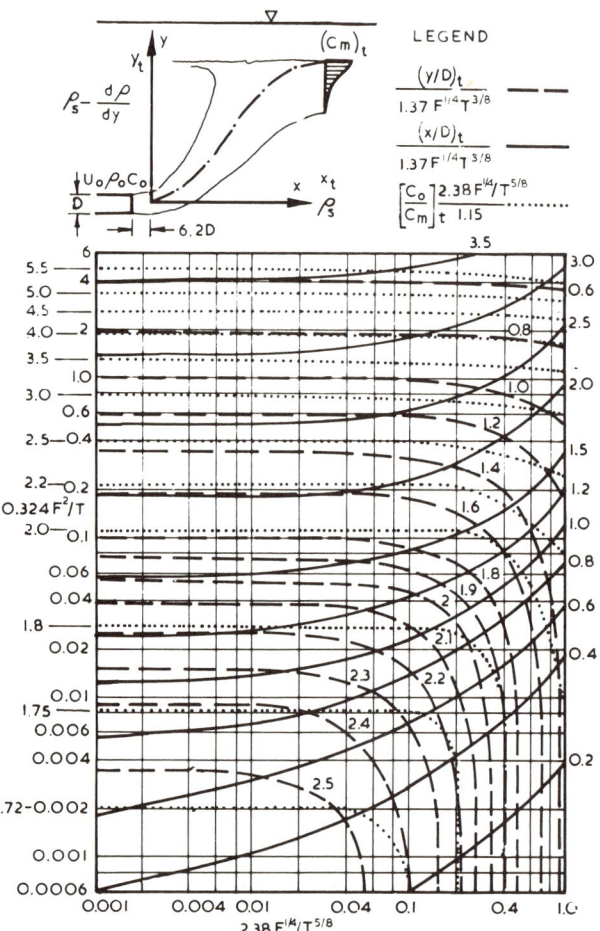

Fig. 6-5. Dilution and height at limit level of horizontal circular jet in ambient fluid of non-uniform density.

Fan and Brooks presented graphs in dimensionless parameters which included α and λ, which have been converted to F by the assumed values noted previously. Isohyets of $(y/D)_t/1.37\ F^{1/4}\ T^{3/8}$ and $(C_o/C_m)_t 2.38\ F^{1/4}/1.15\ T^{5/8}$ have been drawn in Fig. 6-4 for the ordinate and abscissa shown. The factor 1.15 refers the dilution to conditions at the nozzle exit. Accumulation of diluted effluent at y_t will cause a thickness $dp \doteq y_t/5$ to form so restricting the trajectory distance over which dilution can be effected. However, under these neutrally buoyant conditions such accumulation could occur above or below the level y_t so that no allowance is suggested. Note that Fig. 6-4 only supplies terminal values and cannot be used for dilution along the trajectory.

SEWAGE DISPOSAL 249

In Fig. 6-5 is depicted a horizontal circular jet issuing into a similar linearly density-stratified ambient fluid. The trajectory in this case is in the shape of an S rather than the continuously rising curve of the uniform medium. A third series of isohyets are provided, namely $(x/D)_t/1.37\,F^{1/4}T^{3/8}$. In designing for zero pollution at the surface Fan and Brooks [3] questioned whether (y_t) or $(y_t +$ half width of the jet at $y_t)$ should be used as the criterion. However, as the dilution and therefore the density at any radius from the trajectory is less than that at the centre-line it will become neutrally buoyant before reaching the terminal height. No effluent should rise above this limit computed on the basis of the centre-line concentration $(C_m)_t$.

Slotted jet in uniform density medium

In this case the width (B) of the infinitely long slot enters the parameters in place of diameter D and the distance for flow establishment becomes 5.2 B. The Froude number is therefore given by:

$$F = U_o \Big/ \sqrt{\frac{\rho_s - \rho_o}{\rho_s} gB} \qquad (6\text{-}12)$$

The dimensionless momentum flux utilised by Fan and Brooks in this case is related to F by:

$$F = (\lambda/\alpha)^{1/2} M_o^{3/2} \qquad (6\text{-}13)$$

where the assumed values of $\lambda = 0.89$ and $\alpha = 0.16$ [10] have been used in Fig. 6-6.

The vertical height from the nozzle to the point on the trajectory where concentration C_m occurs is obtained by adding to y the vertical component of flow establishment length (L_s) such that:

$$y_s/B = y/B + 5.2 \sin \theta_o \qquad (6\text{-}14)$$

The curves of Fig. 6-6 display the same optima as do those of Fig. 6-3.

Where a row of closely spaced circular ports interact soon after discharge from the outfall an equivalent slot width B_e can be defined as:

$$B_e = \pi D^2/4L \qquad (6\text{-}15)$$

where L is the centre-line spacing of the jets of diameter D. By equating the centreline concentrations from the two- and three-dimensional vertical jets the height to circular jet interaction y_i is given by [2]:

$$y_i = 4.35L \qquad (6\text{-}16)$$

For horizontal ports the jet radius where the concentration is half that of the central value will give an almost uniform concentration along the outfall, after

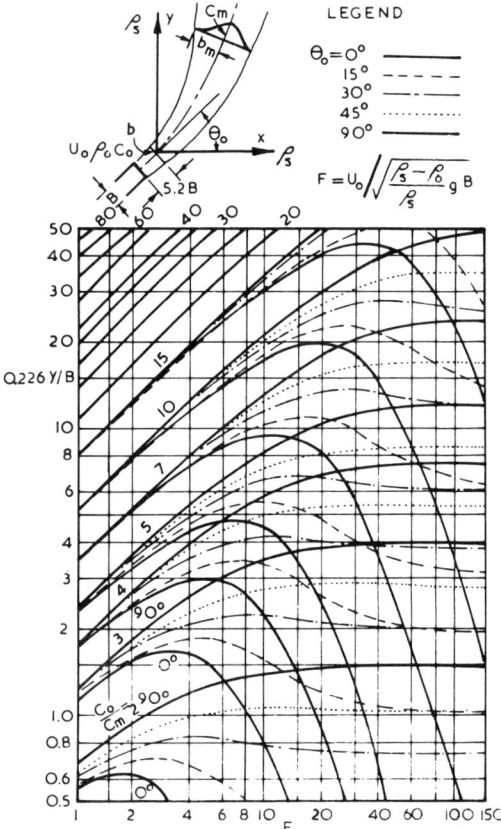

Fig. 6-6. Dilution of lighter slotted jet in ambient fluid of uniform density.

which the effluent could be treated as an equivalent slot. From eq. 6-10 this radius is given by:

$$C_r/C_m = \frac{1}{2} = \exp[-0.74(r/b_m)^2] \quad \text{or} \quad r/b_m = 0.965 \qquad (6\text{-}17)$$

Liseth [12] has carried out model tests on an outfall with ports opposite each other discharging horizontally. He measured the concentration at various distances from the manifold and determined the maximum. This was centred on each jet trajectory until the plumes from each side coalesced, when the maximum concentration occurred directly above the outfall. From various spacings of ports a series of empirical curves were obtained, which are reproduced in Fig. 6-7. The numerical factors in the ordinate and abscissa derive from the fact that Liseth used the diameter of the sharp-edged orifice rather than the diameter of the jet itself. From the curve of Fan and Brooks for plain circular jets it is seen that to heights of

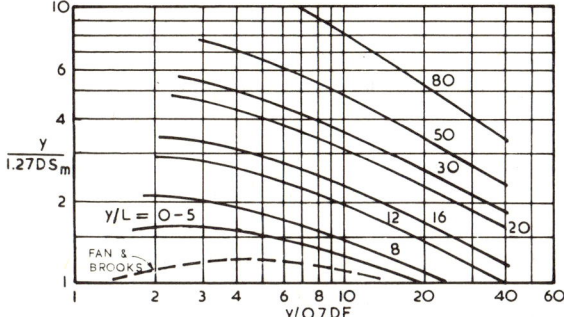

Fig. 6-7. Minimum dilution with height of merging horizontal circular jets in ambient fluid of uniform density.

$y/L = 0$ to 5 the dilution is not quite so good as that for individual jets (theoretical or actual). The reason for this could be the demand in opposite horizontal directions for water to mix with the jets. On the upper half of the jet this water must travel between the jets into the central region above the outfall. An improvement may have been effected by staggering the ports, in which case the water being entrained does not have to turn through 180°. The alternate opposite currents thus established should generate vortices which should aid mixing. Liseth noted the better mixing of jets near the end of the manifold, where water was more readily

TABLE 6-II

Co-ordinates of trajectory and jet width for slotted jet in uniform density medium

θ_0	$0.226\,y/B$	Values of $0.226\,x/B$ (upper) and b_m/b (lower)							
		F 1	2	4	8	16	32	64	128
0	5	0.9	1.9	3.4	6.2	11.6	22.0	38.0	67.5
		5.5	6.0	8.0	12.5	22.0	43.0	75.0	135.0
	10	1.0	2.5	4.5	9.0	15.6	29.0	50.2	–
		11.0	11.5	13.0	18.0	28.0	53.0	98.0	–
	20	1.5	3.0	6.0	11.6	21.0	37.6	66.5	–
		20.5	21.0	23.0	27.5	39.0	67.0	125.0	–
	30	1.7	3.2	7.0	13.5	24.3	44.2	78.5	–
		30.5	31.0	33.0	37.0	48.0	77.0	140.0	–
	40	2.0	3.4	7.3	14.5	27.0	49.3	–	–
		40.5	41.0	43.0	58.0	88.0	148.0	–	–
	50	2.0	3.4	7.5	15.1	29.4	53.7	–	–
		50.5	51.0	53.0	57.0	68.0	100.0	–	–

TABLE 6-II (continued)

θ_o	$0.226\,y/B$	Values of $0.226\,x/B$ (upper) and b_m/b (lower)							
		F 1	2	4	8	16	32	64	128
15	5	0.8	1.8	3.0	5.2	8.5	12.5	15.0	17.5
		5.2	6.0	7.5	10.0	16.0	26.0	31.0	38.0
	10	1.0	2.4	4.2	8.0	13.0	20.0	27.5	32.5
		10.5	11.0	13.0	16.5	25.0	40.0	56.0	68.0
	20	1.5	3.0	5.5	10.5	18.2	29.5	44.6	59.1
		20.5	21.0	23.0	26.5	36.0	56.0	90.0	121.0
	30	1.6	3.2	6.5	12.0	21.6	36.0	56.6	80.0
		30.5	31.0	33.0	36.0	46.0	70.0	110.0	162.0
	40	1.6	3.3	6.8	13.0	24.1	41.2	66.8	–
		40.5	41.0	42.0	46.0	56.0	79.0	129.0	–
	50	1.6	3.4	7.0	13.7	26.0	45.6	75.0	–
		50.0	50.5	52.0	56.0	63.0	90.0	141.0	–
30	5	0.5	1.5	2.7	4.5	6.5	8.0	8.5	9.0
		5.2	6.0	7.5	10.0	15.0	19.0	20.0	21.0
	10	0.7	2.0	3.7	6.5	10.0	13.5	16.0	17.0
		10.5	11.5	13.0	15.5	23.0	30.5	37.0	40.0
	20	1.0	2.5	4.8	9.0	14.6	21.8	29.0	33.0
		20.5	21.0	23.0	26.0	34.0	48.0	65.0	75.0
	30	1.2	2.7	5.5	10.3	17.7	28.0	39.0	47.0
		30.5	31.0	33.0	36.0	44.0	62.0	88.0	108.0
	40	1.3	2.8	5.9	11.4	20.0	32.7	47.5	60.1
		40.5	41.0	43.0	46.0	54.0	74.0	105.0	137.0
	50	1.4	2.9	6.1	12.0	22.0	36.7	55.2	72.0
		50.0	50.5	52.0	56.0	63.0	83.0	120.0	162.0
45	5	0.7	1.4	2.0	3.2	4.5	4.7	4.9	5.1
		5.5	6.5	7.5	9.0	10.0	11.0	12.0	13.0
	10	0.9	1.6	3.0	5.0	7.0	8.8	9.4	10.1
		10.5	11.5	12.5	15.0	19.5	25.0	26.0	28.0
	20	1.0	2.0	3.9	7.0	11.0	15.1	18.0	19.6
		20.5	21.0	22.0	26.0	31.0	42.0	50.0	55.0
	30	1.1	2.2	4.5	8.0	13.5	20.0	25.5	28.8
		30.0	30.5	32.0	36.0	42.0	55.0	72.0	80.0
	40	1.2	2.4	4.7	8.8	15.5	24.0	32.5	37.5
		40.5	41.0	42.0	46.0	52.0	68.0	90.0	105.0
	50	1.2	2.4	5.0	9.5	16.9	27.1	38.0	45.5
		50.0	50.5	52.0	55.0	62.0	78.0	104.0	130.0

SEWAGE DISPOSAL

available for this initial mixing. Fig. 6-7 should serve as a warning not to design marine outfalls on the basis of dilution from a single jet.

The trajectory coordinates with respect to the origin of established flow are given in Table 6-II in terms of x/D and y/D. The width ratio b_m/b (see Fig. 6-6) where $b = (2/\pi)^{1/2}B$ is also contained in the Table. The Gaussian distribution of concentration applies as before (see eq. 6-10).

Slotted jet in non-uniform density medium

The case of the vertical jet of width B is illustrated in Fig. 6-8, issuing with initial values of U_o, ρ_o and C_o into a medium with density ρ_s at the origin of established flow and decreasing at the rate $-d\rho/dy$.

Similar parameters exist as before, except that the terminal height from the slot is $y_t + 5.2\,B$. The dimensionless parameters, for which isolines are provided, are $(y/B)_t/0.908\,F^{1/3}T^{1/2}$ and $(C_o/C_m)_t^2\,2.14\,F^{2/3}/T$. The ordinate and abscissa also

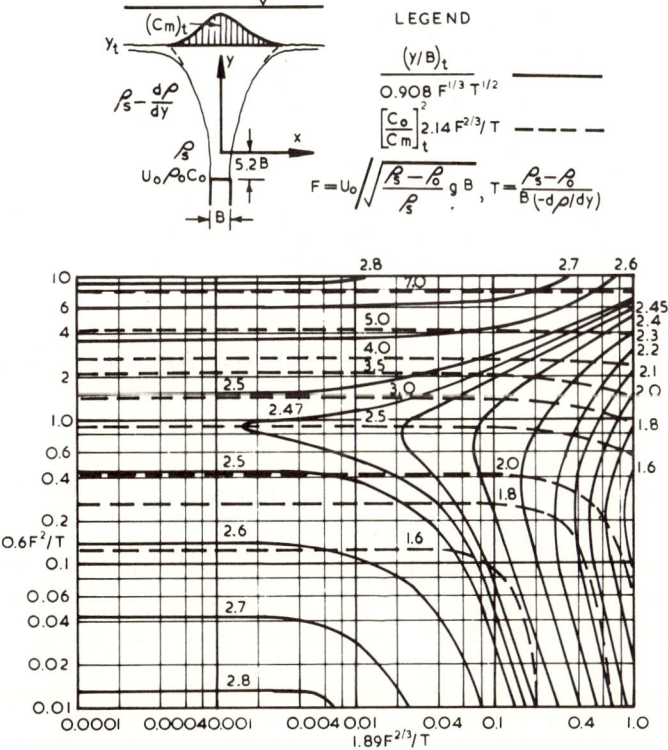

Fig. 6-8. Dilution and height at limit level of vertical slotted jet in ambient fluid of non-uniform density.

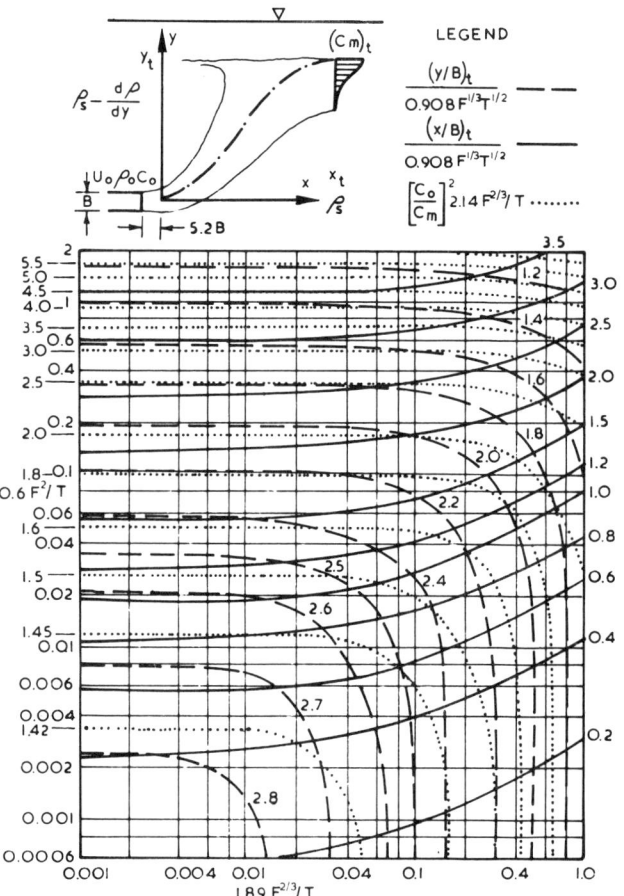

Fig. 6-9. Dilution and height at limit level of horizontal slotted jet in ambient fluid of non-uniform density.

differ from the circular jet case. The multiplication factor referring the dilution to the initial jet conditions is taken as unity. The arguments respecting accumulation of effluent from the nozzle still apply to the slotted outlet.

Fig. 6-9 illustrates the discharge of a slotted jet into a liquid with linearly varying density. The added parameter is that of horizontal distance, or $(x/B)_t/0.908 F^{1/3} T^{1/2}$. Again, no effluent should rise above the terminal limit y_t as computed.

Experimental verification

Cederwall [13] has reported tests on horizontal nozzles, and using a semi-empirical relationship derived by Bousanquet et al. [14] have compared them with

SEWAGE DISPOSAL   255

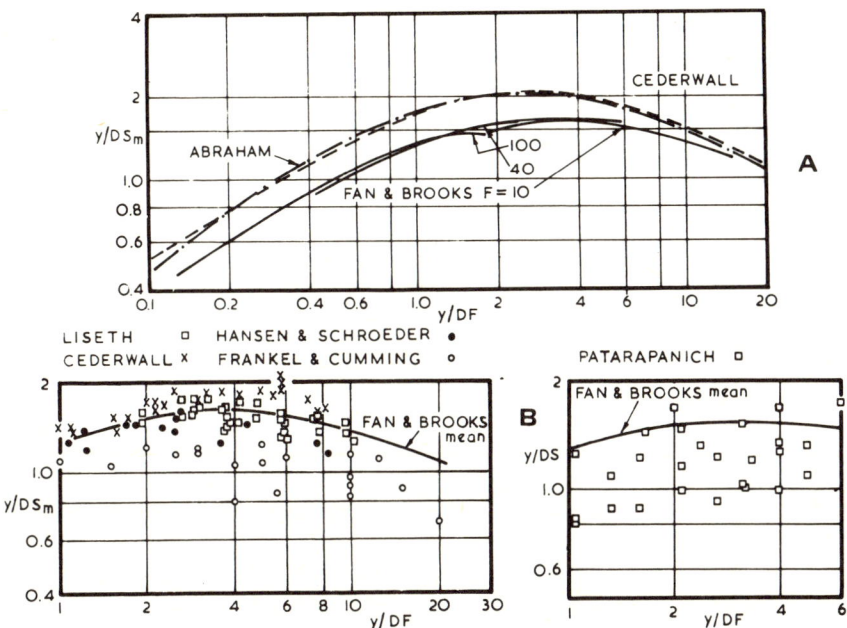

Fig. 6-10. Results for horizontal circular jets in ambient fluid of uniform density; A. theoretical; B. experimental.

an explicit analytical solution:

$$S_m = 0.54 F(y/DF)^{7/16} \tag{6-18}$$

for $y/D < 0.89F$, and:

$$S_m = 0.54 F(0.38 y/DF + 0.68)^{5/3} \tag{6-19}$$

for $y/D > 0.89F$, where:

$$S_m = C_o/C_m \tag{6-20}$$

which has been graphed in Fig. 6-10A, together with a similar presentation of $y/d\,S_m$ versus y/dF for the solutions by Abraham [2] and Fan and Brooks [3], the latter with the dilution factor for flow establishment. This correction could be applied to the curves of Cederwall and Abraham since their solutions apply to established flow only.

The experimental results from a number of workers [12–17] are also incorporated into Fig. 6-10 which indicate that better mixing is obtained than predicted by the theory. In this respect the theoretical solution is conservative.

It has been noted [12,17] that time is required in experiments for the dilutions to reach a steady state. For jets of 1/8–1/4 inch diameter about 10–20 min is required at $y/D \geqslant 50$. Closer to the outlet equilibrium is reached much sooner.

With the aid of further macro-turbulence in prototype conditions it could be expected that even greater dilution could ensue than in the laboratory, so that a mean curve through the experimental data of Fig. 6-10B might well serve for design purposes. Hansen and Schroeder [16] accept the deviation from theory as a "factor of safety". It is worth noting that Daily [18] concluded that for neutrally buoyant suspensions "coarse" rigid particles may enhance diffusion, whilst "fine" rigid particles and fibrous flexible particles could impede diffusion. Effluents containing such fibrous material should therefore have a larger factor of safety.

Mixing tubes

Several devices were used by Hansen and Schroeder [16] to break the jet into parts in order to effect greater dilution. They concluded "Dilution of a jet can be increased by use of a special nozzle design, but the possibilities in this direction are reduced considerably when complicated devices are avoided and little or no additional loss of energy is allowed." In multiport diffusers deflection of one jet into the path of another serves little purpose. Also, such complications must be added to the outfall when it is in place, which can be a costly operation.

However, under certain conditions of insufficient depth, or improving the dilution of an existing outfall, the addition of a mixing tube may be found economical. This consists of a cylinder, larger in diameter than the port or nozzle, which is placed in line with the jet, with its flared entrance in close proximity to the outfall

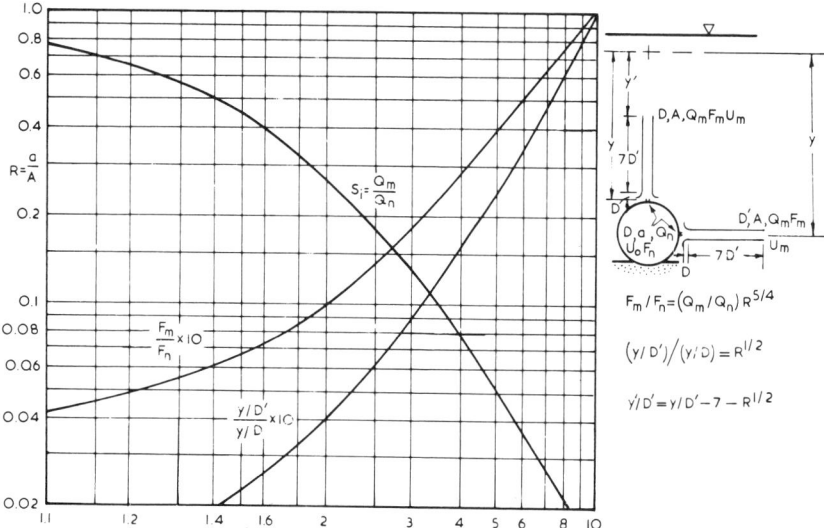

Fig. 6-11. Initial dilution, change of Froude number, and change of dimensionless height by using mixing tube on horizontal or vertical circular jets.

(see Fig. 6-11). The jet then draws seawater into the tube and mixes with it before discharging at a lower velocity at the outlet. Design graphs have been presented [19] for this water-jet-pump action and its application to sewage outfalls presented by Silvester [20].

The initial dilution S_i effected in the mixing tube is in proportion to the water drawn in (Q_s) and the discharge of the driving jet (Q_n), so that:

$$S_i = \frac{Q_n + Q_s}{Q_n} = \frac{Q_m}{Q_n} \qquad (6\text{-}21)$$

The ratio Q_m/Q_n depends upon the ratio (R) of the area of the nozzle $(a = \pi D^2/4)$ to that of the mixing tube $(A = \pi D'^2/4)$. For the specific condition of a plain cylindrical mixing tube with flared entrance a characteristic curve as in Fig. 6-11 can be derived [19]. This is based upon the criterion that the mixing tube length is 7 times its diameter (i.e., $= 7D'$) which has been found to be that for maximum mixing with least friction loss [21].

The discharge velocity from the mixing tube is much lower than that from the nozzle, and inspite of the increased diameter the densimetric Froude number is decreased, such that:

$$F_m/F_n = (Q_m/Q_n)R^{5/4} \qquad (6\text{-}22)$$

where:

$$F_m = U_m/\sqrt{g(\Delta\rho/\rho)D'} \qquad (6\text{-}23)$$

where U_m is the discharge velocity from the mixing tube $(= 4Q_m/\pi D'^2)$. The mixing tube now serves as the nozzle for the further dilution (S'_m) of the jet as it rises to the surface. The height ratio for the design charts (Fig. 6-3 to 6-9) must now be based on the mixing tube diameter D', which can readily be determined from Fig. 6-11 for any chosen R value. When the mixing tube is vertical the remaining height for dilution (y'/D') is the height from the nozzle (y/D') less the height of the end of the mixing tube from the nozzle $(7D' + D = 7D' + D'R^{1/2})$.

There is a multitude of area rations (R) that could be selected, making choice of a mixing tube tedious unless the design criteria limit this choice greatly. A computer study was carried out [17] to optimize the total dilution (S_t) such that:

$$S_t = S_i \times S'_m \qquad (6\text{-}24)$$

The result of this study for the horizontal jet is presented in Fig. 6-12, where the ratio S_t/S_m indicates the improvement of dilution (S_t) with mixing tube over that of the plain jet (S_m). It is seen that greatest gain derives for large y/d, small F and small R. These are rather conflicting demands since smaller F and smaller R may result in a condition where the jet from the nozzle is insufficient to drive the

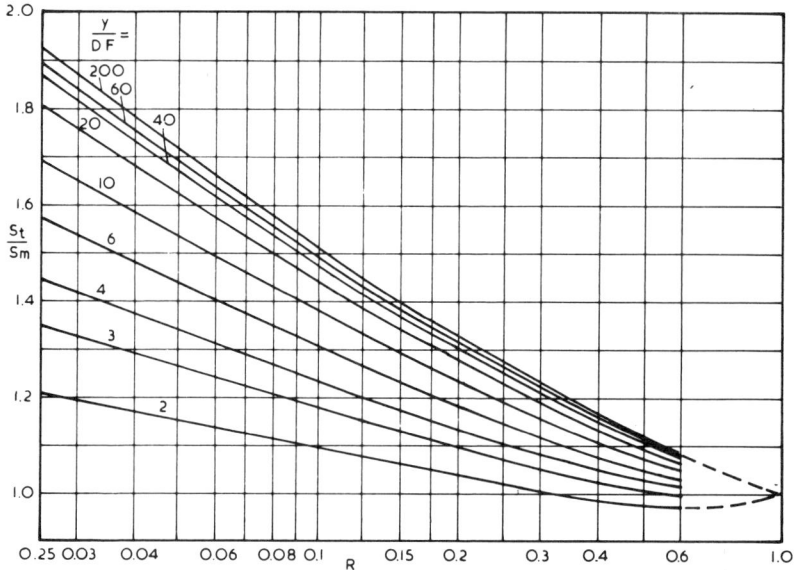

Fig. 6-12. Design chart for selecting size of mixing tube for a horizontal jet.

"jet-pup". The smaller R also implies a relatively large D' and hence length $7D'$. But Fig. 6-12 could serve as a starting point for the design of an outfall, which should then be tested in the laboratory.

A similar optimisation curve is presented in Fig. 6-13 for a vertical jet, in which y/D has been separated from F, because no benefit accrues until $y/D = 120$. This is

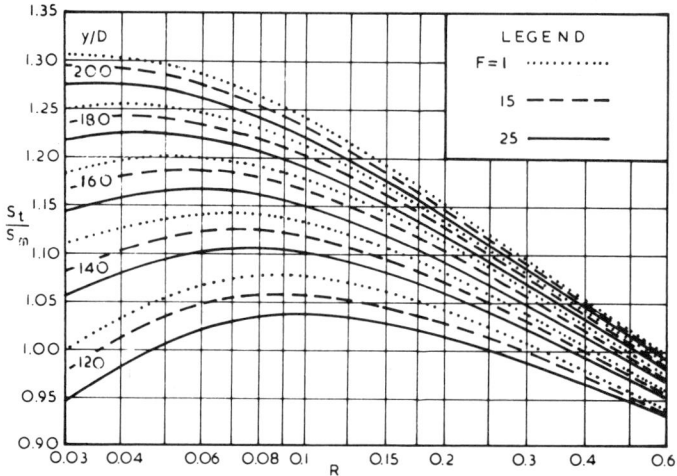

Fig. 6-13. Design chart for selection of mixing tube for a vertical jet.

because the height for dilution S'_m is reduced as the mixing tube length increases vertically. The major benefit from this type of installation is that F is reduced to the stage where "starvation" of the plume occurs and hence mixing promoted. If insufficient height is left for this to take place no increase in dilution over that from the plain jet results.

Laboratory tests [17] have shown that mixing tubes can provide total dilution as indicated by Fig. 6-12 and 6-13. In fact, they gave increased dilution over the theoretical values of the same order as experienced with plain jets. It could be anticipated that this improvement, which probably emanates from the macro-turbulence not accounted for in the theory, might be greater due to the larger turbulons generated in the mixing tube. The tests referred to involved small diameter nozzles 1/8" to 1/5" and small mixing tubes 1/4" to 1/2". Further experiments are warranted, since economical improvements on existing outfalls may be possible, especially where discharge is from a plain end or a few large ports. The addition of such mixing tubes does not impose any extra load on the system, in fact it can increase the discharge because of the slight suction applied at the entrance of the mixing tube. One installation in a river has been reported [22], although the mixing tube in this case was not of correct proportions for optimum mixing [20]. The concept is particularly applicable where an effluent as discharged is particularly obnoxious to marine fauna and flora and where swift initial mixing is called for. It could also serve in density-stratified situations, to reduce the effluent density to a value where it will be retained below the surface.

Pearson [23] made a comprehensive survey of sewage outfalls and provided details of 148 installations throughout the world. This report contains some 250 references on this topic. These outfalls were investigated to see what improvement in dilution could be effected by the addition of mixing tubes. Only 42 could be analyzed due to lack of information or inappropriate nozzle shape (slots, etc.). Of these, 26 could be so improved by percentages ranging from 25 to 245%. The latter were those discharging direct from the end of the outfall into very shallow water, for which an R ratio of 0.1 was used, implying a diameter ratio of about 3, with a length around 20 times the nozzle diameter.

Surface spreading

The spreading of diluted effluent, either at the surface or at some terminal level in a stratified medium, has been analysed on a dimensional basis and verified experimentally by Sharp [24]. For the non-viscous zone of operation, that usually encountered in prototype situations, two dimensionless parameters resulted, namely:

$$Lg^{1/5}(\rho_s-\rho_o)^{1/5}/Q^{2/5}S_m^{1/5}\rho_s^{1/5} \quad \text{and} \quad Tg^{3/5}(\rho_s-\rho_o)^{3/5}/Q^{1/5}S_m^{3/5}\rho_s^{3/5}$$

where L is the distance progressed in time T from the centre-line of the jet at the surface or at some stable intermediate depth, and the dilution (S_m) applies to a continuous discharge Q from the circular nozzle.

The relationship between these two parameters is exhibited in Fig. 6-14, together with a definition sketch. The time T in this form will be expressed in seconds, whilst other variables are dimensionally consistent. The zone verified by Sharp for the smaller values of the parameters is shown distinct from the extrapolated curve, which should serve for design until larger scale or prototype measurements have been made. It permits the computation of spread throughout a period of high- and low-water slack prior to its carriage away from the outfall site. In multiport outfalls, where jets interact to give a longitudinal or two-dimensional plume at the surface the distance travelled by the boundary could be obtained by equating the areas from each condition, or:

$$\frac{\pi}{4}L^2 = 2x_p L' \tag{6-25}$$

where L' is the distance from the outfall centre-line when the port spacing is x_p, in which case the discharge from each single port should be used in Fig. 6-14.

Miscellaneous considerations

The emission of a jet into a cross current for no density difference between effluent and medium has been studied by Pratte and Baines [25] with air. Their results should give a good approximation of a sewage discharge where the mean

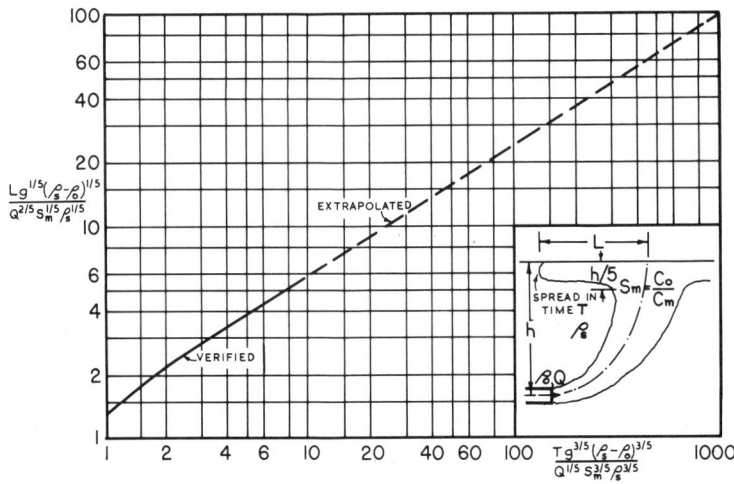

Fig. 6-14. Dimensionless parameters for spread of surface or limit-level layer of buoyant jet.

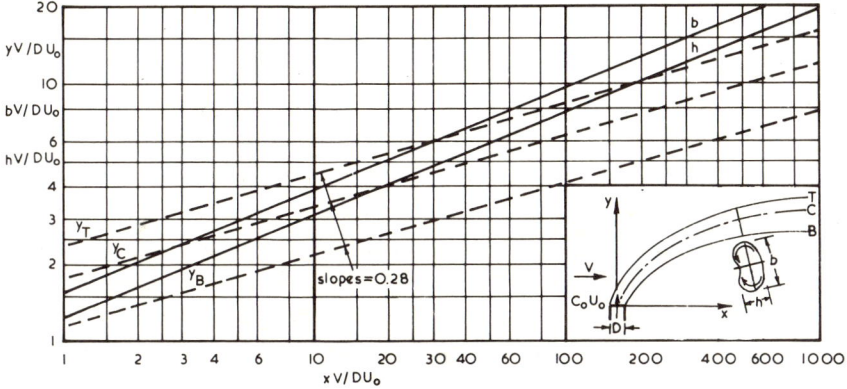

Fig. 6-15. Path and dimensions of jet issuing into cross-current with no density difference.

uniform cross-current velocity V exceeds the jet discharge velocity U_o. Fig. 6-15 displays the definition sketch and curves for the profile of the top, centre-line and bottom of the plume, plus its cross-sectional dimensions (b and h). The dilution of the jet is proportional to the area of the plume, approximately equal to b x h so that it is given by:

$$(S_m)_{ave} = 2.4(U_o/V)^{1.2}(x/D)^{0.8} \tag{6-26}$$

With the elongation of the jet there is a better opportunity for it to be retained within the body of the recipient liquid without reaching the surface, in only slight density stratification.

Waves occurring in a zone where effluent is accumulating at the surface or at some terminal depth will cause some elliptical motion of the water particles. There will be additional macro-turbulence which will aid mixing especially if breaking occurs [26]. Besides this, they generate mass-transport currents which can spread a surface layer in the direction of wave advance or in the opposite direction, depending upon wave steepness as discussed in *Coastal Engineering, I*. Smith [27] has reported a pilot study on waves advancing into a density front. He found that the surface mass-transport first blunted the wedge-shaped layer, slowing down its progress. It then progressively distorted it and turned it down on itself. On the other side of a layer such a net movement of liquid would carry the effluent with it, perhaps breaking the layer into patches if supply was insufficient to keep the layer intact.

The case of a jet issuing vertically in ambient fluid of lesser density is not a usual problem for the coastal engineer to solve, it being more an atmospheric phenomenon. The reader is therefore referred to the work of Abraham [28] who supplies references to other workers.

COOLING WATER STRUCTURES

These can be divided into the two categories of outfalls and inlets, which are somewhat interrelated because any installation must contain both. Outfalls themselves are normally just open channels discharging nearshore into the surface waters. The warm water then spreads laterally, drawing in and mixing with ambient water in the buoyant layer. The horizontal dimensions, temperature distribution, and thickness of the layer should be known. The inlet structure may be constructed nearshore if the bed is rocky and steep, otherwise it is likely to be located some distance offshore in reasonably deep water. It will draw water from some height above the bed, specifically to avoid the induction of warm water from the surface or sand and seaweed from the bed.

Outfalls

Critical conditions for thermal pollution can arise more readily in bays where tidal action or currents from fresh-water flow are negligible. In such stagnant bodies of water the sun can heat up the surface layer, so causing a density difference and strong stratification. When warm water from the condensing plant is poured into this surface layer it will not diffuse with water below the thermocline, where there is a significant change in temperature. Wada [29] has concluded from field measurements that the vertical diffusion coefficient is in the order of 0.01 m^2/sec, whereas the horizontal value is 50 times greater. He found no evidence of turbulent mixing across the thermocline, which in many bays of Japan is 3 to 5 m thick. It varies from bay to bay and from season to season, being most pronounced in the summer months. It can vary across any bay through the influence of wind stress and tidal action. When no stratification occurs, mixing of warm effluent with the body of bay water is easier. In this case wave breaking can generate turbulence at depth, whereas with a thermocline such mixing is effective only in the surface layer.

The mode of dissipating heat in the long term is by radiation from the water surface, convection of sensible heat to the atmosphere and by evaporation. All these depend upon the area of warm water exposed at the surface and the temperature gradient of this plume and the air above it. The overall heat balance must take into account the heat absorbed from the sun and condensation at the surface. The presence of a breeze will aid this cooling process greatly, but design conditions must be the worst possible.

Hayashi and Shuto [30] have conducted experiments in which the overall heat budget was controlled. From their tables Fig. 6-16 has been prepared, which shows the ratio of equilibrium plume area required for no wind (A_o) to that when a wind speed of U exists (A_u). Since the penetration length y is proportional to (area)$^{1/2}$ the ordinate of the figure can serve for the ratio $(y_u/y_o)^2$.

COOLING WATER STRUCTURES

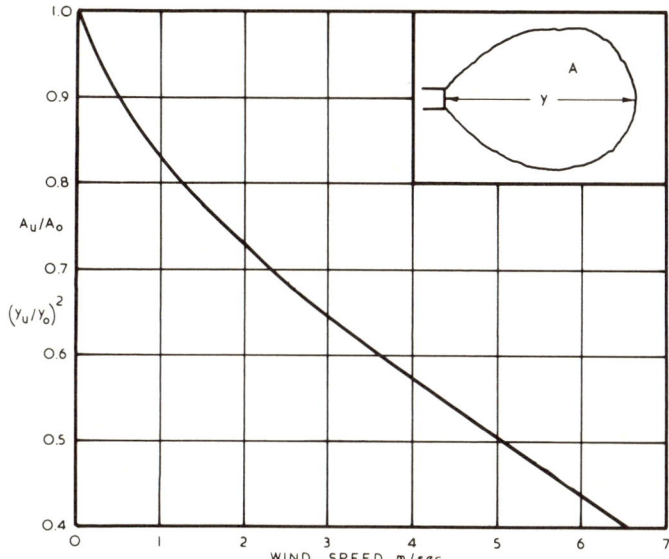

Fig. 6-16. Equilibrium area and centre-line length of plume for continuous dissipation of heat.

Wada [29] measured the plume from a power station when the thermocline was 3 m deep, the atmospheric temperature 28.5°C, the average wind speed 3.3 m/sec and the temperature difference of outlet water and bay water (25°C) was 8°C. His table of results for discharge, area and distance y, have been converted to zero wind conditions, which result in the relationships:

$$A_o = 74Q^2 \tag{6-27}$$

and:

$$y_o = 3Q \tag{6-28}$$

where A_o is the area in m² and y_o the distance in metres for a discharge Q in m³/sec.

Sharp [31] has experimented with discharge of light effluent from a pipe at the water surface, and for non-viscous conditions derived empirical curves of dimensionless ratios similar to those for the temporal spread of diluted sewage horizontally from a submarine outfall. In this case the dilution term S_m can be omitted so that the parameters become:

$$\frac{yg^{1/5}(\rho_s - \rho_o)^{1/5}}{Q^{2/5}\rho_s^{1/5}} \quad \text{and} \quad \frac{Tg^{3/5}(\rho_s - \rho_o)^{3/5}}{Q^{1/5}\rho_s^{3/5}}$$

where centre line distance to the end of the plume y replaces L used previously

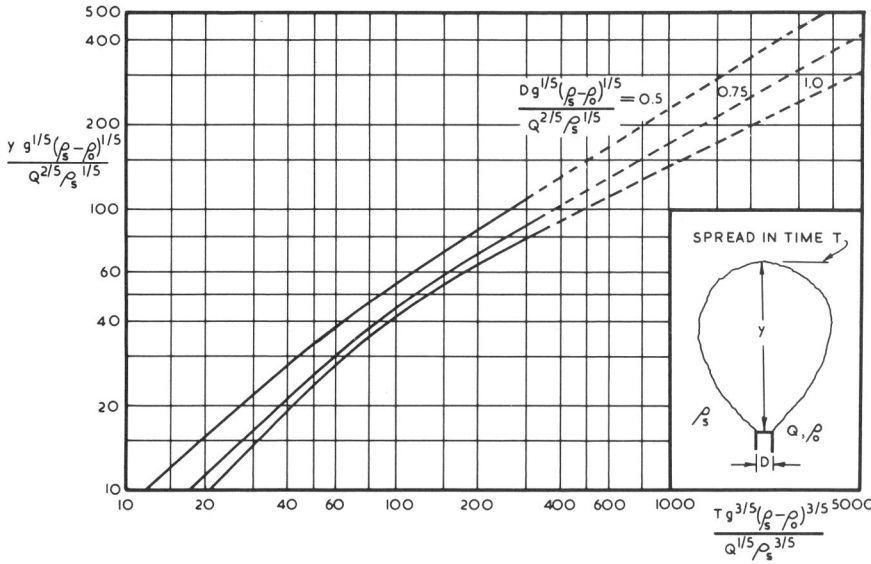

Fig. 6-17. Non-viscous surface spread of buoyant jet from a circular nozzle.

with a third dimensionless variable $\dfrac{Dg^{1/5}(\rho_s - \rho_o)^{1/5}}{Q^{2/5}\rho_s^{1/5}}$ entering the problem. These are presented in Fig. 6-17, where it is seen that for a given discharge the centre-line distance increases with a decrease in diameter D of the outfall. This is due to the initial velocity being greater. For small discharges viscosity can affect the length of the plume. Fig. 6-17 permits the evaluation of plume progress in a given time T. For steady-state mixing, Sharp also derived curves within the viscous affected region. The dotted sections of curve in Fig. 6-17 have been extrapolated from the verified zone at the lower values.

Hayashi and Shuto [30] also provided experimental data on the temperature distribution along the centre-line of the plume from which Fig. 6-18 has been prepared. The y/B in this figure represents a dimensionless distance along the centre-line. The width B can be equated to the diameter D of Fig. 6-17 because in either case the active width is quickly dictated by the thickness of the warm surface layer in which the discharge will be contained. The curve for $T_o - T_s \leqslant 14°$ agrees very well with the relationship derived by Jen et al. [32] as follows:

$$\frac{T_y - T_s}{T_o - T_s} = 7\frac{D}{y}\exp[-3F(x/y)^2] \qquad (6\text{-}29)$$

where T_y is the surface temperature (°C) along the plume at the distance y from the outlet and transverse distance x from the centre-line; T_s is the temperature of

COOLING WATER STRUCTURES

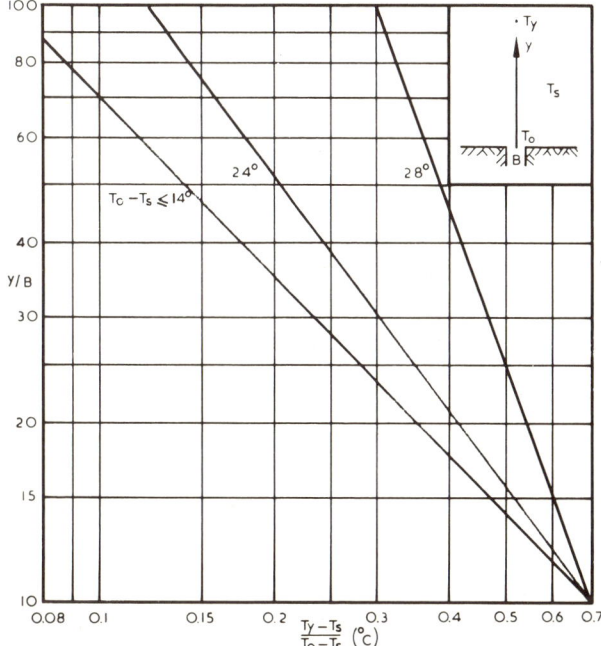

Fig. 6-18. Temperature distribution along the centre-line of a surface plume.

ambient fluid; T_o is the temperature of the effluent; F is the densimetric Froude number $(U_u/\sqrt{(\rho_s-\rho_o)gD/\rho_s}$.

The temperature distribution along the centre-line is given by eq. 6-29 with the exponential term omitted. It is obvious that it can only apply to points more than $7D$ from the outlet since the temperature ratio cannot exceed unity. Sharp also believes eq. 6-29 is inapplicable for $y/D > 100$ since viscosity will start to influence the longitudinal and lateral spread of the plume.

The two-dimensional spread of a buoyant jet discharged at the surface has been examined by Tamai [33]. His theory gave the increased depth along the channel for an ambient liquid of uniform density. Fig. 6-19 combines the results so obtained, which also contains the form of the Froude number.

Such a two-dimensional spread of surface effluent has also been studied by Koh [34], who found that when mixing in a uniform ambient fluid an internal hydraulic jump could form. Depending upon the rate of heat flow to the atmosphere, entrainment beneath the plume, and the densimetric Froude number, so a sudden increase in thickness of the plume may or may not occur. Under certain conditions of little heat dissipation, so the jump will proceed to the outlet and inundation occurs. No experimental verification is available to date on this aspect, which is important because of the possible encroachment of warm water on any submarine cool-water intake nearby.

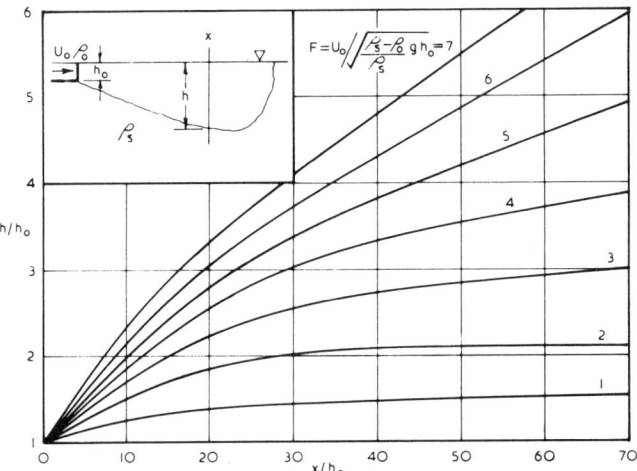

Fig. 6-19. Two-dimensional spread of buoyant surface jet into a homogeneous ambient liquid.

Frazer et al. [35] have reported a comprehensive model study of a cooling water outlet in which congruency diagrams utilising dimensionless parameters already discussed are utilised in the conversion to prototype conditions. Tamai [36] has derived theoretically the density profiles along the centre-line of a plume in the two-dimensional case and compared it to experiments of other workers and his own. This density distribution can be related to temperature variations. Vigander et al. [37] describe a cooling water outfall which rests on a river bed and discharges through multiple ports either 1 or 2 inches in diameter. This design would permit the effluent to meet the strict temperature limits set for the area.

Intakes

With the demand for power doubling every 10 years and the economics of power plants dictating larger installations, sites on inland waterways are giving way to coastal locations because of the greater availability of cooling water. Nuclear power plants consume more water per installed kilowatt than the more conventional steam powered stations. A 200-MW plant must have an intake of 100 m^3/sec when fully loaded. It becomes obvious that the seasonal fluctuations of rivers is too unpredictable for this size of demand.

Besides electric generating stations, cooling water is required for steel works, chemical plants and refineries. For commercial convenience many such enterprises are located together on the coast or within an estuary, generally on reclaimed land. However, this industrialisation promotes pollution of the neighbouring waters in the form of floating debris, oil slicks, sewage effluent, warm water (which promotes marine flora), and siltation if marine structures are badly located.

COOLING WATER STRUCTURES

Since most of the pollution outlined above occurs in the surface layers it is normal practice to induct water at lower levels, but not so low as to draw in water loaded with sediment. In any event, cooling water must be screened to prevent seaweed and other suspended matter from entering condensers. However, if storms bring in large batches of weed such screens must be bypassed temporally in order to keep the cooling water quantities up to the required level. If this happens too often a critical look should be taken at the intake structure. A mid-level induction of water also obviates any surface ice problem in winter.

Curtain-wall intake. Where a complex is located near deep water a curtain-wall intake can be economical. It is suitable for large inputs and can be constructed with flexible or fixed faces to keep out a warm or otherwise polluted surface layer. The entrance must be so designed that induction of water from this layer does not occur even with the draw-down that results with suction.

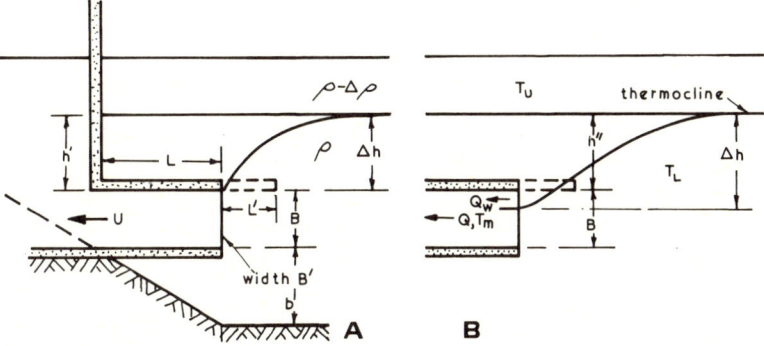

Fig. 6-20. Definition sketch of curtain-wall intake for cooling water.

Consider the curtain wall as in Fig. 6-20 which consists of rectangular openings each of height B, breadth B' and projected length L. They may have a top face projection of L'. The warmer layer of water density $\rho - \Delta\rho$ is of thickness h. It is required to find the induction velocity U which does not include water from the warmer layer, or contains a known proportion of it. The draw-down Δh for no warm-water input must not exceed h', the distance from the interface to the top of the intake. Besides, at this critical discharge, the velocities should not exceed 1 ft./sec (0.3 m/sec) or mixing will occur at the interface.

Senshu [38] has conducted experiments on such a set-up as in Fig. 6-20A and derived the empirical relationship:

$$\Delta h/B = (F/C)^{2/3} \qquad (6\text{-}30)$$

TABLE 6-III

Coefficient C for rectangular openings with B'/B = 2 to 4

b/B	L/B		
	0	3	6
0	1.17	1.44	1.70
0.5	1.30	1.60	1.90
1.0	1.47	1.80	2.15
1.5	1.68	2.02	2.40
2.0	1.89	2.29	2.80
L'/B	C for $b/B = 1$		
0	1.47	1.80	2.15
0.5	2.20	2.50	2.80
1.0	3.00	3.40	3.60
1.5	3.95	4.35	4.60
2.0	5.00	5.40	5.75

where F is densimetric Froude number $(= U/\sqrt{g(\Delta\rho/\rho)B}$ and C is an experimental coefficient dependent upon the geometry of the inlet as given in Table 6-III.

Eq. 6-30 can be expressed as:

$$\Delta h = [UB/C\sqrt{g(\Delta\rho/\rho)}]^{2/3} \qquad (6\text{-}31)$$

and also:

$$Q = UBB' \qquad (6\text{-}32)$$

If $U = 0.1$ to 0.3 m/sec the values of $B \times B'$ for a number of ports can be determined for the flow rate (Q) required. Data are provided for $B'/B = 2$ to 4. With C as determined from Table 6-III eq. 6-31 gives Δh, which for no warm water flow is equated to h' (see Fig. 6-20A) and so the complete dimensions are known.

For the two-dimensional case, termed a "skimmer wall", the relevant equations are:

$$B = q/U \qquad (6\text{-}33)$$

where q is the flow rate per unit length of wall:

$$\Delta h = UB^{1/2}/0.9\sqrt{g(\Delta\rho/\rho)} \qquad (6\text{-}34)$$

In the event of some mixing being permitted, particularly at peak loads on the installation, Δh may exceed h', so introducing an inflow Q_w of warm surface water. Let the ratio of this to the total input (Q) be expressed by:

$$\frac{Q_w}{Q} = \lambda = \frac{B'(\Delta h - h'')U}{B'BU} = \frac{\Delta h - h''}{B} \tag{6-35}$$

assuming the inlet velocity to be uniform across the entrance. Eq. 6-35 also applies to a skimmer wall. For the rectangular ports and skimmer walls Senshu found that for the most economical dimensions:

$$h'' = h'(1-\lambda)^{2/3} - \lambda B \tag{6-36}$$

where h' is the Δh computed by eq. 6-31 for zero influx of warm water, and h'' is the distance from the entrance lip to the thermocline (see Fig. 6-20B).

The similar relationship for the skimmer wall is:

$$h'' = h'(1-\lambda)^{1/2} - \lambda B \tag{6-37}$$

In each case the mean temperature of the cooling water (T_M) is given by:

$$T_M = \lambda T_U + (1-\lambda)T_L \quad \text{or} \quad T_M = T_L + \lambda \Delta T \tag{6-38}$$

where T_U and T_L are the temperatures of upper and lower water bodies, or ΔT is the difference in these.

Offshore intake If an installation is located on a sandy shore, where the bed slope is mild, the cooling water intake must extend seawards in order to draw-in the relatively clear water at the greater depths. This might consist of an open channel formed by sheet piling into which seawater enters through a mid-depth slot at the seaward end. With such structures, which usually run normal to the beach, caution should be exercised in preventing waves from reflecting on the end, thus suspending sand which can be inducted into the channel. Similarly, a knowledge of littoral currents is required, especially those generated during storms when seaweed could be torn from the seafloor and transported to the shore. Sufficient warning cannot be given in respect to such batch incidence of weed, which can clog screens within seconds and so instigate a fire hazard. A comprehensive beach survey should be made for any cooling water site, as carried out by Hom-ma et al. [39].

Where the distance offshore is too great, or the beach processes do not allow for an impediment to littoral drift, a pipe intake is called for. These end in a riser pipe with a flared entrance for minimum entry loss. It is also normal practice to put a velocity cap over the entrance, as depicted in Fig. 6-21 to reduce the induction of fish. The reason for this [40] is that fish can detect and swim against horizontal currents but are not conscious of vertical currents. It was found that the introduction of a velocity cap reduced fish intake by 90%. It will also be seen that the cap permits the intake to be located nearer to the warm surface layer without inducting it, than if a plain flared entrance is used. It also inhibits the formation of vortices [41], which could introduce air to the system at low tidal levels.

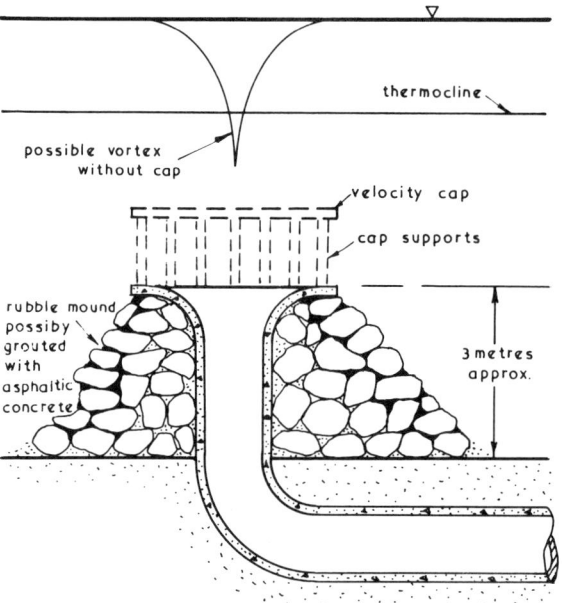

Fig. 6-21. Typical cooling-water intake employing a pipe.

As indicated in Fig. 6-21, the riser must be taken to such a height that suspended sediment is at a minimum, even during storm sequences. Measurements made on the coast of Japan during a typhoon [41] indicated that concentrations decreased up to a height of 3 m from the bed in an 8-m depth. Such a height estimation should take into consideration the change in the level of the floor, for example a build up due to the formation of an offshore bar. In fact, surveys should be made of such transient features so that zones of swift change can be avoided [39].

The intake should be at a sufficient depth below any warm-water surface layer that it will not induct any of it. Nhan [42] has carried out a theoretical and experimental analysis of this problem, for which the definition sketch is provided in Fig. 6-21 of a plain entrance and with a velocity cap.

For the plain entrance (Fig. 6-22A) Nhan obtained:

$$\frac{Q_c^2}{g(\Delta\rho/\rho)y_o^5} = 8\pi^2 K^2 \left(\frac{y_{ac}}{y_o}\right)^2 \left(\frac{y_{ac}}{y_o} + \frac{b}{y_o}\right)^2 \left(1 - \frac{y_{ac}}{y_o}\right) \tag{6-39}$$

where Q_c is the critical input when warm water commences to be inducted; K is an experimental factor; y_{ac} is the height of the interface at the critical flow (Q_c) above the inlet; y_o is the height of the general interface above the inlet; b is the height of the inlet above the seabed.

COOLING WATER STRUCTURES

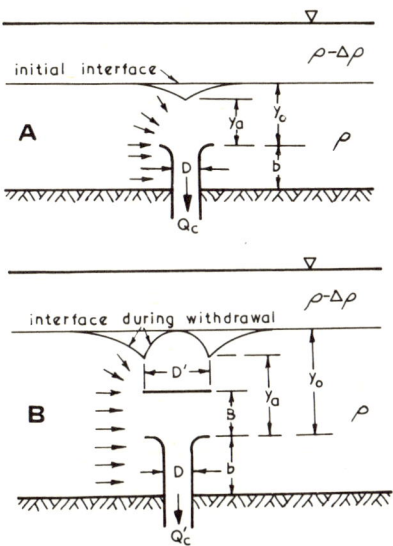

Fig. 6-22. Definition sketch for pipe inlets in a stratified medium for: A. a plain entrance; B. an entrance with velocity cap.

The ratio y_{ac}/y_o is a function of b/y_o such that:

$$\frac{y_{ac}}{y_o} = \frac{1}{10}\left[(4-\frac{3b}{y_o}) + \sqrt{(4-\frac{3b}{y_o})^2 + \frac{40b}{y_o}}\right] \qquad (6\text{-}40)$$

which for a wide range of b/y_o (i.e., ¼ to 1½) is sensably constant, averaging 0.73. The value of K was found by Nhan [41] and also Kato et al. [43] to approximate 0.6 for $0.5 < b/y_o < 1.0$. For $b/y_o \leqslant 0.5$ K could increase to 0.75 [42].

Substitution of $y_{ac}/y_o = 0.73$ and $K = 0.6$ into eq. 6-39 results in:

$$\frac{Q_c^2}{g(\Delta\rho/\rho)y_o^5} = 4.1(0.73 + b/y_o)^2 \qquad (6\text{-}41)$$

For the case of the velocity cap placed above the intake Nhan [42] derived the relationship:

$$\frac{Q_c'^2}{g(\Delta\rho/\rho)y_o^5} = \pi^2 G^2 K'^2/16 \qquad (6\text{-}42)$$

where K' is an experimental factor, and $G = f\left(\frac{D'}{y_o}, \frac{B}{y_o}, \frac{b}{y_o}\right)$, where D' is the diameter of the cap, B is the height of the opening (see Fig. 6-22B) and other dimensions as before.

From a one-dimensional analysis it was found also that for incipient induction of warm water:

$$(y_o)_c = (3B + b)/2 \tag{6-43}$$

when $y_o > (y_o)_c$ withdrawal from the top layer is impossible. When $y_o < (y_o)_c$ induction of warm water may still be prevented by the lack of sufficient flow rate Q'_c as per eq. 6-42.

The ratio of the critical discharges with and without the velocity cap was found by experiment. It varied slightly with y_o/D but mainly with D'/B as in Table 6-IV.

TABLE 6-IV

Critical discharge ratio with and without velocity cap

D'/B	2.5	3.0	3.5	4.0	4.5	5.0	5.5	6.0
Q'_c/Q_c	1.0	1.2	1.5	1.7	1.8	1.9	1.9	2.0

The height ratio of the intake above the floor $b/(b + y_o)$ had little influence on this ratio within the range of 0.4 to 0.65.

The design data provided above for suction of cool denser liquid from beneath a lighter surface layer can be applied also to the suction of a freshwater stratum above a salt-water wedge. In this case it would be advisable to use a velocity cap, so that water is drawn in radially. This will preclude fish induction and impede the uplift of the interface and hence suction of the denser salt water.

Effect of breezes. The first influence of a breeze is to generate short-period waves which will break and help mix water in the warm surface layer. This makes it relatively homogeneous and so stabilizes the boundary between it and the denser liquid below. This permits the surface water, which is having a wind stress applied to it, to move horizontally over the lower liquid and so build up a "surge" against the downwind shoreline. This involves a slight rise in the surface level, but also a manifold deepening of the interface. Such a depression in the warm-water level should be accounted for in the design of intakes outlined above.

Senshu and Wada [44] observed this on Mizushima Bay for prevailing winds of 5–10 m/sec. The analysis is similar to that for storm surge used previously, except that the depth involved is only that to the interface. Also, the depression of this interface (ΔH) is given by:

$$\Delta H/S = \rho/\Delta\rho \tag{6-44}$$

COOLING WATER STRUCTURES

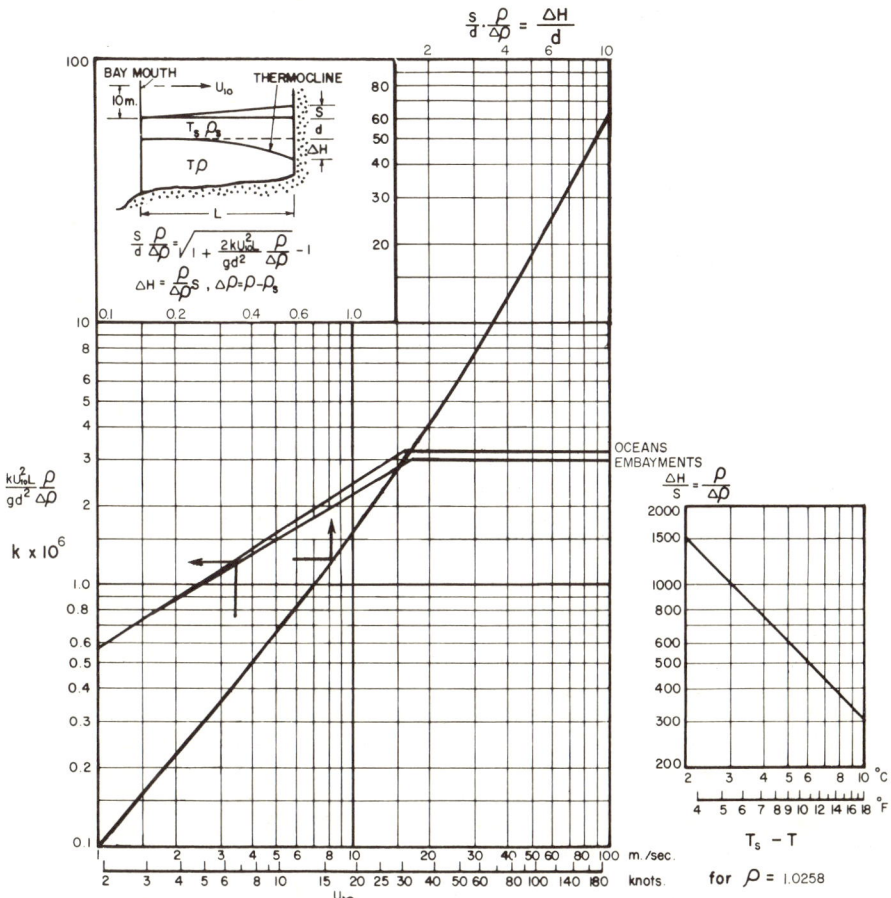

Fig. 6-23. Relationships for warm layer depression due to wind stress.

where S is the rise in surface level; ρ is the density of the lower stratum of water; $\Delta\rho$ is the difference in density of lower and upper strata.

It is obvious that ΔH can be some hundreds of times the surge rise S.

Fig. 6-23 provides the definition sketch, the relevant equations and graphed relationships of $(kU_{10}^2 L/gd^2)(\rho/\Delta\rho)$ versus $\Delta H/d$, k versus U and $\rho/\Delta\rho$ versus $T_S - T$, where U_{10} is the wind velocity (at 10 m above the surface); L is the fetch length of the wind; d is the thickness of the warm surface layer; k is the wind stress coefficient, given by $\tau = k\rho U_{10}^2$ which varies with U_{10} for $U_{10} \leqslant 5$ m/sec; T_S is the temperature of surface layer; T is the temperature of the lower layer.

The graphs are provided assuming that $\rho = 1.0258$. The normal temperature difference of upper and lower layers has been found in Japan to be 5°C and d in the range 3 to 5 m. The velocity distribution in the upper layer must be downwind at the surface and opposite at its base, since no transfer of volume is effected through the interface.

Where winds increase beyond 10 m/sec (20 knots) any reasonable fetch length will produce waves of such a size as to break up the stratification. On these occasions also, lower temperatures could be experienced during the passage of a cyclonic centre. Cooling-water installations must therefore be designed for the critical conditions when the hot summer sun creates a stratification and a sea breeze is generated by the heating of the adjacent land-mass.

EXAMPLES

1

It is required to discharge 140 cfs of sewage effluent into water 52 ft. deep, so that its concentration at the surface is 1/80 that at the pipe. The specific gravity of seawater can be taken as 1.03 and that of the effluent 1.0. Design the outfall for no interaction of the circular jets, which are staggered horizontally along the outfall.

The ultimate object is to dispose of the effluent with the least horsepower demand, which consists of that to discharge a certain quantity through each port and that for pumping through the outfall as a whole. The diameter of the outfall can be calculated on the assumption of a maximum velocity at the landward end of say 8 ft./sec, then $8 d^2 \pi/4 = 140$, or $d = 4.7$ ft. (say 5 ft.). This diameter may be reduced at intervals if this is deemed economical from a construction point of view. Half way along the diffuser section the diameter could be 3.5 ft. for the same pipe velocity. For purposes of calculation the one diameter of 5 ft. will be accepted.

The surface layer thickness $d_p = (52-2.0) \frac{1}{5} = 10$ ft. assuming the pipe penetrates 0.5 ft. into the bed. Then $y = 52-10-2 = 40$ ft. Diameters (D) of 0.67, 0.5, 0.4, 0.33 and 0.267 ft. will be tested for the ports and the calculations are listed in Table 6-V. The specific procedure is:
(a) Compute $0.232 y/D$ from which F is obtained in Fig. 6-3.
(b) Compute U_0 from $F = U_0 / \sqrt{\frac{\rho_s - \rho_0}{\rho_0} gD} = U_0 / \sqrt{0.03 gD}$.
(c) The number of ports (N) is given by $N = Q/(U_0 D^2 \pi/4) = 140/(U_0 D^2 \pi/4)$.
(d) The spacing (L) for no interaction of jets is given by eq. 6-16, so that $L = 40/3.5 = 9.2$ ft. Since ports are located on both sides the total length of diffuser $L_t = 9.2 (N/2)$.

TABLE 6-V

Computations for Example 1

D (ft.)	0.232y/D	F	U_0 (ft./sec)	N	L_t (ft.)	H	h_L	H_T
0.67	14	1.4	1.13	352	1620	1.58	1.7	3.28
0.50	18.6	3.0	2.08	343	1580	1.63	1.66	3.29
0.40	23.2	5.5	3.41	327	1510	1.74	1.59	3.33
0.33	27.9	10	5.62	292	1350	2.05	1.42	3.47
0.267	34.8	30	15.2	165	760	5.15	0.8	5.95

EXAMPLES 275

(e) The pressure head (H_0) in ft. of fresh water to provide velocity U_0 is given by $H_0 - (52-2)1.03 = U_0^2/2g$ (assuming a coefficient of velocity = 1.0), or the head in respect to the sea-surface level $H = 1.56 + U_0^2/2g$.

(f) For a pipe with uniform outlets along its length it can be shown that the equivalent length of pipe is 1/3 full length, so that $h_L = \frac{f}{3} \cdot \frac{L_t}{d} \cdot \frac{Q^2}{2g(d^2\pi/4)^2} = \frac{0.2L(140)^2}{3 \times 5 \times 2g(25\pi/4)^2}$, assuming a pipe friction factor of 0.02.

(g) The total head required $H_T = H + h_L$, which varies for each port diameter. The smallest H_T is chosen from a graph of D versus H_T so that $HP = QH_T w/550$ is minimized.

Although small head differences are involved, the savings are worthwhile over the lifetime of an outfall. For example, a reduction in 1.0 ft. in the present case infers a saving of 12 kw. At 2 cents per kw/h, the saving per annum = $2100, which over 50 years life is $105,000.

(h) Check the concentrations at the surface for the accepted port of 0.60 ft. diameter or 0.232 y/D = 14.0. From Table 6-I for $F = 1.4$, 0.232 x/D = 1.3, $x = 1.3 \times 0.6/0.232 = 3.4$ ft., and $b_m/b = 8 = b_m\sqrt{2}/D$ so that $b_m = 3.3$ ft. This half width of jet at the buoyant surface layer implies a spacing of 6.6 ft. for jets on one side (actually 9.2 ft. has been used). The concentration at this radius is given by eq. 6-10 as $C_r/C_m = e^{-0.74(4.6/3.3)^2} = 0.24$. It is unlikely that the jets either side of the outfall will coalesce and give high concentrations overhead, but checking with Fig. 6-7, y/0.7 DF = 61 and y/1.27 DS_m = 0.6 from which y/L = 8 (extrapolated) could be used or $L = 40/8 = 5$ ft. However, the spacing is limited already by the concentrations along the alignments of the trajectories at the surface layer. If the ports can be considered sharp-edged the actual diameter should be increased by $1/C_c^{1/2}$.

2
A vertical jet of effluent (sg = 1.0) of diameter 0.9 ft. issues with a velocity of 18 ft./sec into a stratified ocean whose density gradient $\left(-\frac{dp}{dy} \cdot \frac{1}{\rho_0}\right) = 1.0 \times 10^{-5}$/ft. Determine the height of the buoyant stratum, and the concentration of this diluted effluent, if the sg of the ambient liquid at the port is 1.026.

From eq. 6-7, $F = U_0/\sqrt{(\rho_s-\rho_0)gD/\rho_s} = 18/\sqrt{\frac{0.026}{1.026} g \times 0.9} = 21$. From eq. 6-11, $T = \frac{\rho_s-\rho_0}{D(-d\rho/dy)} = \frac{0.026}{0.9(1.0 \times 10^{-5})} = 2890$. Parameters to be used in Fig. 6-4 are 0.324 $F^2/T = 0.324(21)^2/2890 = 0.0495$, and 2.38 $F^{1/4}/T^{5/8} = 2.38(21)^{1/4}/(2890)^{5/8} = 0.0343$, giving $(y/D)_t/1.37 \; F^{1/4}/T^{3/8} = 2.42$ and $(C_0/C_m)_t \; 2.38 \; F^{1/4}/T^{5/8} \; 1.15 = 1.775$. Thus $y_t = 142 + 6.2 \times 0.9 = 147.5$ ft. and $(C_0/C_m)_t = 59.5$.

3
Small diameter ports are provided at close intervals along an outfall such that the plume can be considered to issue from a continuous slot of equivalent width 5 cm. at an alignment 45° from the vertical. The sg difference of homogeneous ambient liquid and effluent is 0.027. Find the depth required to effect a dilution of 10 times if the pressure head in the pipe is 0.05 kg/cm² above the surrounding liquid.

$U_0 = \sqrt{2gh} = \sqrt{2gp/w} = \sqrt{2 \times 981 \times 0.05/0.001} = 3.12$ m/sec

$F = U_0/\sqrt{\frac{\rho_s-\rho_0}{\rho_s} gB} = 3.12/\sqrt{\frac{0.027}{1.027} 9.81 \times 0.05} = 27.6$.

From Fig. 6-6, for $\theta_0 = 45°$, 0.226 y/B = 30, $y = 6.6 + 5.2 B \sin 45° = 6.8$ m.

4
An outfall has ports on either side issuing horizontally. When $y = 50$ ft., $F = 14$, $D = 0.5$ ft. and

port spacing $L = 2.5$ ft, the horsepower demanded for nozzle discharge equals that for friction loss in the diffuser section. If the port diameter is doubled what is the relative overal HP demand in the diffuser for the same dilution and the same discharge velocity from each port.

From eq. 6-16 the limit for mixing of circular jets = $4.35 L = 4.35 \times 2.5 = 11$, it is very likely that the two sets of horizontal jets will combine above the outfall, so that Fig.6-7 is applicable, or $\dfrac{y}{0.7 DF} = \dfrac{50}{0.7 \times 0.5 \times 14} = 10.2$, $y/L = 50/2.5 = 20$, giving $\dfrac{y}{1.27 DS_m} = 3$, or $S_m = 26.3$. HP (nozzles) = $QHw/550$ where Q is total discharge, which must remain constant, and $H = U_0^2/2g =$ constant. HP (diffuser) = $Qh_L w/550$ where $h_L = \dfrac{f}{3} \dfrac{(NL)Q^2}{d\, 2g(\pi/4)^2\, d^4}$ or HP (diffuser) $\propto NL$ where $N = Q/q = Q/U_0(\pi/4)D^4$ or $N \propto 1/D^2$.

If diameter is doubled and U_0 remains constant, $F \propto U_0/D^{1/2} \propto 1/D^{1/2}$, $\dfrac{y}{0.7 DF} = \dfrac{50}{0.7 \times 1.0 \times 14/\sqrt{2}} = 7.22$, $\dfrac{y}{1.27 DS_m} = \dfrac{50}{1.27 \times 1.0 \times 26.3} = 1.5$. Fig. 6-7 gives $y/L \doteq 6$, $L = 50/6 = 8.3$, $N_r = \left(\dfrac{0.5}{1.0}\right)^2 = \dfrac{1}{4}$. HP(diffuser)$_r$ = $NL = \dfrac{1}{4} \cdot \dfrac{8.3}{2.5} = 0.83$. Overall HP ratio = $(1 + 0.83)/2 = 0.915$, or a decrease of 8.5%.

5
Compare the horsepower demand for circular ports issuing vertically or horizontally, when the same discharge is to be diluted to the same degree in the same depth of water. Assume F for the horizontal jets is 6 when $C_0/1.15 C_m = 20$.

From Fig. 6-3, for $F = 6$ and $C_0/1.15 C_m = 20$, gives $0.232 y/D = 8.7$. Now $F \propto U_0/D^{1/2} \propto q/D^{5/2}$, so that the trend in Fig. 6-3 is given by a straight line sloping at 0.4:1, modified by the distortion of 1.07 of the ordinate over the abscissa (i.e., slope of 0.43:1). Such a line intersects the vertical jet curve for $C_0/1.15 C_m = 20$ at 15 so that $D_V/D_H = 15/8.7 = 1.725$. Thus, $F_V = 6(1.725)^{5/2} = 23.4$, which matches the value in Fig. 6-3. HP = $wQh/550 \propto U_0^2 \propto D^4 = (1.725)^4 = 8.8$, so that the horsepower demand for vertical jets under these similar conditions is about 9 times that required for horizontal jets.

6
A circular jet with $F = 5$ issues horizontally into a homogeneous ambient fluid whose active $y/D = 100$. Find the dilution at the surface layer. Find the dilution when a mixing tube with area ratio $R = 0.1$ is added.

From Fig. 6-3, for $0.232 y/D = 23.2$ and $F = 5$, $S_m = 75 \times 1.15 = 86$.
From Fig. 6-11, for $R = 0.1$, $S_i = 3.5$, $F_m/F_n = 0.2$, $(y/D')/(y/D) = 0.32$, so that $F_m = 1.0$ and $0.232 y/D' = 23.2 \times 0.32 = 7.4$.
From Fig. 6-3, $S'_m = 35$, so that $S_t = S_i \times S'_m = 35 \times 3.5 = 123$, $S_t/S_m = 123/86 = 1.43$.
Check with design chart (Fig. 6-12) for $y/DF = 100/5 = 20$ and $R = 0.1$, $S_t/S_m = 1.44$. Tests have shown actual dilution to exceed such values of S_t just as $(S_m)_{act} > (S_m)_{th}$.

7
It is desired to test the efficacy of using a mixing tube for a circular effluent jetting horizontally into an ambient fluid with a uniformly decreasing density towards the surface. Assume $F_n = 17$ and $T_n = 2500$ (subscript n referring to nozzle, and m for the mixing tube). Test a mixing tube for which $R = 0.1$.

$0.324 F_n^2/T_n = 0.324 \times 17^2/2500 = 0.0375$, $2.38 F^{1/4}/T^{5/8} = 2.38 (17)^{1/4}/(2500)^{5/8} = 0.0372$.
From Fig. 6-5, $(y/D)_{tn}/1.37 F_n^{1/4}/T_n^{3/8} = 2.12$, and $[S_m]_{tn} 2.38 F_n^{1/4}/T_n^{5/8} = 1.83$.
From Fig. 6-11 for $R = 0.1$, $S_i = 3.5$, $F_m/F_n = 0.2$, $D_n/D_m = 0.33$, $T_m/T_n = 0.33/3.5 = 0.094$. Thus, $0.324 F_m^2/T_m = 0.0375 (F_m/F_n)^2 (T_n/T_m) = 0.0375(0.2)^2/0.094 = 0.016$.
$2.38 F_m^{1/4}/T_m^{5/8} = 0.0372 (F_m/F_n)^{1/4} (T_n/T_m)^{5/8} = 0.0372(0.2)^{1/4}/(0.094)^{5/8} = 0.108$.

EXAMPLES 277

From Fig. 6-5, $(y/D)_{tm}/1.37\ F_m^{1/4}/T_m^{3/8} = 2.18$, and $(S_m)_{tm}\ 2.38\ F_m^{1/4}/1.15\ T_m^{5/8} = 1.78$.

Thus, $\dfrac{y_{tm}}{y_{tn}} = \dfrac{2.18}{2.12}\left(\dfrac{F_m}{F_n}\right)^{1/4}\left(\dfrac{T_m}{T_n}\right)^{3/8}\left(\dfrac{D_m}{D_n}\right) = \dfrac{2.18}{2.12}(0.2)^{1/4}(0.094)^{3/8}/0.33 = 0.85$,

$\dfrac{(S_m)_m}{(S_m)_n} = \dfrac{1.78}{1.83}\left(\dfrac{F_n}{F_m}\right)^{1/4}\left(\dfrac{T_m}{T_n}\right)^{5/8} = \dfrac{1.78}{1.83}\left(\dfrac{1}{0.2}\right)^{1/4}(0.094)^{5/8} = 1.16$.

It is seen in this particular instance that the neutrally buoyant layer of effluent is lower with the mixing tube and that the dilution is increased 16%. Such a gain may not result if values of $0.324\ F^2/T$ are high initially. For design purposes Fig. 6-12 and 6-13 should be inspected to see whether the parameter y/DF is sufficiently high to warrant the use of mixing tubes.

8
The effluent from an outfall reaches the surface with a dilution of 70. If the discharge from each port is 2 cfs and the sg of the ambient liquid is 1.026 and that of the effluent is 1.0, determine the surface spread from each port during the 2 h of relatively stagnant water twice each tidal period.

From Fig. 6-14, for $Tg^{3/5}(\rho_s - \rho_0)^{3/5}/Q^{1/5}S_m^{3/5}\rho_s^{3/5} = 2 \times 3600\ (32.2)^{3/5}(0.026)^{3/5}/2^{1/5}(70)^{3/5}(1.026)^{3/5} = 190$, we have $Lg^{1/5}(\rho_s - \rho_0)^{1/5}/Q^{2/5}S_m^{1/5}\rho_s^{1/5} = 35$ or $L = 35 \times 2^{2/5}(70)^{1/5}(1.026)^{1/5}/(32.2)^{1/5}(0.026)^{1/5} = 112$ ft. This result applies to a single port. If the ports are spaced at 10 ft. intervals the transverse spread of effluent from eq. 6-25 is $L' = L^2\pi/4\ x_p = 112^2\pi/4 \times 10 = 990$ ft. Even at the extremeties of the diffuser the radius of 112 ft. could not be used since there would be pressure from adjacent ports for space in the surface layer. Further research is required on this topic.

9
An effluent is discharged vertically in a river which is 20 ft. deep from a nozzle 0.75 ft. in diameter and at a velocity of 2 ft./sec. Compare the dilution at the surface when there is no river flow and when this flow is 1.0 ft./sec. Find the flow when a mean dilution of 200 occurs at the surface. The density differential of effluent and ambient fluid is 0.025 g/cm^3.

From Fig. 6-3, for $F = 2/\sqrt{0.025\ g\ 0.75} = 2.58$ and $0.232\ y/D = 0.232 \times 20/0.75 = 6.2$, $S_m = 1.15 \times 17 = 19.5$.

When a river current exists treat the mixing as for a homogeneous ambient fluid with zero density difference, and so employ Fig. 6-15, for which $\dfrac{yV}{DU_0} = \dfrac{20 \times 10}{0.75 \times 2.0} = 13.3$. The top surface of the plume (given by curve y_T in Fig. 6-15) thus gives $\dfrac{xV}{DU_0} = 500$ or $x = 500 \times 0.75 \times 2.0/1.0 = 750$ ft. At this value of the abscissa $(bV/DU_0) = 18$, $b = 27$ and $hV/DU_0 = 15$, $h = 22$. The dilution will be in proportion to the area ratio $\dfrac{bh}{\pi D^2/4} = \dfrac{22 \times 27}{\pi(0.75)^2/4} = 1345$. Alternatively from eq. 6-26, $S_m = 2.4(U_0/V)^{1.2}(x/D)^{0.8}$, $S_m = 2.4(2/1.0)^{1.2}(750/0.75)^{0.8} = 1380$. (These are mean values for which the central concentrations will be about double or dilutions half.

To find V to give $(S_m)_{mean} = 200$ (or central dilution about 100), we have from eq. 6-26 $200 = 2.4(2/V)^{1.2}(x/0.75)^{0.8}$ or $x^{0.8}/V^{1.2} = 28.8$. Also the equation of the $y\hat{T}$ curve in Fig. 6-15 is $\dfrac{yV}{DU_0} = 2.35\left(\dfrac{xV}{DU_0}\right)^{0.28}$, so that $\dfrac{20V}{0.75 \times 2} = 2.35\left(\dfrac{xV}{0.75 \times 2}\right)^{0.28}$, or $x^{0.28}/V^{0.72} = 6.35$. Solution of these two equations in x and V gives $V = 0.2$ ft./sec. and $x = 5.6$ ft. Thus, it takes a very small current to effect a significant dilution of effluent, even though the plume is deflected very little. This is no doubt due to the turbulence present and the shearing action exerted on its circumference. This results in the vortices, as illustrated in the inset of Fig. 6-15, which assist the mixing process.

278   MARINE HYDRAULIC WORKS

10
Water is discharged from a power station 8°C above the temperature of the surrounding surface layer of the sea at a rate of 60 m³/sec and a mean velocity of 2 m/sec. Compare the area of the plume and its length along the centre-line of the jet for no wind and a wind of 5 m/sec. In each case find the temperature at the centre of the plume and the time for this steady state to be reached.

From eq. 6-27, $A_0 = 74\, Q^2 = 74 \times 60^2 = 26.7 \times 10^4$ m². From eq. 6-28, $y_0 = 3\,Q = 3 \times 60 = 180$ m. From Fig. 6-16, for wind speed of 5 m/sec, $A_5/A_0 = 0.5$, and $y_5/y_0 = 0.5^{1/2} = 0.7$, so that $A_5 = 13.35 \times 10^4$ and $y_5 = 126$ m.

Pipe diameter is given by $60 = 2.0\, \pi D^2/4$ or $D = 6.5$ m. Length of plume $y_0/D = 27.6$ and $y_5/D = 19.4$. The centre distances are half these, so that from Fig. 6-18 for $T_0 - T_s \leqslant 14°C$ ratio $\dfrac{T_y - T_s}{T_0 - T_s} = 0.5$ and 0.7 respectively, or $T_y - T_s = 4°C$ and $5.6°C$.

The temporal variation of plume length is given by Fig. 6-17, in which $\dfrac{yg^{1/5}(\rho_s - \rho_0)^{1/5}}{Q^{2/5}\rho_s^{1/5}} = \dfrac{180 \times 9.8^{1/5}}{60^{2/5}(380)^{1/5}} = 16.9$, and $\dfrac{Dg^{1/5}(\rho_s - \rho_0)^{1/5}}{Q^{2/5}\rho_s^{1/5}} = \dfrac{6.5(9.8)^{1/5}}{60^{2/5}(380)^{1/5}} = 0.61$.

(Note that the density ratio $\left(\dfrac{\Delta\rho}{\rho}\right)$ is obtainable from the graph in Fig. 6-23 for temperature differential ΔT, i.e., $8°C = 1/380$). So that $\dfrac{Tg^{3/5}(\rho_s - \rho_0)^{3/5}}{Q^{1/5}\rho_s^{3/5}} = 25 = \dfrac{T(9.8)^{3/5}}{60^{1/5}(380)^{3/5}}$ or $T = 512$ sec. The respective time for y_5 to develop is 350 sec.

11
Using the definition sketch in Fig. 6-20 for a curtain wall intake for cooling water, determine the overall depth required for the following conditions: $Q = 12$ m³/sec, $B = 2$m, $B' = 6$ m, $L = 6$ m, $L' = 2$ m, $b = 2$ m, $T_U - T_L = 5°C$, $U = 0.2$ m/sec and surface layer thickness 4 m. If a skimmer wall is used (i.e., $L = L' = b = 0$) determine this overall depth. If 1/5 of the inducted water can be from the surface layer in these conditions what are the respective depths?

From Table 6-III, for $L/B = 6/2 = 3$, $b/B = 2/2 = 1$ and $L'/B = 2/2 = 1$, $C = 3.40$, from Fig. 6-23, for $\Delta T = 5°$, $\rho/\Delta\rho = 600$; eq. 6-31 gives: $h' = \Delta h = [VB/C\sqrt{g\Delta\rho/\rho}]^{2/3} = [0.2 \times 2/3.40 \times \sqrt{9.8/600}]^{2/3} = 1.0$ m.
Total depth = $3 + 1 + 2 + 2 = 8$ m. Number of openings required = $Q/UBB' = 12/0.2 \times 2 \times 6 = 5$. If $Q_w/Q = 0.2 = \lambda$, from eq. 6-36 we have $h'' = h'(1-\lambda)^{2/3} - \lambda B = 1.0(1-0.2)^{2/3} - 0.2 \times 2 = 0.5$ m. Total depth = 7.5 m. Using a skimmer wall of same width = $5 \times 6 = 30$ m, $B = 12/30 \times 0.2 = 2$ m. Eq. 6-33 gives: $h' = \Delta h = UB^{1/2}/0.9\sqrt{g\Delta\rho/\rho} = 0.2 \times 2^{1/2}/0.9\sqrt{9.8/600} = 0.25$.
Total depth = $3 + 0.25 + 2 + 0 = 5.25$ m. If $Q_w/Q = 0.2 = \lambda$, from eq. 6-37, we have: $h'' = h'(1-\lambda)^{1/2} - \lambda B = 0.25(1-0.2)^{1/2} - 0.2 \times 2 = -0.176$ m.
Total depth = $3 - 0.176 + 2 + 0 = 4.82$ m.

12
Using the definition sketch of Fig. 6-22 for a pipe cooling-water intake, compute the critical flow-rate for zero induction of warm surface water for the following conditions. Total water depth 25 ft., thermocline 7 ft. from surface, $\Delta T = 4°C$, $b = 10$ ft., $D = 3.0$ ft. For the case of a velocity cap being employed, where $D' = 6$ ft. and $B = 2$ ft., find the critical depth $(y_0)_c$ for which absolutely no surface-water induction will occur. Also find the critical discharge, with b and y_0 as for the open inlet, when the velocity cap is used.

From eq. 6-41, $Q_c^2/g(\Delta\rho/\rho)y_0^5 = 4.1\left(0.73 + \dfrac{b}{y_0}\right)^2$, where $\Delta\rho/\rho = 1/750$ from Fig. 6-23, and $y_0 =$

25−7−10 = 8 ft. Thus Q_c^2 = 9.8 × 8^5 × 4.1(0.73 + 10/8)^2/750 = 7100, Q_c = 84 cfs. In a 3 ft. diameter pipe this requires a mean velocity of 12 ft./sec.

When a velocity cap is installed, eq. 6-43 gives $(y_o)_c$ = (3B + b)/2 = (3 × 2 + 10)/2 = 8 ft. (same as for open entrance). From Table 6-IV for D'/B = 6/2 = 3, Q_c'/Q_c = 1.2 or Q_c' = 101 cfs. Induction rates below these maxima should be used in design in order to provide a safety factor against particularly hot weather or a breeze thickening the layer of warm surface water.

13

Cooling water is to be taken from a bay 5NM long in which a 3 m thick warm-water layer exists, with a temperature of 5°C above the lower mass. Determine the depth to the thermocline at the cooling-water intake of a power station located at the downwind end of the bay when a breeze of 10 knots blows for a reasonably long time.

From Fig. 6-23, for U = 10 knots, k = 1.5 · 10^{-6}, also, for ΔT = 5°C, $\rho/\Delta\rho$ = 800. Thus:

$$\frac{kU^2L}{gd^2} \cdot \frac{\rho}{\Delta\rho} = \frac{1.5 \cdot 10^{-6}(10 \times 6080/3600)^2 \, 5 \times 6080}{32.2(3 \times 3.28)^2/600} = 2.5.$$

From Fig. 6-23, $\Delta H/d$ = 1.45, ΔH = 3 × 1.45 = 4.35 m. It is important, therefore, to consider such mild-wind possibilities during summer periods.

PROBLEMS

1

A 60-inch diameter sewage outfall discharges 180 cfs at a depth of 108 ft. through its open end. Compare the dilution at the surface under these conditions and when the line is plugged and 45 horizontal ports 9 inches diameter are drilled into both sides of the line at 18 ft. centres. The ports should be considered as sharp edged and $\Delta\rho/\rho$ = 0.027.

2

A 90-inch diameter outfall 7900 ft. long divides into two 60-inch lines, which are each 1200 ft. in length, each containing 34 ports 6½ inch diameter, 16 ports 7½ inch diameter and 1 port at the end 15 inches diameter, all discharging horizontally. Determine the minimum dilution at the surface if the depth of water is 203 ft. and the total discharge is 520 cfs. Compute the HP consumption for the outfall if the Darcy Weisbach f = 0.024 for all pipes. The ports are at 24 ft. spacing and are rounded (take $\Delta\rho/\rho$ = 0.03).

3

A buoyant jet 9 inches diameter issues vertically at 4 ft./sec into seawater which has a density of 0.027 g/cm^3 at the bottom, but which decreases at the rate of $-(d\rho/dy)/\rho$ = 0.9·10^{-5} ft. Find the depth of water required for the neutrally-buoyant layer to be located at 1/3 depth from the surface. What is the maximum concentration of this layer if that at the outlet is taken as unity.

4

If the jet in Problem 3 issues horizontally determine the depth of water for similar conditions of buoyancy and the concentration as before.

5

An outfall is in 80 ft. of water and contains ports on both sides and along its top surface. These are 0.5 ft. diameter at 6 ft. spacing. What minimum dilution can be expected above the outfall at the surface if the densimetric Froude number of each port is 5 and the ambient liquid is homogeneous?

6
What spacing should 1 ft. diameter ports have if they issue horizontally with a velocity of 3 ft./sec into ambient fluid with a densimetric ratio of 0.025 and are not expected to give concentrations greater than those on the centre-lines of the trajectories? The water depth is 70 ft. and ports should be considered sharp edged. What pressure should be maintained in the outfall for these conditions?

7
An outfall contains a series of slotted ports where equivalent width is 2 inches. They discharge at 15° to the horizontal on one side only. If the velocity of discharge is 6 ft./sec and the density ratio $\Delta\rho/\rho = 0.023$ find the dilution at the surface. What combination of circular ports and spacing with similar angle θ_0 will give the same dilution?

8
An existing outfall contains 16 inch-diameter, rounded nozzles which discharge horizontally with a densimetric Froude number of 3.56 into 40 ft. of water. What effect on dilution at the surface will the addition of mixing tubes 48 inches diameter have? What length should these tubes be and what is the best spacing from the nozzle to the commencement of the cylindrical section?

9
What maximum improvement in dilution can be expected when a 6 inch-diameter vertical jet, issuing with a densimetric Froude number of 5 into water 70 ft. deep, is supplied with a mixing tube. What diameter and length should such an appurtenance be? If the jet is reduced to 4½ inch-diameter, but the discharge through each retained at the same rate, what improvement in dilution can be expected from the same mixing tube? What increase in horsepower for jetting is required? Is this HP the sole criterion for judging the situation?

10
A sewage effluent is ejected vertically into a fresh-water river at the bed, through circular nozzles in an outfall running across its bed. If the nozzles are 6 inch diameter, the discharge from each 0.7 cfs, the effective river depth 20 ft., and its mean velocity is 0.5 ft./sec, find the spacing which will preclude interaction of the jets as they reach the surface. At what distance downstream will the whole surface be contaminated and what will be the dilution of the effluent.

11
An effluent is to be disposed of through a port 1 ft. diameter at the end of an outfall in 30 ft. of water. If the discharge is 4 cfs, what distance from shore should the outlet be taken so that the diluted surface effluent does not come within 50 yards of the bank during the 30 min of slack water four times daily? The ambient fluid has a density of 1.027 g/cm^3 whilst that of the effluent is unity.

12
Determine the area and dimensions of a cooling-water plume which forms from the discharge of 20 m^3/sec of seawater 7°C above the temperature of the surface layer. If the outlet is 3 m across, compute this equilibrium area for zero wind and for a breeze of 3 m/sec. What effect can the breeze have besides helping to transfer heat from the water to the atmosphere? If the outlet temperature is 28°C, what is the temperature of the plume at its outermost extremity and at its centre?

13
A cooling-water intake is to be designed for a power station which demands 300 cfs. Deep

PROBLEMS 281

water near shore over a length of 25 m permits a proposal for a curtain-wall installation. If a thermocline exists 4 m below the surface, determine the depth of water at which a skimmer wall must be installed if the induction velocity is not to exceed 0.3 m/sec. Calculate for no induction of warm surface water and then for a proportion of 0.25 of the total intake. If the surface layer is 27.5°C and the lower water is 21°C, assess the temperature of the water inducted in each case.

14

Consider the power station in Problem 13 to be located in an embayment, and the surface layer referred to as existing in calm conditions. What allowance should be made for lowering of the thermocline in the presence of an 8 knot breeze along the 15 km length of the bay.

15

A refinery is located on a sandy shoreline so that its cooling-water intake must be located offshore. A warm surface layer 8 ft. thick exists, whose temperature is 6°C above the bottom water. If the flared entrance of the intake pipe must be set 12 ft. up from the bed, to impede sand induction during storms, to what depth of water must the inlet be taken in order that only cool bottom water is inducted at the rate of 100 cfs. Provide an answer for a plain entrance and one containing a velocity cap with a diameter of 8 ft. and a vertical opening of 2 ft.

16

Show that for a given discharge of sewage, to be diluted 50 times as it reaches the surface of the sea, where the depth is 40 ft., the number of vertical jets required is approximately 50% greater than those inserted horizontally, if the exit velocity is constant at 4 ft./sec. Besides the lesser number, what is an added advantage of the horizontal jets? Base your calculations on the three-dimensional case.

17

The cooling-water discharge from a power station is contained within floating booms 20 ft. apart in order to direct it 300 ft. seawards away from a nearby intake. The discharge is 300 cfs, through a 5 ft. deep channel. If this water is 8°C above ambient water temperature, to what depth should the booms be taken?

18

Treated sewage discharges from a 3-ft. diameter outfall under a pressure of 20 p.s.i. into water 40 ft. deep. If the 4 inch-diameter ports are located only at the top of the pipe, which sits on the seabed, determine their spacing in order that the concentration between the jets is equal to that on the jet centrelines at the surface. Assume the density of seawater is 1.03 g/cm^3.

19

A cooling-water intake is to be designed for a power station at the head of an embayment, whose opening to the sea precludes swell from entering. The embayment is 1.7 km in length and during the summer months a warm layer 3 m thick forms at the surface, which is 7°C warmer than the lower water. A skimmer wall intake at the rocky site is to be designed, so that no warm surface water is inducted when the input is 100 m^3/sec over the 130 m effective length of the wall. Allowance is to be made for a surface breeze along the embayment towards the site of 3 m/sec. Determine the depth to the head of the opening from the still-water level and the size of the opening if a maximum velocity of 0.3 m/sec is contemplated.

20

Cooling-water intakes, comprising a pipeline and flared inlet, normally have a velocity cap placed over them. What purposes does this serve? Under what conditions is the pipe intake preferred to the curtain wall structure?

21

As a consultant you are required to advise on the design of cooling-water intake and outlet structures for a power station. List the topics you would need to investigate when reaching a decision on these matters.

22

Whilst the computation of dilution is of prime importance in the design of a marine outfall, list the topics to be discussed in any proposal of such an installation.

REFERENCES

[1] A.M. Rawn, F. Bowerman and N.H. Brooks, 1960. Diffusers for disposal of sewage in seawater. *Proc. ASCE*, 86(SA2): 65–106.

[2] G. Abraham, 1963. Jet diffusion in stagnant ambient fluid. *Delft Hydr. Lab., Publ.*, 29.

[3] L.N. Fan and N.H. Brooks, 1969. Numerical solutions of turbulent buoyant jet problems. *Calif. Inst. Tech., Rep.*, KH-R-18.

[4] A.M. Rawn and H.K. Palmer, 1930. Pre-determining the extent of a sewage field in seawater. *Trans. ASCE*, 94: 1036–1060.

[5] W.E. Hart, 1960. Jet discharge into a fluid with a density gradient. Univ. Calif., I.E.R. Ser., 138(4).

[6] P.S. Bulson, 1961. Currents produced by an air curtain in deep water. *Dock Harbour Authority*, 42: 15–22.

[7] G. Abraham and P. van der Burgh, 1962. Reduction of salt water intrusion through locks by pneumatic barriers. *Delft Hydr. Lab., Publ.*, 28.

[8] L.N. Fan, 1967. Turbulent buoyant jets in stratified or flowing ambient fluids. *Calif. Inst. Tech., Rep.*, KH-R-15.

[9] M.L. Albertson, Y.B. Dai, R.A. Jensen and H. Rouse, 1950. Diffusion of submerged jets. *Trans. ASCE*, 115: 639–697.

[10] H. Rouse, C.S. Yih and H.W. Humphreys, 1952. Gravitational convection from a boundary source. *Tellus*, 4: 201–210.

[11] J.J. Sharp, 1968. Physical interpretation of jet dilution parameters. *Proc. ASCE*, 94(SA1): 55–64.

[12] P. Liseth, 1970. Mixing of merging buoyant jets from a manifold in stagnant receiving water of uniform density. *Univ. Calif., Rep.*, HEL 23-1.

[13] K. Cederwall, 1968. Hydraulics of marine waste water disposal. *Chalmers Inst. Tech., Goteborg, Rep.*, 42.

[14] C.H. Bosanquet, G. Horn and N.W. Thring, 1961. The effect of density differences on the path of jets. *Proc. R. Soc.*, A263: 340–352.

[15] R.J. Frankel and J.D. Cumming, 1965. Turbulent mixing phenomena of ocean outfalls. *Proc. ASCE*, 91(SA2): 33–59.

[16] J. Hansen and H. Schroeder, 1968. Horizontal jet dilution studies by use of radioactive isotopes. *Acta Polytech. Scand., Civil Build. Constr. Ser.*, 49.

[17] M. Patarapanich, 1971. *Use of Mixing Tube in Marine Sewage Disposal*. Thesis Asian Inst. Tech., Bangkok.

[18] J.W. Daily, 1965. Aspects of the role of suspensions as related to waste water diffusion. *Proc. IAHR Congr., 11th*, 2: Pap. 2.12.

[19] R. Silvester and N.H.G. Mueller, 1968. Design data for the liquid-liquid jet pump. *J. Hydr. Res.*, 6: 129–162.

[20] R. Silvester, 1967. Jet mixers in sewage outfalls. *J. Inst. Eng. Austr.*, 39: 33–37, 95.

[21] H.N.G. Mueller, 1964. Water jet pump. *Proc. ASCE*, 90(HY3): 83–113.

[22] Anonymous, 1966. Jet mixer on effluent main for Surface Paradise sewerage. *J. Inst. Eng. Austr.*, 38: 74–76.
[23] E.A. Pearson, 1956. Investigation of the efficacy of submarine outfall disposal of sewage and sludge. *Calif. Water Pollution Control Board, Publ.*, 14.
[24] J.J. Sharp, 1969. Spread of buoyant jets at the free surface. *Proc. ASCE*, 95(HY3): 811–825.
[25] B.D. Pratte and W.D. Baines, 1967. Profiles of the round turbulent jet in a cross flow. *Proc. ASCE*, 92(HY6): 53–64.
[26] F.D. Masch, 1963. Mixing and dispersion of wastes by wind and wave action. *Air Water Pollution*, 7: 697.
[27] A.A. Smith, 1965. The effect of gravity waves on the spread of an effluent. *Proc. IAHR Congr., 11th*, 2: Pap. 2.7.
[28] G. Abraham, 1967. Jets with negative buoyancy in homogeneous fluid. *J. Hydr. Res.*, 5: 235–248.
[29] A. Wada, 1967. Study of recirculation of cooling water of power station sited on a bay. *Coastal Eng. Japan*, 10: 143–170.
[30] T. Hayashi and H. Shuto, 1967. Diffusion of warm water jets discharged horizontally at the water surface. *Proc. IAHR Congr., 12th*, 4: 47–59.
[31] J.J. Sharp, 1971. Unsteady spread of buoyant surface discharge *Proc. ASCE*, 97(HY9): 1471–1492.
[32] Y. Jen, R.L. Wiegel and I. Mobarek, 1966. Surface discharge of horizontal warm water jet. *Proc. ASCE*, 92(PO2): 1–30.
[33] N. Tamai, 1969. Diffusion of horizontal buoyant jet discharged at water surface. *Proc. IAHR Congr., 13th*, 3: 215–222.
[34] R.C.Y. Koh, 1971. Two-dimensional surface warm jets. *Proc. ASCE*, 97(HY6): 819–836.
[35] W. Frazer, D.I.H. Barr and A.A. Smith, 1968. A hydraulic model study of heat dissipation at Longannet Power Station. *Proc. Inst., Civil Eng.*, 39: 23–44.
[36] N. Tamai. 1969. Surface discharge of horizontal buoyant jets. *Coastal Eng. Japan*, 12: 159–177.
[37] S. Vigander, R.A. Elder and N.H. Brooks, 1970. Internal hydraulics of thermal discharge diffusers. *Proc. ASCE*, 96(HY1): 509–527.
[38] S. Senshu, 1968. Some considerations on hydraulic design of bottom water intake. *Proc. 11th Conf. Coastal Eng.*, 2: 1442–1452.
[39] M. Hom-ma, K. Horikawa and C. Sonu, 1962. Field investigation at Tokai, Japan, conducted by combined procedure of macroscopic and microscopic approaches. *Coastal Eng. Japan*, 5: 1–18.
[40] R.H. Weight, 1958. Ocean cooling water system for 800 MW power station. *Proc. ASCE*, 84(PO6): Pap. 1888.
[41] M. Hom-ma and K. Horikawa, 1962. Experimental studies on the cooling water intake system for the atomic power plant at Tokai, Japan. *Coastal Eng. Japan*, 5: 19–42.
[42] H.K. Nhan, 1970. *Cooling Water Intake*. Thesis Asian Inst. Technol., Bangkok.
[43] M. Kato, Y. Mizutani and T. Kishi, 1969. Study on a vertical circular intake for selective withdrawal. *Proc. Ann. Congr. Japan. Soc. Civil Eng.*, 1969: 191–194.
[44] S. Senshu and A. Wada, 1966. Study on bottom water intake for condenser cooling system of power stations sited on a bay. *Proc. 10th Conf. Coastal Eng.*, 2: 1493–1512.

Chapter 7

HYDRAULIC MODELLING

New structures or machines involving fluid motion can be designed from: (*a*) pure reasoning or theory; (*b*) past experience of similar structures; (*c*) hydraulic or fluid modelling.

Few practical problems can be treated purely from theory. The one case in which no experimental coefficients arise is that of laminar flow in pipes, in which case:

$$h/L = 32\nu v/gd^2 \tag{7-1}$$

where h is the head loss over a length L of pipe with diameter d when a mean velocity v exists of a fluid with kinematic viscosity ν. But this equation applies only in sections of the pipe where the central velocity is twice the mean value, so that in the entrance and exit sections experimental coefficients would be necessary.

Past experience from one prototype situation to another can be used in design of simple structures, for example conduits, open channels and weirs, where variables are related by coefficients which remain reasonably constant. But where accelerated flow, non-uniform depths or velocities, or complicated networks are involved resort must be made to models.

Hydraulic models are generally smaller than the natural phenomenon they depict, although this is not necessarily the case. In fact, some processes must be exaggerated in order to allow for proper observation, for example, the flow of blood in arteries. Not all engineers become involved in such laboratory investigations, but they should be aware of the possibilities and the limitations of hydraulic or any other models that purport to explain or quantify certain phenomena. The possibility of reproducing actions to scale should not inhibit a thorough examination in the field. In fact, the search for input data to a model can widen the observation of the natural setting.

The advantages of modelling are many. It is cheap compared to any empirical approach on a full scale. If the design is faulty the model can be readily dismantled or altered, whereas in nature costly remedial measures may be required because of the experiment. Models can be built and tested relatively quickly. In fact, they are generally requested under these conditions, since they are in many cases the last resort, so are left to the last minute in the planning of an installation. Models provide ease in measurement as input variables can be controlled and instrumentation can be highly sensitive in a laboratory atmosphere. Experiments can be repeated and multiple recordings made simultaneously. Turbulence and other flow phenomena can be visually observed in a model when it may be impossible in a prototype situation.

SIMILITUDE

Any action or reaction in a model can predict similar action in a model of different size (even full-scale) if certain conditions are met. This similarity can be geometric, kinematic or dynamic. For geometrical similarity corresponding dimensions must be in the same proportion. However, the scale used for the horizontal magnitudes may differ from those for the vertical, so introducing the concept of geometric distortion. Kinematic similarity implies that motion and paths of homologous particles are in proportion. Dynamic similarity requires that the masses and forces on fluid particles in the model and prototype are in proportion. For dynamic similarity to be achieved both kinematic and geometric similarity are also required. This complete similarity is not often achieved as there are many forces present to produce acceleration of fluid particles. However, one force generally predominates, so that by a judicious choice of scales the others can be minimised.

The forces entering hydraulic models are gravity, pressure, viscosity, surface tension, elasticity and inertia. The forces that may thus act on a mass M of fluid are:

gravity $F_g = Mg = \rho L^3 g$ (7-2)

where ρ is the density of fluid, and L is a dimension of length;

pressure $F_p = pA = pL^2$ (7-3)

where p is mean pressure, and A is area over which this pressure acts;

viscosity $F_v = \tau A = \mu \dfrac{dV}{dy} L^2 = \mu \dfrac{V}{L} L^2 = \mu VL$ (7-4)

where τ is shear stress, μ is dynamic viscosity, and dV/dy is the differential velocity across a width (dy) of fluid;

surface tension $F = \sigma L$ (7-5)

where σ is the surface tension;

elasticity $F_e = KA = KL^2$ (7-6)

where K is the bulk modulus of elasticity of the fluid;

inertia $F_i = Ma = \rho L^3 V^2/L = \rho V^2 L^2$ (7-7)

where a is the acceleration of fluid particle.

SIMILITUDE 287

The most universal of the above forces is the inertial force (F_i) since it must be present no matter what remaining force is predominant. Thus, for dynamic similarity the ratio of this to the other forces must remain constant, so that for predominance of:

$$\text{gravity } \frac{F_i}{F_g} = \frac{\rho V^2 L^2}{\rho L^3 g} = \frac{V^2}{Lg} \text{ (Froude No.)} \tag{7-8}$$

$$\text{pressure } \frac{F_i}{F_p} = \frac{\rho V^2 L^2}{pL^2} = \frac{\rho V^2}{p} = \text{(Euler No.)} = \frac{V^2}{Lg} \tag{7-9}$$

$$\text{viscosity } \frac{F_i}{F_v} = \frac{\rho V^2 L^2}{\mu V L} = \frac{V L \rho}{\mu} \text{ (Reynolds No.)} \tag{7-10}$$

$$\text{surface tension } \frac{F_i}{F} = \frac{\rho V^2 L^2}{\sigma L} = \frac{\rho V^2 L}{\sigma} \text{ (Weber No.)} \tag{7-11}$$

$$\text{elasticity } \frac{F_i}{F_e} = \frac{\rho V^2 L^2}{K L^2} = \frac{V^2}{K/\rho} \text{ (Cauchy or Mach No.)} \tag{7-12}$$

The dimensionless numbers above are well known to hydraulic engineers, but if two or more forces enter a problem it is difficult to model, because equality of the respective parameters must be maintained. The exception is the equality of the Froude and Euler numbers which are essentially the same. Whilst suitable scaling might minimise the effect of one parameter, such as that for viscosity predominance, a check should be made that critical Reynolds numbers are not approached, when laminar flow may erroneously occur in the model.

WAVES

Surface tension enters as a force when a liquid surface suffers extreme curvature, such as in droplets or in the steepening and breaking of waves. The celerity of a wave is given by:

$$C = \sqrt{\left(\frac{gL}{2\pi} + \frac{2\pi\sigma}{\rho L}\right) \tanh 2\pi d/L} \tag{7-13}$$

where the second term in the bracket constitutes the surface tension effect. The ratio of this to the gravity force is thus $(2\pi\sigma/\rho L)/(gL/2\pi)$, which, on substitution of $\sigma = 0.005$ lb./ft. and $\rho = 2$, becomes $0.0032/L^2$. Thus, if surface tension is to be restricted to 1% the effect of gravity then $L = 0.5$ ft. Waves of this order are termed "ripples" and should be avoided in normal model work concerned with coastal engineering.

Long waves such as surges, or more-so tides, suffer from viscous forces, but normal wave action is controlled predominantly by gravity. Swift changes from velocity to pressure (i.e., $V^2/2g = H$) infer an equivalence of Froude number or:

$$\frac{V_m^2}{L_m g_m} = \frac{V_p^2}{L_p g_p} \quad \text{or} \quad \frac{V_r^2}{L_r g_r} = 1 \tag{7-14}$$

where subscripts m, p and r refer to model, prototype and ratio of model/prototype, respectively.

From eq. 7-14 it is seen that $V_r = (L_r)^{1/2}$, since gravity (g) is similar in model and prototype. On some occasions gravity has been exaggerated in order to study the influence of Coriolis force in some tidal phenomena.

Since $t_r = L_r/V_r$, $t_r = L_r^{1/2}$, wave periods are thus proportioned to the square root of a length scale. The length ratio in this case must be the vertical scale of the model because, for either shallow or deep-water waves, the depth enters into the celerity (e.g., $C = \sqrt{gd}$ or $C_0 = \sqrt{gL/2\pi}$). The wave length ratio $(L_w)_r = C_r T_r = L_r$, so that it is proportional to the vertical scale. Where geometric distortion occurs (normally the case for coastal models because of the large areas involved) there will be less waves per unit horizontal length $(L_h)_r$, the ratio being in proportion to the distortion $(L_v/L_h)_r$ where L_{v_r} is the vertical scale ratio. However, all wave processes dependent upon depth, such as steepening, refraction, breaking height and depth, and breaking angle will be reproduced correctly in the model at the respective depth contours. This will be so unless the distortion is greater than about 3, or some areas of the bed are steeply sloped, in which case bottom reflections may occur in the model and not in the prototype. Wave reflection from sloping surfaces and those containing rough or permeable surfaces (rubble-mound breakwaters, for example) are difficult to reproduce to scale unless no distortion is used, as for example in flume tests where properly scaled stone is employed in the model structure. Even then viscous forces can influence results [1]. The phenomenon of seiching in harbour basins, which is dictated by travel time of waves from one reflecting wall to another should be carried out in non-distorted models, whose depths are great enough to preclude frictional effects if these are not apparent in the prototype.

Wave breaking is somewhat dependent upon beach slope or breakwater slope so that distortion of these can influence this phenomenon. It may be possible to distort the major bed zone and revert to nearly zero distortion at boundaries where wave breaking is of greater importance.

The usual scales for wave action are: flume studies (no distortion) — 1:20 to 1:80; basin studies — vertical 1:60 horizontal 1:180; seiching and surges (no distortion) — 1:200 to 1:1000. Sometimes a pilot study is necessary to a small scale in order to obtain a picture of the complete problem. From this important aspects can be identified and separated into larger scale models.

WAVES 289

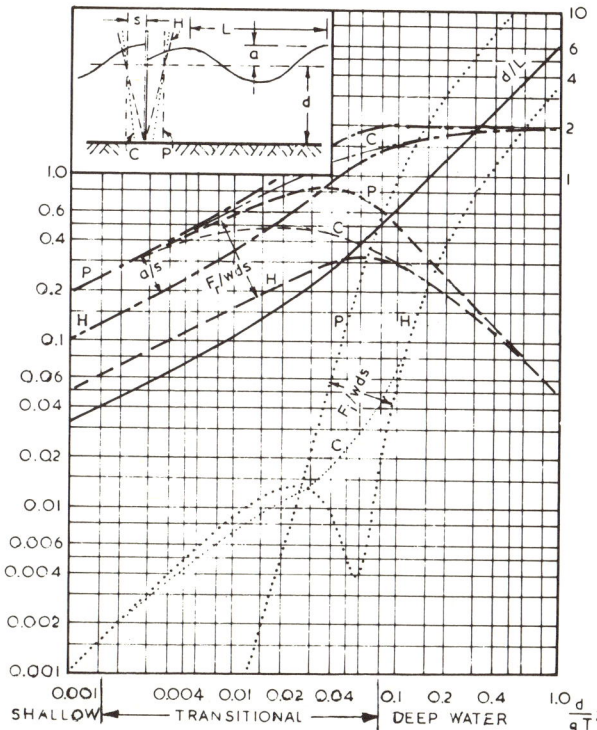

Fig. 7-1. Design parameters for wave generators of hinged (*H*), piston (*P*) or combined (*C*) type.

The generation of waves in models is a highly sophisticated procedure particularly in flumes where larger-scale oscillations are required. The various types of generator have been described by Biésel and Suquet [2]. Theoretical design charts have been derived by Gilbert et al. [3] for hinged and piston-type machines. The curves covering regular waves have been combined in Fig. 7-1, the reader is referred to the original article for relationships in the production of a wave spectrum. Fig. 7-1 presents curves of a/s, d/L, F_r/wds and F_i/wds versus d/gT^2, where a is the amplitude of wave with period T and wave length L when SWL amplitude of the blade is s, and where F_r and F_i are the forces per unit length of blade in phase with the velocity (resistive) and the acceleration (inertia) respectively, for specific weight of water w and depth d.

These forces (F_r and F_i) account for only the hydraulic forces on one side of the blade so they should be doubled. The inertia of the blade itself must be added to F_i, so that the resultant inertia force F_I is given by:

$$F_I = 2F_i + 4\pi^2 sM/9T^2 \qquad (7\text{-}15)$$

where M is the mass of blade per unit length. The total force acting through the SWL is then:

$$F_T = [4F_r^2 + (2F_i + 4\pi^2 sM/9T^2)^2]^{1/2} \qquad (7\text{-}16)$$

The power carried away from both sides of the blade is:

$$P = 4\pi s F_r/T \qquad (7\text{-}17)$$

but the power to drive the blade will be greater than this due to mechanical and electrical losses. Gilbert et al. [3] also treated the case of an oscillating wedge generator.

Between the curves H (hinged) and P (piston) are those marked C (combination), in which both the surface and bottom edges of the blade are oscillated horizontally, the former having a larger amplitude than the latter, so as to more truly represent the orbital motion of water particles in transitional depths. A method of achieving this in a wide basin generator has been described by Silvester [4], the results from which confirm the design curves of Fig. 7-1.

For normal operation of models employing waves, attenuation from the generator to the site of the structure or beach under study need not be considered. However, if small differences in height are being considered, as in reflection studies from submerged objects, the decrease of energy from the measuring probe to the object and back to the probe must be taken into account. Van Dorn [5] has provided the following solution for laminar losses in flumes:

$$\frac{H}{H_o} = \exp\left[-\frac{2x}{LB}(\pi\nu T)^{1/2}\frac{(2\pi B/L)\cosh^2 2\pi d/L + \sinh 4\pi d/L}{4\pi d/L + \sinh 4\pi d/L}\right] \qquad (7\text{-}18)$$

where H is the height of the wave reduced from H_o after travelling distance x in a flume of depth d and width B when it has a period T and length L, and ν is the kinematic viscosity of the liquid.

Treloar and Brebner [6] have conducted tests in different widths of channel to differentiate side and bottom frictional effects. They derived an empirical relationship similar to eq. 7-18, namely:

$$\frac{H}{H_o} = \exp\left[-\frac{2x}{LB}(\pi\nu T)^{1/2}\frac{(2.96\pi B/L) + 0.94\sinh 4\pi d/L}{4\pi d/L + \sinh 4\pi d/L}\right] \qquad (7\text{-}19)$$

which is seen to differ only in the numerator involving the hyperbolic functions. By substitution it would be seen that greater attenuation is predicted in shallow water, but less in deep water, than by the theoretical value of eq. 7-18. At $d/L = 0.1$ similar ratios of H/H_o are obtained.

Since waves must be modelled on the Froudian law the case of an undistorted model permits model-to-prototype ratios of any variables to be derived explicitly as in Table 7-I. The similitude proportions involving length ratio (L_r), gravity ratio (g_r)

TABLE 7-I

Froudian similitude scales for wave motion in undistorted models with length ratio L_r

Variable	Ratio model/proto	Similitude
Water depth	d_r	L_r
Wave height	H_r	L_r
Wave length	λ_r	L_r
Wave period	T_r	$L_r^{1/2} g_r^{-1/2}$
Orbital velocity	u_r or v_r	$L_r^{1/2} g_r^{1/2}$
Refraction angle	α_r	1
Breaking angle	$(\alpha_b)_r$	1
Diffraction angle	β_r	1
Diffraction height	$(H_d)_r$	1
Reflection	$(H_R)_r$	1
Mass transport	$(U_b)_r$	$L_r^{1/2} g_r^{1/2}$
Littoral current	U_r	$L_r^{1/2} g_r^{1/2}$
Time scale	t_r	$L_r^{1/2} g_r^{-1/2}$
Acceleration	a_r, g_r	g_r
Density	ρ_r	ρ_r
Force	F_r	$L_r^3 \rho_r g_r$
Mass	m_r	$L_r^3 \rho_r$
Specific weight	γ_r, w_r	w_r
Dynamic viscosity	μ_r	$L_r^{3/2} g_r^{1/2} \rho_r$
Kinematic viscosity	ν_r	$L_r^{3/2} g_r^{1/2}$
Surface tension	σ_r	$L_r^2 g_r \rho_r$
Bulk elasticity modulus	K_r	$L_r g_r \rho_r$
Pressure	p_r	$L_r g_r \rho_r$
Momentum	$(mv)_r$	$L_r^{7/2} g_r^{1/2} \rho_r$
Energy-work	E_r	$L_r^4 g_r \rho_r$
Power	P_r	$L_r^{7/2} g_r^{3/2} \rho_r$
Discharge	Q_r	$L_r^{5/2} g_r^{1/2}$

and density ratio (ρ_r) will revert to a length ratio only if $g_r = \rho_r = 1$, as is generally the case. The derivation of these parameters has been discussed by Fan and Le Méhauté [7].

TIDES AND CURRENTS

Where flow in channels and across coastal shallows is involved in a coastal phenomenon, distortion of the vertical to horizontal scale is necessary for several reasons. One is the need to have vertical variations in water level which are measurable. If a natural scale is adopted for a wide expanse of coastline or estuary the depths become extremely small because of the limited area generally available for a model. Also the vertical scale must be such as to minimize viscosity effects in the bottom boundary layer. Keulegan [8] has derived the lower bound of the vertical scale (L_v) as:

$$(L_v)^{3/4} = 3.5(\nu T_p)^{1/2}/d_p \qquad (7\text{-}20)$$

where T_p is the period of the prototype fluctuation, say of a tide, and d_p is a vertical measurement in the prototype, say depth of water. Substituting in d_p = 30 ft., $\nu = 11 \cdot 10^{-6}$ ft.^{2}/sec and T_p = 44700 sec, gives L_v = 1/28. Since this scale cannot normally be applied to the horizontal dimensions distortions up to 5 times have to be countenanced, with 3 being the normal value.

As noted previously, velocities associated with wave motion must be scaled to the vertical dimension, so that $u_r = (L_v)^{1/2}$. Since $L_H = t_r u_r$, so $t_r = L_H/(L_v)^{1/2}$. By similar reasoning the variables listed in the first two sections of Table 7-II can be derived, based upon the assumption that g_r and ρ_r are both unity [7]. When stratification is introduced into a model the resulting currents must be based on similar densimetric Froude numbers ($U/\sqrt{g(\Delta\rho/\rho)d}$), in which the density ratio $(\Delta\rho/\rho)_r$ = 1. Sharp [9] has discussed requirements for models of spreading buoyant jets in which the main criterion is similarity in densimetric Froude number which can be expressed as:

$$D_r g_r^{1/5} (\Delta\rho/\rho)_r^{1/5}/Q_r^{2/5} = 1 \qquad (7\text{-}21)$$

Generation of tidal cycles in models consists mainly of discharging water into the basin and out of it at predetermined rates. The level at all times must be controlled to the stages of the tide in the prototype sequence and timing. A mechanism is shown diagrammatically in Fig. 7-2, where it is seen that a float moves with the water surface to make or break the contact with a metal strip shaped in the form of the tidal fluctuation to be followed in the model. This electrical contact raises or lowers the weir through the electric motor. An intermittent cut-out is necessary in the circuit to prevent overcompensation, otherwise the lag in hydraulic response would cause fluctuations about the desired tidal curve. Many sophistications have been developed in such tide generators to which the reader should refer [10–12].

It will be seen from Table 7-II that the roughness of the model surface must be

TABLE 7-II

Froudian similitude for tide and current action in distorted models with $g_r = \rho_r = 1$

Variable	Ratio model/proto	Similitude
Horizontal length	L_H	—
Vertical length	L_V	—
Distortion	$L_V/L_H = L_d$	—
Beach slope	m_r	L_d
Water depth	d_r	L_V
Velocity	V_r	$L_V^{1/2}$
Friction factor	f_r	L_d
Chezy coefficient	C_r	$L_d^{-1/2}$
Shear velocity	$(u_*)_r$	$L_V L_H^{-1/2}$
Roughness	K_r	$L_V^4 L_H^{-3}$
Energy slope	S_r	L_d
Head loss	$(h_L)_r$	L_V
Time scale	t_r	$L_H L_V^{-1/2}$
Discharge	Q_r	$L_H L_V^{3/2}$
Thickness of viscous sublayer	δ_V	$(L_V L_d)^{-1/2}$
Density difference	$(\Delta\rho/\rho)_r$	1
Intrusion length	$(L_A)_r$	$(L_V)^{11/8}$
Diffusion coefficient	$(D')_r$	$L_H L_V^{1/2}$
Sediment size	D_r	$(L_V L_d)^{-1/2}$
Sediment distribution	$(\sigma_D)_r$	1
Apparent specific weight	$(\gamma')_T$	$(L_V L_d)^{3/2}$
R_n (b layer)	$(R_n)_r = (u_* D/\nu)_r$	1 for $(R_n)_m \geq 500$
F_n (b layer)	$(F_n)_r = (u_*/\sqrt{g\gamma'D})_r$	1
Volume moved	$(V_s)_r$	$L_H^2 L_V$
Transport rate	$(Q_s)_r$	$L_H^{5.61} L_V^{-4.35}$
Time for bed evolution	$(t_b)_r$	$L_V^{5.35} L_H^{-3.61}$

Note. All wave characteristics in Table 7-I apply in the case of distorted models.

exaggerated in order to maintain similarity. It will be found that model tides normally pass through an estuary or river zone too quickly. From knowledge of the respective high tides at two or three sites within the prototype area the equivalent

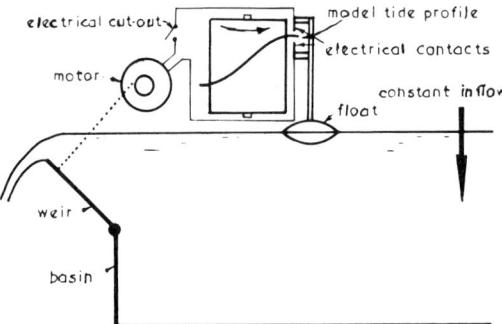

Fig. 7-2. Essential features of a model tide generator.

model times must be checked by gauges. Roughness is added, in the form of vertical metal strips which penetrate the flow from support beams laid across the waterway, until corresponding times are observed.

MOVABLE BED MODELS

Many hydraulic phenomena can be studied in models which have a fixed bed profiled to the existing shape of the prototype, distorted or undistorted. But where the bed of the prototype consists of sediment which can be transported by the waves and currents, it may be found necessary to replicate this in the model. As can readily be imagined, this immediately introduces many new distortions into the action. Such problems have been encountered in fluvial hydraulics over some decades, whereas use of movable beds in coastal problems has had a much shorter history.

Le Méhauté [13], whilst equating an engineer's belief or disbelief in movable-bed scale model technology to a religion, contends that reproduction of sedimentary events on the coast can be more successful than those in river transport. To support this he gives the following reasons:

(*a*) bottom roughness and hydraulic head loss do not enter the coastal model;

(*b*) a shoreline evaluation can be produced by adjusting wave and tidal current input, whereas in fluvial hydraulics the flood histograms must be adhered to;

(*c*) except for rip currents the flow pattern along a coast is steady and predictable, whereas a river bed can meander and shoal in unexpected places;

(*d*) the bed material of a beach is reasonably uniform, any change in size taking place slowly across the profile seawards, as distinct from the large spectrum of particle sizes and shapes moving along a river bed, where even cohesive material may interrupt erosive forces.

Perhaps the writer tends to be an agnostic in respect to this "religious" issue, but the following observations are worthy of note. The empirical approach as embodied in reason (b) above must be based upon a certain knowledge as to how the erosional or accretional feature has evolved. A sandy protuberance on a coast can be caused by offshore shoals, abundant supply from a river, diffraction around an island some distance at sea, or a batch of sediment being accreted upcoast and eroded on the downcoast side. All these demand a different solution, which can be reproduced in a model only if the correct duration of wave direction is used in a model, together with a correct supply of sediment in terms of volume and source of entry to the system.

Differentiation must be made between transient features and the longer-term ones. The former result from batch supply of material from rivers, the non-uniform distribution of wave energy along the coast as storms pass across it, and the interference of littoral drift by man-made structures or dredging operations. On a well known area of coastal margin historic data may be readily available, but on remote sites information can only be gathered from more recent hydrographic surveys.

Le Méhauté [13] distinguishes between conditions and criteria of similitude. The conditions are the basic laws of sediment motion which are relatively well known in fluvial hydraulics. Even so, a choice must be made of the most predominant mechanism and this choice is still a topic for much discussion [14]. Criteria of similitude, on the other hand, are the simplifications in the model procedure that must be specified by the experimenter. In coastal processes this will involve a determination of types of wave, the distribution and duration of directional energy, the size of tide and the location of its generator in the model. The area of sea to be covered by the model is also a criterion of similitude which is in the hands of the experimenter. Cognisance should be taken of such choices made when drawing conclusions from a model, so that nothing can replace a sound knowledge of the physical processes involved. Before a model is commenced a thorough report should be prepared on these processes, which may obviate the need for a model.

From the discussion in Chapter 2 it will be appreciated that crenulate shaped bays provide different littoral conditions around their peripheries. Whilst some sand may travel around the beachline, more may traverse the bay from the downcoast headland. The beach profiles in the sheltered zone, curved in plan, will be steeper than those on the exposed straight tangent section downcoast even though the waves are higher there. The reason for this is that the wave energy is more normal to the tangent section and therefore the longshore component is small. This requires a greater width of shoal water to transport a given quantity of sand downcoast. If it is not transported downcoast it accumulates until the excess can be removed. Such variations in conditions around a bay must be considered when perhaps only a portion of it is being reproduced in a model. There may even be a tendency to introduce rates of sediment transport in a model of one section of bay which has been measured in the field in another section of the same bay. Even if these assessments have encompassed a reasonable width of profile, to include transport well beyond the breaker line, the volumes involved in the curved sheltered zone may be vastly different from those in the tangential downcoast section, because of its reception of material direct from the downcoast headland.

Le Méhauté [13] also distinguishes between short and long models, the former dissipating energy by turbulence, such as wave breaking or hydraulic jump mechanisms, and the latter dissipating it by viscous friction as in river or tidal currents. A coastal model cannot readily replicate frictional processes, as may occur in a littoral current turning into a rip. Models of beaches which involve river or estuary outlets are difficult to conduct when both wave and current action are involved. Bijker [15] has dealt with this subject in depth, and derived certain conditions of similitude which can only be applied with many subjective decisions.

Even where waves are the sole source of energy for sediment transport, their mode of action differs inside and outside the surf zone. The conditions of similitude therefore differ for these onshore and offshore areas, and different scales and modes of modelling are called for. The discussion to follow will be divided into these two categories.

Onshore wave action

The criterion for similitude in models involving beaches is the equilibrium beach slope for similar wave input, in fact a slope that is similar for a reasonable range of wave heights and periods. As discussed in previous chapters, swell and storm-wave profiles can occur, the latter containing the offshore bar. Such shapes and comparative slopes should be reproducible in the model. As will be seen, it may be necessary to distort the model in order that the model sediment can assume the prototype profile with the scaled wave motion.

As noted by Noda [16], a relationship is needed between the wave and sediment characteristics for identical beach profiles. For this purpose Noda carried out extensive flume tests on a variety of sediments which differed in density and size. These were then compared with larger scale flume tests, conducted by Watts [17], Rector [18] and Saville [19], by maintaining wave characteristics scaled from these tests. From the profiles so measured a similarity existed between the SWL distance normal to the shoreline to the point of beach discontinuity after equilibrium had been established. From this a direct relationship was devised between the ratios L_H, L_V, D_r and $(sg-1)_r$ where D and sg refer to characteristic dimension and specific gravity of the sediment respectively, and subscripts H and V refer to horizontal and vertical scales of the model. The reader is referred to original report [16] for details of the test procedure and mathematical manipulations, the resulting equations being:

$$D_r(sg-1)_r^{1.46} = (L_v)^{0.55} \tag{7-22}$$

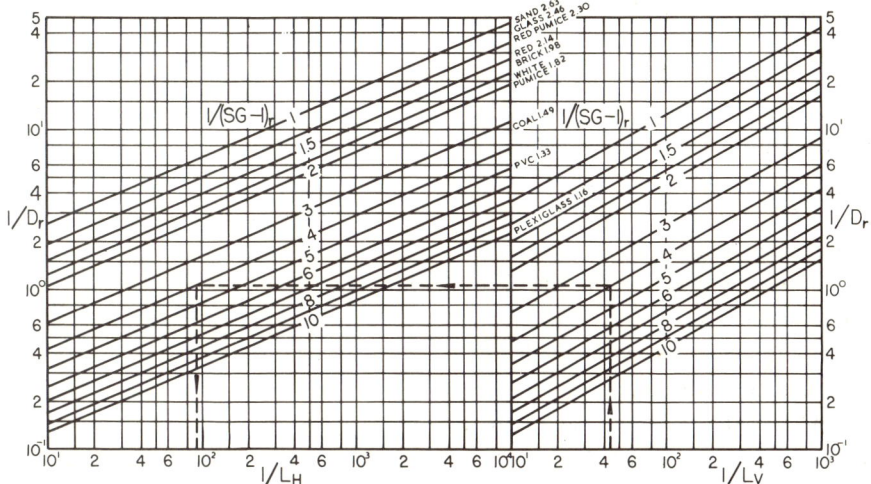

Fig. 7-3. Characteristics of sediment to be used in distorted movable bed models.

and:

$$(L_H)(sg-1)_r^{0.35} = (L_V)^{1.32} \qquad (7\text{-}23)$$

Eq. 7-22 and 7-23 have been graphed in Fig. 7-3, where the reciprocals of the variables are presented which represent prototype/model ratios. The dotted line in the figure indicates how scales L_V and L_H can be obtained through a common D_r ordinate, using similar values of $(sg-1)_r$ in each section of the figure. It will readily be seen that distortion is normally necessary in order to achieve similitude, as noted by Le Méhauté. Since four unknowns exist in 7-22 and 7-23 the experimenter has freedom to choose two, the remainder becoming constrained.

Other conditions of similitude for distorted movable-bed models in which wave action is predominant are listed in the lower portion of Table 7-II. The last two variables of transport rate $(Q_s)_r$ and time of bed changes $(t_b)_r$ have been derived through modifying ratios given by Le Méhauté [13] from the equalities of eq. 7-22 and 7-23. By the substantial differences in some of these ratios [13] given by various workers, there is a clear indication for more research.

The other aspect of beach reproducability is the formation of the offshore bar. Criteria for this have been proposed by Johnson [20] who suggested H_o/L_o alone, Iwagaki and Noda [21] who used H_o/L_o and H_o/D_{50}, and Nayak [22], who changed this latter parameter to $H_o/D_{50}(sg-1)$, where D_{50} represents the median diameter of the sediment particle. Noda [16] contends that eq. 7-22 provides a more consistent parameter in the form $H_o^{0.55}/D_{50}(sg-1)^{1.46}$. If this ratio is the same in model and prototype, not only will the swell beach profile be reproduced with similitude, but also the offshore bar will form at the appropriate wave steepness. Comparison with the test results of the several workers [17-22] indicated that, up to a value of $H_o^{0.55}/D_{50}(sg-1)^{1.46} = 400$ in ft. units for sand, the critical steepness $H_o/L_o = 0.048$. Beyond this value offshore bars were formed at all wave steepnesses. Thus, for a wave 6 ft. deep-water height, sand smaller than 1 mm will always provide an offshore bar. In a model a wave with $H_o = 2$ inches will require sand 0.14 mm diameter or less to reproduce this condition.

From his experience with the beach-model tests Noda [16] made the following recommendations:

(1) Use sand wherever possible because of its cheapness, ease of handling and its good shape factor.

(2) The minimum sand size should be approximately 0.1 mm so as not to include cohesive material.

(3) When the prototype sand is about 0.2 mm lighter material must generally be used in the model. If specific gravity differs little from that of the sand the grains should be spheroidal so that the beach profile will be properly reproduced.

(4) Specific gravities should exceed 1.3, since lighter material attenuates waves and so affects the beach profile.

(5) Light material will smooth out any prototype irregularities whilst fine sand will form ripples which are exaggerations of prototype roughness.

(6) The beach profile is sensitive to changes in wave height, especially for low steepness conditions.

(7) Median diameter is a suitable measure for model sediment, and size distribution similar to that in the prototype is unnecessary unless sorting effects are to be examined.

(8) Where sediment distribution must be modelled it could be done on the bases of $(D_{90})_r = (D_{50})_r = (D_{10})_r$, etc.

(9) The model should be as large as economically feasible.

Offshore wave action

Waves moving into shore across the continental shelf will be travelling at angles to each other, even though they may be almost aligned close to the beach (see *Coastal Engineering, I,* chapter 5). The resulting short-crestedness has an important influence on the transport of bed material beyond the surf zone. Silvester [23,24] has emphasized the need for more research into this aspect of the overall longshore movement of sediment. This condition of angled wave trains offshore is far more prevalent than the uni-directional trains usually replicated in models.

Even though the ability of waves to suspend sand from the seabed is reduced substantially in the deeper water offshore, the transport mechanism depends equally upon any horizontal current present. As emphasized in the chapters on sediment movement, the current of paramount importance is that of the mass-transport due to wave propagation. This is optimum in the direction of wave advance within the boundary layer at the bed. Thus sand grains need be lifted very little in order to be caught in this current and shifted a step towards or along the shore as each wave passes overhead. In conditions of great turbulence, where material may be thrown to great heights above the bed, the net movement across the sea floor may be minimal. It is therefore the small waves which can be the more effective in the transport mechanism offshore.

Incipient motion of bed particles involves near laminar conditions, so that reproduction in models should be based upon the Reynolds number as much as the Froude number. For this reason it is preferable to use full-scale amplitudes of water motion and full-scale sediment. The sand found offshore will be much finer than that close to the beach so that scaling down of its size is the more difficult. The phenomena of incipient bed motion, ripple formation, dune development, and final smooth bed from large water oscillations, has been reproduced in models, in which either the water moved back and forth or the bed so moved in stationary water. The major difficulty is in trying to quantify the resulting transport.

It is possible to conduct flume or basin tests on offshore transport but, because the scales for such operations must be about 1:20 and 1:50, respectively, great distortions are introduced. Also, the wave boundary layer is exaggerated and the dunes are excessive compared to prototype conditions. Since direct comparison of transport rates is only possible from such small scale experiments, it can safely be said that reliable relationships are not yet possible for transport in the offshore zone.

The alternative method of producing oscillatory flow at the bed, by moving either the bed or the water, comes closer to representing prototype conditions. Even so, distortions can occur, due to the lack of water mass in which the macro-turbulence structure can exist, and the lack of transient pressure fluctuation in correct phase with the wave cycle. In this type of model the mass-transport velocity can be introduced, but only from some theoretical or experimental value otherwise obtained. The velocity within the boundary layer may be incorporated into the total oscillating mass of water, but the correct vertical distribution close to the bed cannot be duplicated.

All attempts at reproducing oscillatory flow to full scale have so far been concerned with two-dimensional motion, and then mainly for sinusoidal oscillations as pertain to a single train of waves. From previous discussion it is realised that such conditions are relatively rare over the deeper zones of the continental shelf. At present it can only be surmised what influence the cumulative accelerations and decelerations of a more complex wave system has on sediment suspension, to which must be added the macro-vortex generation of the short-crested system.

Flumes. Information needs collating from the various flume studies [25–28] even for data on incipient motion and bed forms. The determination of transport rate is more complicated since the mode of measurement influences the results so greatly. The normal procedure is to record movement of sand to or from a tray inserted in the bed. When sand from either side moves into a tray the step projection of the tray edge generates vortices [29,30], which greatly increase the turbulence intensity near the sand bed. When material is removed from a tray, the smooth flow either side influences the velocities on the section of the bed which is of prime importance. As soon as some sand has been removed, the tray edge then starts to exert an influence. A suggested method to overcome this problem is to alternate batches of dyed sand so that a long length of uniform sedimentary bed exists without end effects.

The net movement of sediment in a flume is by mass-transport of the water, which is established quickly throughout the depth of a flume. In the ocean it may take many weeks for this net motion to be convected or diffused from the bed or the surface into the mid-depths, as discussed in *Coastal Engineering, I,* chapter 4. This introduces errors when the distribution of sediment throughout the depth of the flume differs from that in the ocean, even for the same two-dimensional wave conditions. The mass-transport distribution also varies with bed roughness, which in a model is exaggerated. Because the net movement of water and of sediment within the boundary layer is so much more important than in upper layers it is preferable to concentrate on this bed load in any quantitative studies.

Yalin [31,32] has derived dimensionless parameters involving amplitude of water motion alone. These necessarily exclude velocity and period of the wave, so are not likely to be so universal as those which incorporate these variables. Carstens' sediment number [33] may be restricted because of its exclusion of fluid viscosity, the apparent value of which alters with sediment concentration.

Basins. The major benefit of studying sediment transport in a basin is that the combined action of waves and currents can be assessed. The work of Bijker [15] in this respect has already been discussed, in which he observed an increase in shear stress from a current due to the wave action. This involved an evaluation of bed roughness or friction factor. Jonsson [34] has summarised research on boundary layers as they affect scale models. Changes of wave intensity, as occurs continually in nature, can significantly change the bed forms, and even the presence of small obstacles on the seabed can generate dunes on an otherwise smooth floor. Application of model results must therefore be undertaken with care. Bijker concludes: "It will, however, always be necessary to carry out some calibration tests for the determination of the transport and time scale."

Such calibration would include the measurement of suspended and bed load in these deeper zones of the sea. Besides the sand transport itself, the waves and currents producing it would also have to be recorded over lengthy periods. In these zones also, the waves will comprise trains angled to each other, whose influence at

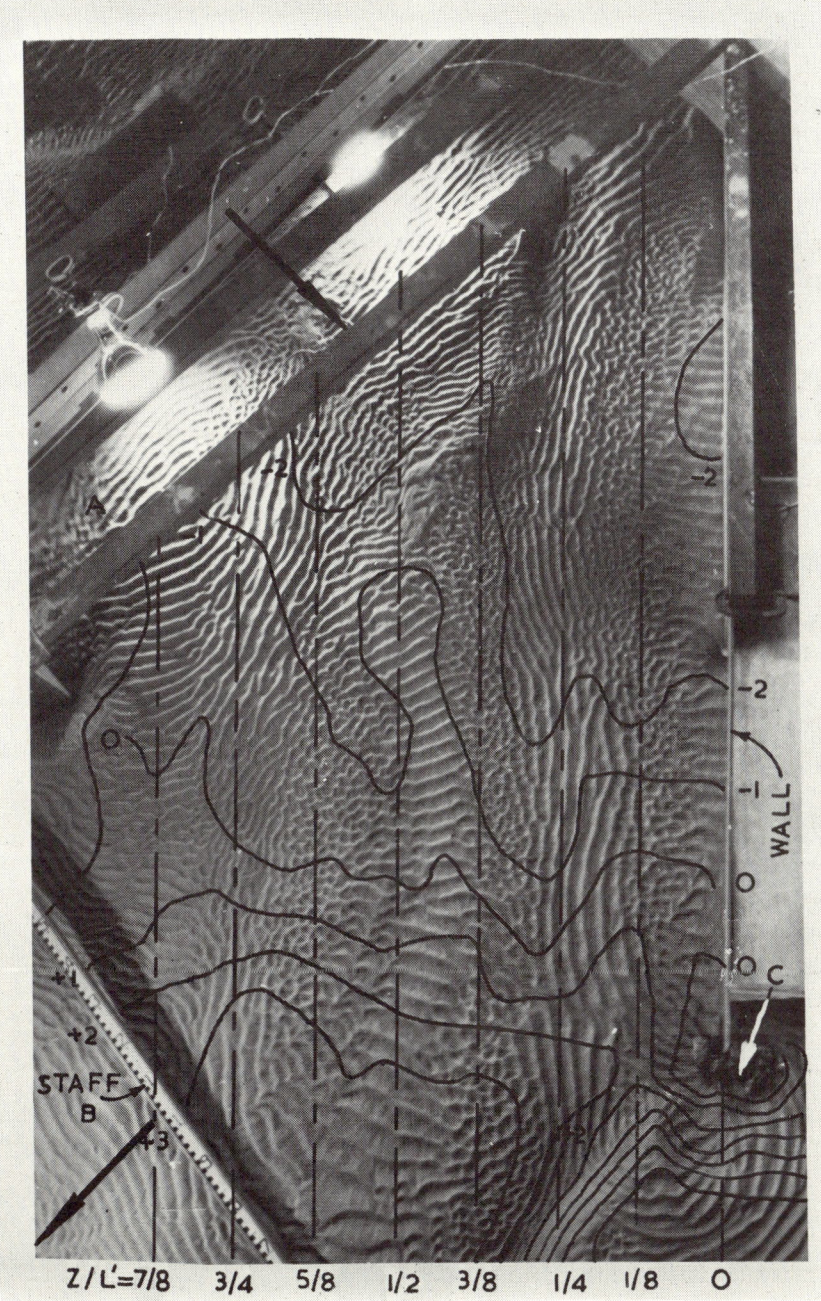

Fig. 7-4. Sand-bed forms emanating from short-crested waves produced by 45° reflection from a wall.

the bottom is dictated by their height and period. Perhaps a more macroscopic view needs to be taken in which threshold bed movement is related to wave power, as suggested by Bagnold [35]. Into this might be introduced energy vectors from which longshore transport rates might be determined across the continental shelf. If correlation coefficients between model and prototype could be determined, then the relationship between wave power and transporting capacity might be evaluated.

The influence of short-crested waves on the incipient motion and bed forms of sediment particles must first be evaluated. A pilot study by Silvester [36] in a wave basin confirmed the orbital water-particle motions predicted by theory. By reflecting waves at 45° from a vertical wall the sedimentary dune pattern formed under angled-wave trains was observed. Fig. 7-4 is a photograph of the model bed after some 60 h duration of waves 1 sec in period, 4 cm high, in water 12.5 cm deep. Also identified in the figure are contours of the bed marked in centimetres from the original flat bed. Erosion has occurred in the zone of entry of the waves and accretion in the zone of exit of the reflected waves. Also marked are the alignments of the crests in the short-crested system. These run parallel to the reflecting wall at the right side of the figure, and are marked as fractions of the crest length (L'). A staff laid on the bed in the left-hand bottom corner of Fig. 7-4 has alternate centimetres painted black to give a scale to the bed forms.

Distinctive ripple patterns can be observed. Along the crest alignments, of $Z/L' = 0$, 1/2 and 1, the crests of the ripples are normal to the wall, or normal to axis of rectilinear oscillation of the water particles in these planes. Similarly, on alignments $Z/L' = 1/4$ and 3/4, the crests are parallel to the wall or normal to the transverse water oscillations, which are also rectilinear. At the remaining alignments, of $Z/L' = 1/8$, 3/8 and 5/8, a speckled appearance derives from the spherical depressions caused by the vortices of the water particles in these regions. This swirling motion was closely observed in the model and grains of sand were seen to be plucked from the centre and deposited at the edge, to form a circular mound. These have coalesced as seen in the figure, to give a snakelike appearance.

The dark patch at the end of the wall, marked C, was a zone filled with stone to prevent erosion from the strong vortex formed at this point. Even so, a swift accumulation of material took place just beyond it as indicated by the several contours. As already noted, this experiment was in the nature of a pilot study, but indicates many possibilities on the use and abuse of wave reflection, plus knowledge of sand particles under short-crested wave action. A comparison needs to be made in a basin of the erosion capacity of uni-directional waves with that of angled trains. This would provide some indication of the differences between the action normally studied in the laboratory and that existing on the continental shelf, where waves are propagating in many directions.

Oscillating bed. Since basin experiments must necessarily be small-scale, attempts have been made to reproduce the water-particle oscillations at the seabed to full scale. In this case the sediment can also be of natural size. The first to tackle this problem was Bagnold in 1946 [37], who pendulated a narrow curved bed containing sediment in still water. This concept has been carried on by workers at the University of California, Berkeley [38–40] who have oscillated a flat bed horizontally in a narrow tank. At the threshold of movement and initial ripple formation, the forces on sand grains are similar to those in wave flumes. However, as the velocities increase so do the rotational speeds of the eddies. It is relevant, therefore, to examine the Magnus effect on both sand particles and water turbulons for the two modes of generation, still water or still bed.

In Fig. 7-5 are depicted the motions of a rotating mass, which can represent a sand grain or a body of water of either spherical or cylindrical shape. Motions in A refer to the oscillating bed and in B to the oscillating water case. Stages throughout the wave cycle are recorded in the figure as follows: *1* = commencement of motion; *2* = acceleration; *3* = deceleration; *4* = zero motion; *5* = reverse acceleration.

The forces exerted on any mass result from its relative rectilinear motion in respect to the surrounding medium and the speed and direction of its rotation. The velocity profiles of the water are shown approximately for cases A and B. The flat bed condition is assumed, but the arguments would also hold for initial ripple development. The case of a duned bed is discussed later.

It is seen from Fig. 7-5 that for stages *3*, *4* and *5* an upward force is exerted on the particle or eddy when the bed oscillates and a downward force in the case of water oscillation. Thus, in the former case vortices tend to move away from the bed

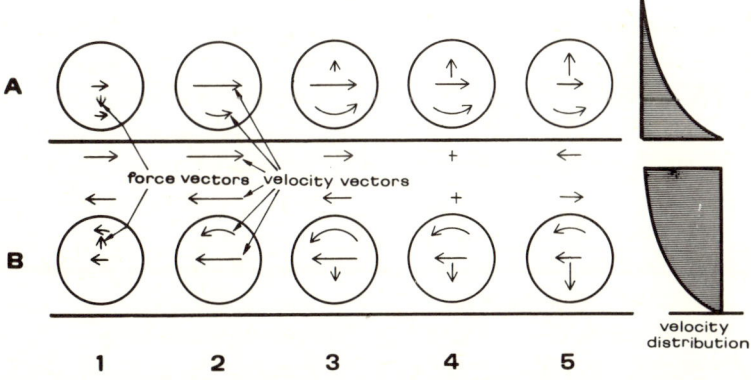

Fig. 7-5. Force vectors applying to rotating masses from oscillatory flow produced by: A. oscillating bed; B. oscillating water. (For explanation of stages 1 to 5, see text.)

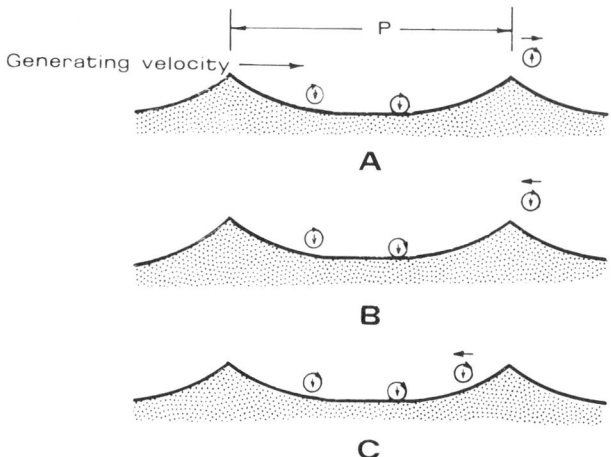

Fig. 7-6. Forces exerted on vortices near a duned bed produced by oscillatory flow, when: A. $x \gg P$; B. $x \geqslant P$; C. $x < P$.

or sand grains remain suspended longer. This contention should be tested in a facility where both water and bed oscillation can be effected separately.

When dunes have formed on the bed Magnus influences are similar for the two types of oscillation. Fig. 7-6 shows vortices in different locations with respect to the dunes with a pitch of P. Those in the trough zone hug the bed. Those ejected from the crest of the dunes are forced upward whilst the generating velocity is still high, but are deflected downwards when deceleration or zero velocity occurs. The relative amplitude of water oscillation (x) with respect to the spacing of the dunes (P) thus assumes importance, as was noted by Bagnold [37]. This is illustrated by cases A, B and C in Fig. 7-6.

Vincent [41] has questioned the validity of the oscillating bed concept on the grounds of certain phenomena being incorrectly reproduced, "such as entrainment currents, acceleration of fluid particles, pressure fluctuations in the immediate vicinity of the sea bed, development of turbulence, etc." L'Hermitte [42] has queried the effect of inertia on the sand grains, which would be small whilst incipient motion was taking place, but could become significant as acceleration increased. However, the similarities of dimensionless parameters of Carstens [38] and Manohar [38] as discussed in Chapter 1, would appear to indicate that the motions are not dissimilar, unless some compensatory influences have operated.

The presence of trays in an oscillating bed will introduce secondary effects, perhaps more so than in flume tests, because the motions are to full scale. As illustrated in Fig. 7-7, trays into which sediment is carried and measured, would generate eddies which are carried across the adjacent sand bottom, and so influence the turbulence intensity there. It could also be presumed that the proportion of bed

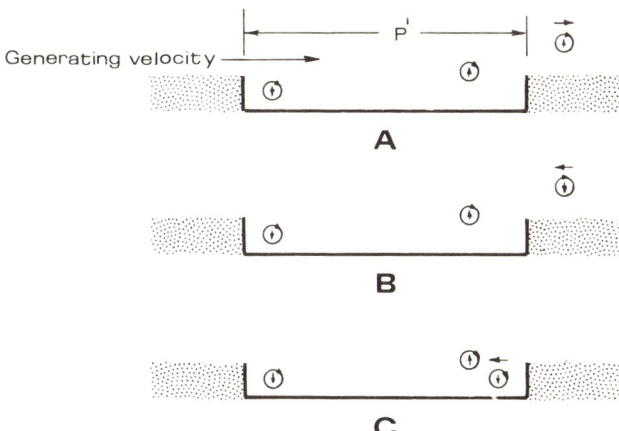

Fig. 7-7. Vortex generation by trays where: A. $x \gg P$; B. $x \geqslant P$; C. $x < P$.

oscillation (x) to the width of the train (P') could have a significant effect on the volume of "transport" as computed from the transfer into the tray. Where the still-water tanks are narrow, the movement of the bed can generate vortices with vertical axes, which could persist longer than the horizontal ones.

Water tunnels. The most recent development for testing the hydrodynamic effects of oscillatory flow over sedimentary beds and other submerged obstacles is the water tunnel. These have now been constructed in Denmark [43], the United Kingdom [44] and the United States [45]. Their relative size and shape are illustrated in Fig. 7-8 (A, B and C). Units depicted in A and C are U-shaped tubes in which a water mass is oscillated continuously with the assistance of an air blower. These are necessarily limited to sinusoidal oscillations, the period of which can be varied by altering the volume of water in the system. The United Kingdom facility, in B of the figure, is a sophisticated unit in which the driving piston can be made to oscillate in a complex manner. A mass transport velocity can be introduced over the full cross-section by means of the pump provided.

All three installations suffer certain limitations respecting ocean conditions. These mainly concern the replication of boundary-layer effects from walls and ceiling of the working section. These, in turn, affect the macro-turbulence in the vicinity of the bed. Such distortions increase as the velocities grow, but should not be greatly influential during incipient motion of bed particles or even early stages of dune formation.

The cross-section of unit A in Fig. 7-8 is 16 inches by 12 inches high and is therefore limited in the velocities employed for dune studies, because of the ceiling and wall effects. Although C has the same height its width is 48 inches, so mini-

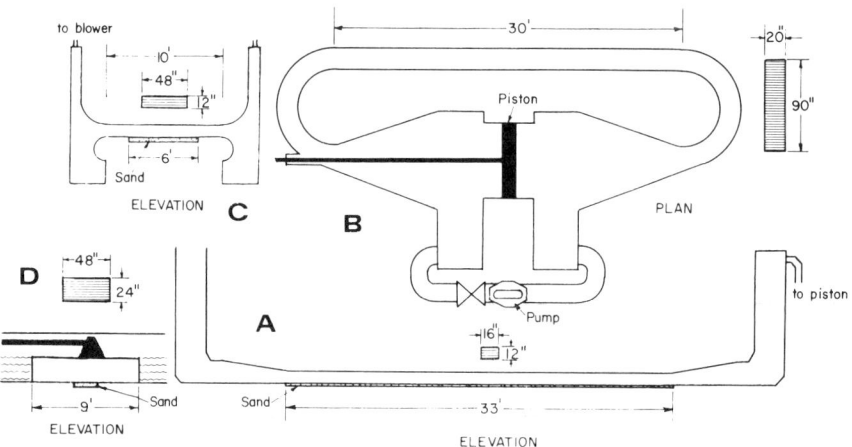

Fig. 7-8. Installations for providing oscillatory water flow constructed in: A. Denmark; B. United Kingdom; C. United States; and D. Australia.

mizing the influence of the walls. Unit B has a working height of 90 inches and width 20 inches. Whilst more depth is provided for macro-turbulence the width still restricts the dimension of turbulons that can be developed or sustained. The 180° bends at the extremity of the working section could introduce uneven velocity distributions across the width of the waterway, which could generate strong vortices with vertical axes. Flow straighteners would not overcome this secondary effect. Future designs might well utilise Tee- or elbow-shaped conduits with honey-comb filters to dissipate the large-scale eddies likely to be generated at these locations.

Water block. To permit reasonable development of macro-turbulence, emanating from the bed boundary layer, it is advisable to maximize the area of the working section. But to accomplish this with the techniques outlined would involve massive inertia forces. The United Kingdom unit is designed for a maximum piston thrust of 230 tons, because of the large water hammer effect [44]. A more economical approach might be to create a "block" of water which can be oscillated over a fixed bed. Such an attempt has been made in Australia [46] where a 9 ft. by 2 ft. high steel box (ends and top only) has been oscillated in a 4 ft. wide flume. This is illustrated diagrammatically in Fig. 7-8D. The forces to overcome were those of inertia of the water in the box and of the box itself, wave making in the flume, and friction in the wheel supports and driving gears. This installation appears to have potential for introducing complex oscillations of one or a number of concurrent wave trains. Mass-transport could be reproduced by a net movement of the box over a number of wave cycles, followed by a slow return to its original position. By maintaining the mean water level in the flume the same as the box height, pressure differentials inside and outside the box are minimized. With adequate sealing, cyclic pressures could be introduced if deemed necessary. The synchronous motion of the top with the water obviates any ceiling boundary effects as occur in the tunnels previously described.

The water-block concept seems to be the only one to lend itself to the study of angled wave trains with full-scale water particle motions. A box could be oscillated in a basin by cranks in two directions normal to each other. By a suitable choice of amplitudes and phases, the rectilinear or circular motions at any point across the short-crested pattern being modelled could be reproduced. Again, mass transport and pressure fluctuations could be introduced without great complication. If windows were provided and lights inserted the macro-turbulence of the water and its effect on the sand particles could be viewed and filmed.

Short-crested wave systems, which are so prevalent in the deeper water offshore, should receive more attention from the theoretician and experimentalist alike. Even closer inshore, reflected waves can create erosive conditions of great concern to the coastal engineer. The use of this double application of wave energy to the bed might solve many problems of silting at harbour and river mouths. Whilst some aspects of this phenomenon must be studied in wave basins, the actual mechanism of sediment suspension should be studied full-scale in rigs as suggested above or similar full-scale concepts [49].

EXAMPLES

1

A beach phenomenon is to be modelled with a vertical scale of 1:40 and a horizontal scale of 1:100. The wave train involved has prototype characteristics of $T = 10$ sec, $H_o = 6$ ft. The depth at the boundary of the model, where the waves are to be generated, has a prototype value of 30 ft. Find the height length and period of the model waves and the SWL horizontal amplitude of a hinged, piston or combined type of wave generator. In the last case suggest an amplitude for the base of the blade. If the fixed bed is to represent a beach slope of 1:30, what scale effects can be expected at the shoreline? What is this effect for a prototype beach slope of 1:10?

$d = 30$ ft., $L_o = 5.12 \cdot 10^2 = 512$, $d/L_o = 30/512 = 0.0585$, $d/L = 0.1028$, $\tanh 2\pi d/L = 0.57$, $\sinh 2\pi d/L = 0.69$, $H/H_o = 0.997$.

Wave height at point of generation $\doteq 6$ ft. From Table 7-I, $H_m = 72/40 = 1.8$ inch, $d_m = 30/40 = 0.75$ ft., $T_m = 10/\sqrt{40} = 1.58$ sec.

From Fig. 7-1, for $d/gT^2 = 30/32.2 \cdot 10^2 = 0.0093$, $(a/S)_H = 0.34$, $S_H = 1.8/(2 \times 0.34) = 2.65$ inch, $(a/S)_P = 0.62$, $S_p = 1.8/(2 \times 0.62) = 1.45$ inch, $(a/S)_C = 0.60$, $S_c = 1.8/(2 \times 0.60) = 1.5$ inch, x at SWL $= H/2 \tanh 2\pi d/L = 1.8/(2 \times 0.57) = 1.58$ inch, x at bed $= H/2 \sinh 2\pi d/L = 1.8/(2 \times 0.69) = 1.3$ inch, S at bed for combination blade $= 1.5 \times 1.3/1.58 = 1.235$ inch.

From Fig. 7-18, for beach slope of 1:30 = 3% and $H_o/L_o = 6/512 = 0.0118$, $K_r = 0$. Slope with distortion of 100/40 = 2.5 is 1:12 = 8.3%, so that $K_r = 0.05$. Thus reflection in model is minimal as for prototype. When prototype slope is 1:10, model slope is 1:4 or 25%. For $H_o/L_o = 0.0118$, $K_r = 0.5$, so that reflection could produce substantial scale effect. This might be over-

308   HYDRAULIC MODELLING

come by flattening the beach zone in the model, so long as no tides were being reproduced to change the water level.

2
What force per unit length would be expected against a sea wall if a 1:36 fresh-water model wall 3 ft. long experiences a force of 27 lb? If the model waves are of 2.0 sec period and 0.3 ft. height, what are the prototype values?

From Table 7-I, $F_r = \rho_r L_r^3 = \gamma_r L_r^3$ so that $F_p/F_m = 64(36)^3/62.4(1)^3 = 47,900$, $F_p = 47,900 \times 27/(3 \times 36) = 11,950$ lb./ft. From Table 7-I, $T_r = L_r^{1/2}$, $T_p = 2\sqrt{36} = 12$ sec, $H = 0.3 \times 36 = 10.8$ ft. Such a model must be undistorted.

3
A tidal river is to be modelled to a vertical scale of 1:50 and horizontal scale of 1:1000. Its mean depth is 25 ft. and mean width 2000 ft. at a point where the maximum tidal current is 3 ft./sec. Find the respective model values and also the model discharge, comparative friction factor and tidal period.

$d_m = 25/50 = 0.5$ ft., $W_m = 2000/1000 = 2$ ft., $v_r = L_r^{1/2} = 1/\sqrt{50}$, $v_m = 3/\sqrt{50} = 0.424$ ft./sec.

From Table 7-II, $Q_r = L_H L_v^{3/2} = \frac{1}{1000}\left(\frac{1}{50}\right)^{3/2}$, $Q_m = 2000 \times 25 \times 3/1000(50)^{3/2} = 0.435$ cfs, $f_r = L_d = \frac{1}{50}\bigg/\frac{1}{1000} = 20$, model should be 20 times rougher than prototype; $t_r = L_H L_v^{-1/2} = \frac{(50)^{1/2}}{1000} = 0.00707$, if $t_p = 44,700$ sec, $t_m = 317$ sec $= 5.28$ min. Such a model would be classed as a pilot model because of its large distortion and short tidal period. It could serve to obtain the overall picture of large section of coastal zone with a view to selecting specific areas for more intensive study.

4
An undistorted model of a cooling-water outlet is to be built to a scale of 1:80. The prototype discharge is 3,000 cfs through an opening whose equivalent diameter is 35 ft. The temperature difference in both prototype and model will be 5°C. Find the model diameter and discharge for dynamic similarity. Determine the likely travel of the buoyant jet in the model over 10 min and assess the prototype values of this distance and time.

From Fig. 6-23, for $\Delta T = 5°C$, $\Delta\rho/\rho = 1/600$. Eq. 7-21 gives $\left[\frac{D(g\Delta\rho/\rho)^{1/5}}{Q^{2/5}}\right]_r = 1$, which results also from equivalence of densimetric Froude numbers in model and prototype. With $(g\Delta\rho/\rho)_r = 1$, $Q_r^{2/5} = D_r = 1/80$, $Q_m = 3000/80^{5/2} = 0.0517$ cfs, $D_m = 35/80 = 0.438$ ft., $\left[\frac{D(g\Delta\rho/\rho)^{1/5}}{Q^{2/5}}\right]_p = \frac{35(32.2/600)^{1/5}}{3000^{2/5}} = 0.79$.

From Fig. 6-17, for $\frac{T_m(g\Delta\rho/\rho)_m^{3/5}}{Q_m^{1/5}} = \frac{600(32.2/600)^{3/5}}{(0.0517)^{1/5}} = 187.5$, $\frac{y_m(g\Delta\rho/\rho)_m^{1/5}}{Q_m^{2/5}} = 65$, $y_m = 35.8$ ft. Since $y_r = Q_r^{2/5} = (L_v^{5/2})^{2/5} = L_v$, so $y_p = 35.8 \times 80 = 2870$ ft. Also, $T_r = Q_r^{1/5} = (L_v^{5/2})^{1/5} = L_v^{1/2}$, so $T_p = 10 \cdot 80^{1/2} \doteq 90$ min.

5
Design a movable-bed model to represent conditions in a prototype beach where the median sediment size in the surf zone is 0.5 mm, if the horizontal scale must be less than 1:50. What wave characteristics must be introduced to allow for swell and storm-wave beach profiles?

From Fig. 7-3, for specific values of L_H and $1/(sg-1)_r = 1$ (i.e., sand to be used in model), the relative values of L_v, d_m, L_v/L_H and D_m are given in Table 7-III.

TABLE 7-III

Values of model parameters for example 5

L_H	D_r	L_v	d_m (ft.)	L_v/L_H	D_m (mm)
50	5.0	19	1.58	2.64	0.100
60	5.4	23	1.30	2.61	0.093
70	5.7	25	1.20	2.80	0.088
80	6.05	28	1.07	2.85	0.0825
90	6.3	30	1.00	3.00	0.080
100	6.6	33	0.91	3.00	0.076
150	7.0	45	0.67	3.33	0.071

From Table 7-III it is seen that as L_H increases so does L_v in such a way that model distortion increases. The L_H = 50 model is the limit of 0.1 mm sand recommended for use in models, so that the use of this size of sand at any of the other length scales will introduce a scale effect in the sedimentation process. However, for small variations this should not be serious, so that the scales could be chosen on the basis of a reasonable area and depth of model. Also scales should be chosen that can readily convert prototype measurements to model sizes. For example, if fps units are employed, vertical scales of 24, 48, 60 and 72 are convenient to transpose feet or fathoms to inches in the model. A reasonable compromise in this case would be L_H = 72, L_v = 24 giving a distortion of 3 and D_r = 5.8 or sediment distortion of 0.1/(0.5/5.8) = 1.16. This demands a depth at the boundary of the model of 1.25 ft. which is reasonable.

To obtain a winter profile either $H_o/L_o > 0.048$ or $H_o^{0.55}/D_{50}(sg - 1)^{1.46} > 400$. Testing the latter criterion first, $(H_o)_m^{0.55}/(0.1/305)(2.63-1)^{1.46} > 400$ or $(H_o)_m > 0.09$ ft. say $H_o = 0.15$ ft. Thus for $H_o/L_o = 0.06$ (say), $(L_o)_m = 0.15/0.06$ or $T_m = 0.7$ sec representing $T_p = 0.7\sqrt{24} = 3.4$ sec, which is not realistic. Waves of at least 7 sec prototype should be introduced, so that the storm profile must be based upon the sediment size criterion. For the swell profile $(H_o)_m$ must therefore be less than 0.09, say 0.07 ft. or 0.84 inch. Periods can then be selected from the most predominant in the winter and summer situation in order to have a reasonable refraction pattern from the wave generator to shore.

PROBLEMS

1
It is required to test a model breakwater whose undistorted scale is 1:20. The prototype waves have a height of 10 ft. and a period of 12 sec in water 30 ft. deep. Find the SWL amplitude for a combination wave generator and also the amplitude of the blade at the bed. Calculate also the force in the link arm connected at the SWL and the horsepower carried away by the waves. What would be the order of power required for a machine in a 4 ft. wide flume?

2
The reflection from a submerged mound is being tested in a flume. The distance from the wave probe to the mound is 10 ft. Calculate the correction to be applied to the input and the reflected wave if the water depth is 2.5 ft., the wave period 2 sec, the flume width 4 ft. Assume $\nu = 12 \cdot 10^{-6}$ ft.^{2}/sec. Show first that K_r (actual) = K_r (experimental)/$(H/H_o)^2$ where H and H_o refer to attenuation in each 10 ft. distance.

3

A section of coast one mile in extent is to be studied in a laboratory where 40 ft. is available for the model. The offshore zone to be included has a maximum depth of 35 ft. The waves to be included at various stages range from 6 to 12 ft. in height and 11 to 7 sec in period, respectively. Compute the SWL sway of a hinged wave generator required. If the bed is fixed and the prototype beachline has a slope of 1:15, what scale effects could be expected?

4

If the model beach in Problem 3 is movable and consists of sand 0.2 mm diameter, what is the appropriate prototype median sediment size for the scales chosen. Would the waves chosen produce swell and storm wave profiles?

5

It has been suggested in the literature [47] that waves generated by wind in flumes and basins can simulate natural conditions very well. Discuss such a claim and the difficulties related thereto.

6

The problem of testing wind and wave loads on offshore structures has been discussed [48], in which the Cauchy number of the structural response must be equated in prototype and model. Is it feasible to test wind and wave loads simultaneously in a model?

7

A semi-diurnal tide is to be modelled in a waterway 40 ft. deep. What is the smallest vertical scale that could be used if viscosity effects at the bed are to be avoided?

8

A model is to reproduce both waves and tides. The prototype variables are wave height 8 ft., wave period 10 sec, maximum water depth 35 ft., tidal range 5 ft., tidal period 44,700 sec. Select vertical and horizontal scales such that a distortion of 3 is not exceeded and the water surface fluctuations are measurable. An accuracy in the laboratory of 0.005 ft. can be accepted. Determine the model values for wave and tide characteristics.

9

An undistorted model of a cooling-water outlet is to be built. The prototype fluctuations in discharge from the power station is 500 to 2500 cfs. from an outlet 70 ft. wide by 20 ft. depth from the sea surface. Concern is felt for stagnant water which may occur for an hour in the prototype when warm water can accumulate. What size of basin is necessary to study the spreading of this plume during this period. The temperature differential of discharge and ambient water is $6°C$ and the outlet is to be simulated by a 4 inches diameter pipe.

10

A flume study is to be made of a beach profile so that the erosion and offshore bar formation can be studied in connection with a sewage outfall running beneath the area. The sand in the prototype is 1 mm median diameter and the deep-water wave heights 12 ft. and period 9 sec. Derive suitable scales of wave and sand when the maximum water depth in the flume is 2 ft. What height and period of wave should be produced for the study? What comments would you offer in a report of such a study to a client, in order to direct his attention to the possible variable nature of the prototype conditions?

11

Eq. 7-22 and 7-23 are based upon the use of fresh water in the model, the specific gravity term should read $(\gamma_s - \gamma_f)/\gamma_f$. What is the possibility of using a liquid which is denser than prototype seawater but which has a similar kinematic viscosity? The range of sg recommended is 2.63 to 1.33 so that the discussion should be restricted to this equivalent density difference.

12

Discuss the limitations in the various rigs described for reproducing oscillatory water motion at the seabed full scale. Differentiate between reproduction of single wave trains, complex but co-linear waves and short-crested systems.

13

In what way do you think that wave power might be related to sediment transport beyond the breaker zone? Besides power as such, what other variables do you consider must enter the problem?

14

Models can replicate conditions over a very modest length of coastline. Discuss the need for a macroscopic view of an area in respect to sediment supply, wave energy, tidal currents etc. List the types of survey you would have carried out in order to collect data for a model dealing with longshore sediment transport.

15

Why are all models that reproduce wave action based upon the Froude law? Which length scale in a distorted model is so related, and why?

16

What determines the greatest scale ratio (i.e., the smallest model) that can be used for any specific problem. Differentiate between "long" and "short" models.

17

Although diffracted waves cannot be simulated in respect to wave height in a distorted model, whereas refracted waves can, why does this not matter much in sedimentation processes where both refraction and diffraction are occurring together?

18

In a phenomenon of wave reflection from a wall in relatively shallow water, on what basis of similitude should a model be based, the beach slope or offshore zone criterion? Since a wave basin would be involved in any such study, full-scale reproduction would be impossible. Discuss the distortions you would contemplate.

19

Discuss the types of problems likely to be studied in estuarine models. Comment on the peculiar difficulties related thereto.

REFERENCES

[1] J.W. Johnson, H. Kondo and R. Wallihan, 1966. Scale effects in wave action through porous structures. *Proc. 10th Conf. Coastal Eng.*, 2: 1022–1024.

[2] F. Biésel and F. Suquet, 1951–1952. Les appareils générateurs de houle en laboratoire. *Houille Blanche*, 6: 147–155, 475–496, 723–727; 7: 372–376.

[3] G. Gilbert, D.M. Thompson and A.J. Brewer, 1971. Design curves for regular and random wave generators. *J. Hydr. Res.*, 9: 163–196.

[4] R. Silvester, 1956. A mobile multi-purpose wave making machine. *Houille Blanche*, 11: 854–863.

[5] W.G. van Dorn, 1966. Boundary dissipation of oscillatory waves. *J. Fluid Mech.*, 24: 769–779.

[6] P.D. Treloar and A. Brebner, 1970. Energy losses under wave action. *Proc. 12th Conf. Coastal Eng.*, 1: 257–268.

[7] L.N. Fan and B. Le Méhauté, 1969. Coastal movable bed scale model technology. *Tetra Tech. Rep.*, TC-131.

[8] G.H. Keulegan, 1966. Model laws for coastal and estuarine models. In: A.T. Ippen (Editor), *Estuary and Coastline Hydrodynamics*. McGraw-Hill, New York, N.Y., pp.691–710.

[9] J.J. Sharp, 1971. Unsteady spread of buoyant surface discharge. *Proc. ASCE*, 97(HY9): 1471–1492.

[10] M.J. Wilkie and G.A.J. Young, 1952. Pneumatic tide generator. *Engineer*, 194: 133–137.

[11] H.R.A. Dedow, 1965. The control of hydraulic models. *Engineer*, 219: 259–263, 301–304.

[12] J.A. Zwamborn, 1965. Simple apparatus for producing model tides. *Proc. ASCE*, 91(HY2): 39–49.

[13] B. Le Méhauté, 1970. Comparison of fluvial and coastal similitude. *Proc. 12th Conf. Coastal Eng.*, 2: 1077–1096.

[14] J.G. Herbertson, 1969. Scaling procedures for mobile bed hydraulic models in terms of similitude theory. *J. Hydr. Res.*, 7: 315–353.

[15] E.W. Bijker, 1967. Some considerations about scales for coastal models with movable bed. *Delft Hydraul. Lab., Rep.*, 50.

[16] E.K. Noda, 1971. Coastal movable-bed scale-model relationships. *Tetra Tech. Rep.*, Tetrat-P-71-191-1.

[17] G.M. Watts, 1954. Laboratory study of varying wave periods on beach profiles. *Beach Erosion Board, Tech. Mem.*, 53.

[18] L.R. Rector, 1954. Laboratory study of equilibrium profiles of beaches. *Beach Erosion Board, Tech. Mem.*, 41.

[19] T. Saville Jr., 1957. Scale effects in two-dimensional beach studies. *Proc. IAHR Congr., 7th*, Pap. A-3.

[20] J.W. Johnson, 1949. Scale effects in hydraulic models involving wave motion. *Trans. Am. Geophys. Union*, 30: 517–525.

[21] Y. Iwagaki and H. Noda, 1962. Laboratory study of scale effects in two-dimensional beach processes. *Proc. 8th Conf. Coastal Eng.*, 194–210.

[22] I.V. Nayak, 1970. Equilibrium profiles of model beaches. *Univ. Calif. Rep.*, HEL-2-25.

[23] R. Silvester, 1965. Coastal sediment movement - some fundamental problems with discussion of research support. *J. Inst. Eng. Austr.*, 37: 311–323.

[24] R. Silvester, 1970. Sediment movement beyond the breaker zone. *Trans. Inst. Eng. Austr.*, CE12: 63–71.

[25] T. Ishihara and T. Sawaragi, 1962. Fundamental studies of sand drifts. *Coastal Eng. Japan*, 5: 59–65.

[26] M. Hom-ma, K. Horikawa and R. Kajima, 1965. Study of suspended sediment due to wave action. *Coastal Eng. Japan*, 8: 85–103.

[27] D.L. Inmann and A.J. Bowen, 1962. Flume experiments on sand transport by waves and currents. *Proc. 8th Conf. Coastal Eng.*, 137–150.

[28] M. Hom-ma and K. Horikawa, 1963. A laboratory study on suspended sediment due to wave action. *Proc. IAHR Congr., 10th*, 1: 213–220.

[29] F.A. Locher and E. Naudascher, 1967. Some characteristics of macroturbulence in flow past a normal wall. *Proc. IAHR Congr., 12th*, 2: 298–307.

REFERENCES

[30] T.K. Hung, 1967. A computational investigation of impulsively-generated eddies. *Proc. IAHR Congr., 12th,* 2: 315–322.

[31] M.S. Yalin and R.C.H. Russell, 1962. Similarity in sediment transport due to waves. *Proc. 8th Conf. Coastal Eng.,* 151–167.

[32] M.S. Yalin, 1963. Method for selecting scales for models with movable bed involving wave motion and tidal currents. *Proc. IAHR Congr., 10th,* 1: Pap. I-30,

[33] M.R. Carstens, 1966. Similarity laws for localised scour. *Proc. ASCE,* 92(HY3): 13–36.

[34] I.G. Jonsson, 1967. Wave boundary layers and friction factors. *Proc. 10th Conf. Coastal Eng.,* 1: 127–148.

[35] R.A. Bagnold, 1963. Beach and nearshore processes, 1.Mechanics of marine sedimentation. In: M.N. Hill (Editor), *The Sea,* 2 Wiley, New York, N.Y. pp.507–528.

[36] R. Silvester, 1972. Wave reflection at seawalls and breakwaters. *Proc. Inst. Civil Eng.,* 51: 123–131.

[37] R.A. Bagnold, 1946. Motion of waves in shallow water – interaction between waves and sand bottoms. *Proc. R. Soc.,* A187: 1–18.

[38] M. Manohar, 1955. Mechanics of bottom sediment movement due to wave action. *Beach Erosion Board, Tech. Mem.,* 75.

[39] M.M. Abou-Seida, 1965. Bed load function due to wave action. *Univ. Calif. IER Tech., Rep.,* HEL-2-11.

[40] G. Kalkanis, 1966. Transportation of bed material due to wave action. *Coastal Eng. Res. Centre, Tech. Mem.,* 2.

[41] G.E. Vincent, 1958. Contribution to the study of sediment transport on a horizontal bed due to wave action. *Proc. 6th Conf. Coastal Eng.,* 326–355.

[42] P. l'Hermitte, 1961, Mouvements des matériaux de fond sous l'action de la houle. *Proc. 7th Conf. Coastal Eng.,* 211–261.

[43] H. Lundgren and T. Sorensen, 1957. A puisating water tunnel. *Proc. 6th Conf. Coastal Eng.,* 356–358.

[44] H.R.A. Dedow, 1966. A pulsating water tunnel for research in reversing flow. *Houille Blanche,* 21: 837–843.

[45] M.R. Carstens and F.M. Neilson, 1967. Evolution of a duned bed under oscillatory flow. *J. Geophys. Res.,* 72: 3053–3059.

[46] G.R. Mogridge, 1970. Testing sediment movement due to wave action. *Proc. ASCE,* 96(HY7): 1587–1604.

[47] J.M. Colonell and B. Perry, 1968. Laboratory simulation of sea waves. *Proc. ASCE,* 94(WW2): 159–174.

[48] E.J. Plate and J.H. Nath, 1969. Modelling of structures subjected to wind waves. *Proc. ASCE,* 95(WW4): 491–511.

[49] R. Silvester, 1970. Modelling of sediment motion offshore. *J. Hydr. Res.,* 8: 227–259.

APPENDIX

TABLE OF FUNCTIONS OF d/L_o

d/L_o	d/L	$2\pi d/L$	tanh $2\pi d/L$	sinh $2\pi d/L$	cosh $2\pi d/L$	$4\pi d/L$	sinh $4\pi d/L$	cosh $4\pi d/L$	H/H'_o
0	0	0	0	0	1	0	0	1	
.0001000	.003990	.02507	.02506	.02507	1.0003	.05014	.05016	1.001	4.467
.0002000	.005643	.03546	.03544	.03547	1.0006	.07091	.07097	1.003	3.757
.0003000	.006912	.04343	.04340	.04344	1.0009	.08686	.08697	1.004	3.395
.0004000	.007982	.05015	.05011	.05018	1.0013	.1003	.1005	1.005	3.160
.0005000	.008925	.05608	.05602	.05611	1.0016	.1122	.1124	1.006	2.989
.0006000	.009778	.06144	.06136	.06148	1.0019	.1229	.1232	1.008	2.856
.0007000	.01056	.06637	.06627	.06642	1.0022	.1327	.1331	1.009	2.749
.0008000	.01129	.07096	.07084	.07102	1.0025	.1419	.1424	1.010	2.659
.0009000	.01198	.07527	.07513	.07534	1.0028	.1505	.1511	1.011	2.582
.001000	.01263	.07935	.07918	.07943	1.0032	.1587	.1594	1.013	2.515
.001100	.01325	.08323	.08304	.08333	1.0035	.1665	.1672	1.014	2.456
.001200	.01384	.08694	.08672	.08705	1.0038	.1739	.1748	1.015	2.404
.001300	.01440	.09050	.09026	.09063	1.0041	.1810	.1820	1.016	2.357
.001400	.01495	.09393	.09365	.09407	1.0044	.1879	.1890	1.018	2.314
.001500	.01548	.09723	.09693	.09739	1.0047	.1945	.1957	1.019	2.275
.001600	.01598	.1004	.1001	.1006	1.0051	.2009	.2022	1.020	2.239
.001700	.01648	.1035	.1032	.1037	1.0054	.2071	.2086	1.022	2.205
.001800	.01696	.1066	.1062	.1068	1.0057	.2131	.2147	1.023	2.174
.001900	.01743	.1095	.1091	.1097	1.0060	.2190	.2207	1.024	2.145
.002000	.01788	.1123	.1119	.1125	1.0063	.2247	.2266	1.025	2.119
.002100	.01832	.1151	.1146	.1154	1.0066	.2303	.2323	1.027	2.094
.002200	.01876	.1178	.1173	.1181	1.0069	.2357	.2379	1.028	2.070
.002300	.01918	.1205	.1199	.1208	1.0073	.2410	.2433	1.029	2.047
.002400	.01959	.1231	.1225	.1234	1.0076	.2462	.2487	1.031	2.025
.002500	.02000	.1257	.1250	.1260	1.0079	.2513	.2540	1.032	2.005
.002600	.02040	.1282	.1275	.1285	1.0082	.2563	.2592	1.033	1.986
.002700	.02079	.1306	.1299	.1310	1.0085	.2612	.2642	1.034	1.967
.002800	.02117	.1330	.1323	.1334	1.0089	.2661	.2692	1.036	1.950
.002900	.02155	.1354	.1346	.1358	1.0092	.2708	.2741	1.037	1.933
.003000	.02192	.1377	.1369	.1382	1.0095	.2755	.2790	1.038	1.917
.003100	.02228	.1400	.1391	.1405	1.0098	.2800	.2837	1.040	1.902
.003200	.02264	.1423	.1413	.1427	1.0101	.2845	.2884	1.041	1.887
.003300	.02300	.1445	.1435	.1449	1.0104	.2890	.2930	1.042	1.873
.003400	.02335	.1467	.1456	.1472	1.0108	.2934	.2976	1.043	1.860
.003500	.02369	.1488	.1477	.1494	1.0111	.2977	.3021	1.045	1.847
.003600	.02403	.1510	.1498	.1515	1.0114	.3020	.3065	1.046	1.834
.003700	.02436	.1531	.1519	.1537	1.0117	.3061	.3109	1.047	1.822
.003800	.02469	.1551	.1539	.1558	1.0121	.3103	.3153	1.049	1.810
.003900	.02502	.1572	.1559	.1579	1.0124	.3144	.3196	1.050	1.799
.004000	.02534	.1592	.1579	.1599	1.0127	.3184	.3238	1.051	1.788
.004100	.02566	.1612	.1598	.1619	1.0130	.3224	.3280	1.052	1.777
.004200	.02597	.1632	.1617	.1639	1.0133	.3263	.3322	1.054	1.767
.004300	.02628	.1651	.1636	.1659	1.0137	.3302	.3362	1.055	1.756
.004400	.02659	.1671	.1655	.1678	1.0140	.3341	.3403	1.056	1.746

APPENDIX (Continued)

d/L_o	d/L	$2\pi d/L$	tanh $2\pi d/L$	sinh $2\pi d/L$	cosh $2\pi d/L$	$4\pi d/L$	sinh $4\pi d/L$	cosh $4\pi d/L$	H/H_o'
.004500	.02689	.1690	.1674	.1698	1.0143	.3380	.3444	1.058	1.737
.004600	.02719	.1708	.1692	.1717	1.0146	.3417	.3483	1.059	1.727
.004700	.02749	.1727	.1710	.1736	1.0149	.3454	.3523	1.060	1.718
.004800	.02778	.1745	.1728	.1754	1.0153	.3491	.3562	1.062	1.709
.004900	.02807	.1764	.1746	.1773	1.0156	.3527	.3601	1.063	1.701
.005000	.02836	.1782	.1764	.1791	1.0159	.3564	.3640	1.064	1.692
.005100	.02864	.1800	.1781	.1809	1.0162	.3599	.3678	1.066	1.684
.005200	.02893	.1818	.1798	.1827	1.0166	.3635	.3715	1.067	1.676
.005300	.02921	.1835	.1815	.1845	1.0169	.3670	.3753	1.068	1.669
.005400	.02948	.1852	.1832	.1863	1.0172	.3705	.3790	1.069	1.662
.005500	.02976	.1870	.1848	.1880	1.0175	.3739	.3827	1.071	1.654
.005600	.03003	.1887	.1865	.1898	1.0178	.3774	.3864	1.072	1.647
.005700	.03030	.1904	.1881	.1915	1.0182	.3808	.3900	1.073	1.640
.005800	.03057	.1921	.1897	.1932	1.0185	.3841	.3937	1.075	1.633
.005900	.03083	.1937	.1913	.1949	1.0188	.3875	.3972	1.076	1.626
.006000	.03110	.1954	.1929	.1967	1.0192	.3908	.4008	1.077	1.620
.006100	.03136	.1970	.1945	.1983	1.0195	.3941	.4044	1.079	1.614
.006200	.03162	.1987	.1961	.2000	1.0198	.3973	.4079	1.080	1.607
.006300	.03188	.2003	.1976	.2016	1.0201	.4006	.4114	1.081	1.601
.006400	.03213	.2019	.1992	.2033	1.0205	.4038	.4148	1.083	1.595
.006500	.03238	.2035	.2007	.2049	1.0208	.4070	.4183	1.084	1.589
.006600	.03264	.2051	.2022	.2065	1.0211	.4101	.4217	1.085	1.583
.006700	.03289	.2066	.2037	.2081	1.0214	.4133	.4251	1.087	1.578
.006800	.03313	.2082	.2052	.2097	1.0217	.4164	.4285	1.088	1.572
.006900	.03338	.2097	.2067	.2113	1.0221	.4195	.4319	1.089	1.567
.007000	.03362	.2113	.2082	.2128	1.0224	.4225	.4352	1.091	1.561
.007100	.03387	.2128	.2096	.2144	1.0227	.4256	.4386	1.092	1.556
.007200	.03411	.2143	.2111	.2160	1.0231	.4286	.4419	1.093	1.551
.007300	.03435	.2158	.2125	.2175	1.0234	.4316	.4452	1.095	1.546
.007400	.03459	.2173	.2139	.2190	1.0237	.4346	.4484	1.096	1.541
.007500	.03482	.2188	.2154	.2205	1.0240	.4376	.4517	1.097	1.536
.007600	.03506	.2203	.2168	.2221	1.0244	.4406	.4549	1.099	1.531
.007700	.03529	.2218	.2182	.2236	1.0247	.4435	.4582	1.100	1.526
.007800	.03552	.2232	.2196	.2251	1.0250	.4464	.4614	1.101	1.521
.007900	.03576	.2247	.2209	.2265	1.0253	.4493	.4646	1.103	1.517
.008000	.03598	.2261	.2223	.2280	1.0257	.4522	.4678	1.104	1.512
.008100	.03621	.2275	.2237	.2295	1.0260	.4551	.4709	1.105	1.508
.008200	.03644	.2290	.2250	.2310	1.0263	.4579	.4741	1.107	1.503
.008300	.03666	.2304	.2264	.2324	1.0266	.4607	.4772	1.108	1.499
.008400	.03689	.2318	.2277	.2338	1.0270	.4636	.4803	1.109	1.495
.008500	.03711	.2332	.2290	.2353	1.0273	.4664	.4834	1.111	1.491
.008600	.03733	.2346	.2303	.2367	1.0276	.4691	.4865	1.112	1.487
.008700	.03755	.2360	.2317	.2381	1.0280	.4719	.4896	1.113	1.482
.008800	.03777	.2373	.2330	.2396	1.0283	.4747	.4927	1.115	1.478
.008900	.03799	.2387	.2343	.2410	1.0286	.4774	.4957	1.116	1.474
.009000	.03821	.2401	.2356	.2424	1.0290	.4801	.4988	1.118	1.471
.009100	.03842	.2414	.2368	.2438	1.0293	.4828	.5018	1.119	1.467
.009200	.03864	.2428	.2381	.2452	1.0296	.4855	.5049	1.120	1.463
.009300	.03885	.2441	.2394	.2465	1.0299	.4882	.5079	1.122	1.459
.009400	.03906	.2455	.2407	.2479	1.0303	.4909	.5109	1.123	1.456
.009500	.03928	.2468	.2419	.2493	1.0306	.4936	.5138	1.124	1.452
.009600	.03949	.2481	.2431	.2507	1.0309	.4962	.5168	1.126	1.448
.009700	.03970	.2494	.2443	.2520	1.0313	.4988	.5198	1.127	1.445
.009800	.03990	.2507	.2456	.2534	1.0316	.5014	.5227	1.128	1.442
.009900	.04011	.2520	.2468	.2547	1.0319	.5040	.5257	1.130	1.438

Table of functions of d/L_0

APPENDIX (Continued)

d/L_0	d/L	$2\pi d/L$	tanh $2\pi d/L$	sinh $2\pi d/L$	cosh $2\pi d/L$	$4\pi d/L$	sinh $4\pi d/L$	cosh $4\pi d/L$	H/H_0'
.01000	.04032	.2533	.2480	.2560	1.0322	.5066	.5286	1.131	1.435
.01100	.04233	.2660	.2598	.2691	1.0356	.5319	.5574	1.145	1.403
.01200	.04426	.2781	.2711	.2817	1.0389	.5562	.5853	1.159	1.375
.01300	.04612	.2898	.2820	.2938	1.0423	.5795	.6125	1.173	1.350
.01400	.04791	.3010	.2924	.3056	1.0456	.6020	.6391	1.187	1.327
.01500	.04964	.3119	.3022	.3170	1.0490	.6238	.6651	1.201	1.307
.01600	.05132	.3225	.3117	.3281	1.0524	.6450	.6906	1.215	1.288
.01700	.05296	.3328	.3209	.3389	1.0559	.6655	.7158	1.230	1.271
.01800	.05455	.3428	.3298	.3495	1.0593	.6856	.7405	1.244	1.255
.01900	.05611	.3525	.3386	.3599	1.0628	.7051	.7650	1.259	1.240
.02000	.05763	.3621	.3470	.3701	1.0663	.7242	.7891	1.274	1.226
.02100	.05912	.3714	.3552	.3800	1.0698	.7429	.8131	1.289	1.213
.02200	.06057	.3806	.3632	.3898	1.0733	.7612	.8368	1.304	1.201
.02300	.06200	.3896	.3710	.3995	1.0768	.7791	.8603	1.319	1.189
.02400	.06340	.3984	.3786	.4090	1.0804	.7967	.8837	1.335	1.178
.02500	.06478	.4070	.3860	.4184	1.0840	.8140	.9069	1.350	1.168
.02600	.06613	.4155	.3932	.4276	1.0876	.8310	.9310	1.366	1.159
.02700	.06747	.4239	.4002	.4367	1.0912	.8478	.9530	1.381	1.150
.02800	.06878	.4322	.4071	.4457	1.0949	.8643	.9760	1.397	1.141
.02900	.07007	.4403	.4138	.4546	1.0985	.8805	.9988	1.413	1.133
.03000	.07135	.4483	.4205	.4634	1.1021	.8966	1.022	1.430	1.125
.03100	.07260	.4562	.4269	.4721	1.1059	.9124	1.044	1.446	1.118
.03200	.07385	.4640	.4333	.4808	1.1096	.9280	1.067	1.462	1.111
.03300	.07507	.4717	.4395	.4894	1.1133	.9434	1.090	1.479	1.104
.03400	.07630	.4794	.4457	.4980	1.1171	.9588	1.113	1.496	1.098
.03500	.07748	.4868	.4517	.5064	1.1209	.9737	1.135	1.513	1.092
.03600	.07867	.4943	.4577	.5147	1.1247	.9886	1.158	1.530	1.086
.03700	.07984	.5017	.4635	.5230	1.1285	1.0033	1.180	1.547	1.080
.03800	.08100	.5090	.4691	.5312	1.1324	1.018	1.203	1.564	1.075
.03900	.08215	.5162	.4747	.5394	1.1362	1.032	1.226	1.582	1.069
.04000	.08329	.5233	.4802	.5475	1.1401	1.047	1.248	1.600	1.064
.04100	.08442	.5304	.4857	.5556	1.1440	1.061	1.271	1.617	1.059
.04200	.08553	.5374	.4911	.5637	1.1479	1.075	1.294	1.636	1.055
.04300	.08664	.5444	.4964	.5717	1.1518	1.089	1.317	1.654	1.050
.04400	.08774	.5513	.5015	.5796	1.1558	1.103	1.340	1.672	1.046
.04500	.08883	.5581	.5066	.5876	1.1599	1.116	1.363	1.691	1.042
.04600	.08991	.5649	.5116	.5954	1.1639	1.130	1.386	1.709	1.038
.04700	.09098	.5717	.5166	.6033	1.1679	1.143	1.409	1.728	1.034
.04800	.09205	.5784	.5215	.6111	1.1720	1.157	1.433	1.747	1.030
.04900	.09311	.5850	.5263	.6189	1.1760	1.170	1.456	1.766	1.026
.05000	.09416	.5916	.5310	.6267	1.1802	1.183	1.479	1.786	1.023
.05100	.09520	.5981	.5357	.6344	1.1843	1.196	1.503	1.805	1.019
.05200	.09623	.6046	.5403	.6421	1.1884	1.209	1.526	1.825	1.016
.05300	.09726	.6111	.5449	.6499	1.1926	1.222	1.550	1.845	1.013
.05400	.09829	.6176	.5494	.6575	1.1968	1.235	1.574	1.865	1.010
.05500	.09930	.6239	.5538	.6652	1.2011	1.248	1.598	1.885	1.007
.05600	.1003	.6303	.5582	.6729	1.2053	1.261	1.622	1.906	1.004
.05700	.1013	.6366	.5626	.6805	1.2096	1.273	1.646	1.926	1.001
.05800	.1023	.6428	.5668	.6880	1.2138	1.286	1.670	1.947	.9985
.05900	.1033	.6491	.5711	.6956	1.2181	1.298	1.695	1.968	.9958
.06000	.1043	.6553	.5753	.7033	1.2225	1.311	1.719	1.989	.9932
.06100	.1053	.6616	.5794	.7110	1.2270	1.3231	1.744	2.011	.9907
.06200	.1063	.6678	.5834	.7187	1.2315	1.336	1.770	2.033	.9883
.06300	.1073	.6739	.5874	.7256	1.2355	1.348	1.795	2.055	.9860
.06400	.1082	.6799	.5914	.7335	1.2402	1.360	1.819	2.076	.9837

APPENDIX (Continued)

d/L_o	d/L	$2\pi d/L$	tanh $2\pi d/L$	sinh $2\pi d/L$	cosh $2\pi d/L$	$4\pi d/L$	sinh $4\pi d/L$	cosh $4\pi d/L$	H/H_o'
.06500	.1092	.6860	.5954	.7411	1.2447	1.372	1.845	2.098	.9815
.06600	.1101	.6920	.5993	.7486	1.2492	1.384	1.870	2.121	.9793
.06700	.1111	.6981	.6031	.7561	1.2537	1.396	1.896	2.144	.9772
.06800	.1120	.7037	.6069	.7633	1.2580	1.408	1.921	2.166	.9752
.06900	.1130	.7099	.6106	.7711	1.2628	1.420	1.948	2.189	.9732
.07000	.1139	.7157	.6144	.7783	1.2672	1.432	1.974	2.213	.9713
.07100	.1149	.7219	.6181	.7863	1.2721	1.444	2.000	2.236	.9694
.07200	.1158	.7277	.6217	.7937	1.2767	1.455	2.026	2.260	.9676
.07300	.1168	.7336	.6252	.8011	1.2813	1.467	2.053	2.284	.9658
.07400	.1177	.7395	.6289	.8088	1.2861	1.479	2.080	2.308	.9641
.07500	.1186	.7453	.6324	.8162	1.2908	1.490	2.107	2.332	.9624
.07600	.1195	.7511	.6359	.8237	1.2956	1.502	2.135	2.357	.9607
.07700	.1205	.7569	.6392	.8312	1.3004	1.514	2.162	2.382	.9591
.07800	.1214	.7625	.6427	.8386	1.3051	1.525	2.189	2.407	.9576
.07900	.1223	.7683	.6460	.8462	1.3100	1.537	2.217	2.432	.9562
.08000	.1232	.7741	.6493	.8538	1.3149	1.548	2.245	2.458	.9548
.08100	.1241	.7799	.6526	.8614	1.3198	1.560	2.274	2.484	.9534
.08200	.1251	.7854	.6558	.8687	1.3246	1.571	2.303	2.511	.9520
.08300	.1259	.7911	.6590	.8762	1.3295	1.583	2.331	2.537	.9506
.08400	.1268	.7967	.6622	.8837	1.3345	1.594	2.360	2.563	.9493
.08500	.1277	.8026	.6655	.8915	1.3397	1.605	2.389	2.590	.9481
.08600	.1286	.8080	.6685	.8989	1.3446	1.616	2.418	2.617	.9469
.08700	.1295	.8137	.6716	.9064	1.3497	1.628	2.448	2.644	.9457
.08800	.1304	.8193	.6747	.9141	1.3548	1.639	2.478	2.672	.9445
.08900	.1313	.8250	.6778	.9218	1.3600	1.650	2.508	2.700	.9433
.09000	.1322	.8306	.6808	.9295	1.3653	1.661	2.538	2.728	.9422
.09100	.1331	.8363	.6838	.9372	1.3706	1.672	2.568	2.756	.9411
.09200	.1340	.8420	.6868	.9450	1.3759	1.684	2.599	2.785	.9401
.09300	.1349	.8474	.6897	.9525	1.3810	1.695	2.630	2.814	.9391
.09400	.1357	.8528	.6925	.9600	1.3862	1.706	2.662	2.843	.9381
.09500	.1366	.8583	.6953	.9677	1.3917	1.717	2.693	2.873	.9371
.09600	.1375	.8639	.6982	.9755	1.3970	1.728	2.726	2.903	.9362
.09700	.1384	.8694	.7011	.9832	1.4023	1.739	2.757	2.933	.9353
.09800	.1392	.8749	.7039	.9908	1.4077	1.750	2.790	2.963	.9344
.09900	.1401	.8803	.7066	.9985	1.4131	1.761	2.822	2.994	.9335
.1000	.1410	.8858	.7093	1.006	1.4187	1.772	2.855	3.025	.9327
.1010	.1419	.8913	.7120	1.014	1.4242	1.783	2.888	3.057	.9319
.1020	.1427	.8967	.7147	1.022	1.4297	1.793	2.922	3.088	.9311
.1030	.1436	.9023	.7173	1.030	1.4354	1.805	2.956	3.121	.9304
.1040	.1445	.9076	.7200	1.037	1.4410	1.815	2.990	3.153	.9297
.1050	.1453	.9130	.7226	1.045	1.4465	1.826	3.024	3.185	.9290
.1060	.1462	.9184	.7252	1.053	1.4523	1.837	3.059	3.218	.9282
.1070	.1470	.9239	.7277	1.061	1.4580	1.848	3.094	3.251	.9276
.1080	.1479	.9293	.7303	1.069	1.4638	1.858	3.128	3.284	.9269
.1090	.1488	.9343	.7327	1.076	1.4692	1.869	3.164	3.319	.9263
.1100	.1496	.9400	.7352	1.085	1.4752	1.880	3.201	3.353	.9257
.1110	.1505	.9456	.7377	1.093	1.4814	1.891	3.237	3.388	.9251
.1120	.1513	.9508	.7402	1.101	1.4871	1.902	3.274	3.423	.9245
.1130	.1522	.9563	.7426	1.109	1.4932	1.913	3.312	3.459	.9239
.1140	.1530	.9616	.7450	1.117	1.4990	1.923	3.348	3.494	.9234
.1150	.1539	.9670	.7474	1.125	1.5051	1.934	3.385	3.530	.9228
.1160	.1547	.9720	.7497	1.133	1.5108	1.944	3.423	3.566	.9223
.1170	.1556	.9775	.7520	1.141	1.5171	1.955	3.462	3.603	.9218
.1180	.1564	.9827	7543	1.149	1.5230	1.966	3.501	3.641	.9214
.1190	.1573	.9882	.7566	1.157	1.5293	1.977	3.540	3.678	.9209

Table of functions of d/L_0

APPENDIX (Continued)

d/L_o	d/L	$2\pi d/L$	tanh $2\pi d/L$	sinh $2\pi d/L$	cosh $2\pi d/L$	$4\pi d/L$	sinh $4\pi d/L$	cosh $4\pi d/L$	H/H'_o
.1200	.1581	.9936	.7589	1.165	1.5356	1.987	3.579	3.716	.9204
.1210	.1590	.9989	.7612	1.174	1.5418	1.998	3.620	3.755	.9200
.1220	.1598	1.004	.7634	1.182	1.5479	2.008	3.659	3.793	.9196
.1230	.1607	1.010	.7656	1.190	1.5546	2.019	3.699	3.832	.9192
.1240	.1615	1.015	.7678	1.198	1.5605	2.030	3.740	3.871	.9189
.1250	.1624	1.020	.7700	1.207	1.5674	2.041	3.782	3.912	.9186
.1260	.1632	1.025	.7721	1.215	1.5734	2.051	3.824	3.952	.9182
.1270	.1640	1.030	.7742	1.223	1.5795	2.061	3.865	3.992	.9178
.1280	.1649	1.036	.7763	1.231	1.5862	2.072	3.907	4.033	.9175
.1290	.1657	1.041	.7783	1.240	1.5927	2.082	3.950	4.074	.9172
.1300	.1665	1.046	.7804	1.248	1.5990	2.093	3.992	4.115	.9169
.1310	.1674	1.052	.7824	1.257	1.6060	2.104	4.036	4.158	.9166
.1320	.1682	1.057	.7844	1.265	1.6124	2.114	4.080	4.201	.9164
.1330	.1691	1.062	.7865	1.273	1.6191	2.125	4.125	4.245	.9161
.1340	.1699	1.068	.7885	1.282	1.6260	2.135	4.169	4.288	.9158
.1350	.1708	1.073	.7905	1.291	1.633	2.146	4.217	4.334	.9156
.1360	.1716	1.078	.7925	1.300	1.640	2.156	4.262	4.378	.9154
.1370	.1724	1.084	.7945	1.308	1.647	2.167	4.309	4.423	.9152
.1380	.1733	1.089	.7964	1.317	1.654	2.177	4.355	4.468	.9150
.1390	.1741	1.094	.7983	1.326	1.660	2.188	4.402	4.514	.9148
.1400	.1749	1.099	.8002	1.334	1.667	2.198	4.450	4.561	.9146
.1410	.1758	1.105	.8021	1.343	1.675	2.209	4.498	4.607	.9144
.1420	.1766	1.110	.8039	1.352	1.681	2.219	4.546	4.654	.9142
.1430	.1774	1.115	.8057	1.360	1.688	2.230	4.595	4.663	.9141
.1440	.1783	1.120	.8076	1.369	1.696	2.240	4.644	4.751	.9140
.1450	.1791	1.125	.8094	1.378	1.703	2.251	4.695	4.800	.9139
.1460	.1800	1.131	.8112	1.388	1.710	2.261	4.746	4.850	.9137
.1470	.1808	1.136	.8131	1.397	1.718	2.272	4.798	4.901	.9136
.1480	.1816	1.141	.8149	1.405	1.725	2.282	4.847	4.951	.9135
.1490	.1825	1.146	.8166	1.415	1.732	2.293	4.901	5.001	.9134
.1500	.1833	1.152	.8183	1.424	1.740	2.303	4.954	5.054	.9133
.1510	.1841	1.157	.8200	1.433	1.747	2.314	5.007	5.106	.9133
.1520	.1850	1.162	.8217	1.442	1.755	2.324	5.061	5.159	.9132
.1530	.1858	1.167	.8234	1.451	1.762	2.335	5.115	5.212	.9132
.1540	.1866	1.173	.8250	1.460	1.770	2.345	5.169	5.265	.9132
.1550	.1875	1.178	.8267	1.469	1.777	2.356	5.225	5.320	.9131
.1560	.1883	1.183	.8284	1.479	1.785	2.366	5.283	5.376	.9130
.1570	.1891	1.188	.8301	1.488	1.793	2.377	5.339	5.432	.9129
.1580	.1900	1.194	.8317	1.498	1.801	2.387	5.398	5.490	.9130
.1590	.1908	1.199	.8333	1.507	1.809	2.398	5.454	5.544	.9130
.1600	.1917	1.204	.8349	1.517	1.817	2.408	5.513	5.603	.9130
.1610	.1925	1.209	.8365	1.527	1.825	2.419	5.571	5.660	.9130
.1620	.1933	1.215	.8381	1.536	1.833	2.429	5.630	5.718	.9130
.1630	.1941	1.220	.8396	1.546	1.841	2.440	5.690	5.777	.9130
.1640	.1950	1.225	.8411	1.555	1.849	2.450	5.751	5.837	.9130
.1650	.1958	1.230	.8427	1.565	1.857	2.461	5.813	5.898	.9131
.1660	.1966	1.235	.8442	1.574	1.865	2.471	5.874	5.959	.9132
.1670	.1975	1.240	.8457	1.584	1.873	2.482	5.938	6.021	.9132
.1680	.1983	1.246	.8472	1.594	1.882	2.492	6.003	6.085	.9133
.1690	.1992	1.251	.8486	1.604	1.890	2.503	6.066	6.148	.9133
.1700	.2000	1.257	.8501	1.614	1.899	2.513	6.130	6.212	.9134
.1710	.2008	1.262	.8515	1.624	1.907	2.523	6.197	6.275	.9135
.1720	.2017	1.267	.8529	1.634	1.915	2.534	6.262	6.342	.9136
.1730	.2025	1.272	.8544	1.644	1.924	2.544	6.329	6.407	.9137
.1740	.2033	1.277	.8558	1.654	1.933	2.555	6.395	6.473	.9138

Appendix

APPENDIX (Continued)

d/L_o	d/L	$2\pi d/L$	tanh $2\pi d/L$	sinh $2\pi d/L$	cosh $2\pi d/L$	$4\pi d/L$	sinh $4\pi d/L$	cosh $4\pi d/L$	H/H_o
.1750	.2042	1.282	.8572	1.664	1.941	2.565	6.465	6.541	.9139
.1760	.2050	1.288	.8586	1.675	1.951	2.576	6.534	6.610	.9140
.1770	.2058	1.293	.8600	1.685	1.959	3.586	6.603	6.679	.9141
.1780	.2066	1.298	.8614	1.695	1.968	2.597	6.672	6.747	.9142
.1790	.2075	1.304	.8627	1.706	1.977	2.607	6.744	6.818	.9144
.1800	.2083	1.309	.8640	1.716	1.986	2.618	6.818	6.891	.9145
.1810	.2092	1.314	.8653	1.727	1.995	2.629	6.890	6.963	.9146
.1820	.2100	1.320	.8666	1.737	2.004	2.639	6.963	7.035	.9148
.1830	.2108	1.325	.8680	1.748	2.013	2.650	7.038	7.109	.9149
.1840	.2117	1.330	.8693	1.758	2.022	2.660	7.113	7.183	.9150
.1850	.2125	1.335	.8706	1.769	2.032	2.671	7.191	7.260	.9152
.1860	.2134	1.341	.8718	1.780	2.041	2.681	7.267	7.336	.9154
.1870	.2142	1.346	.8731	1.791	2.051	2.692	7.345	7.412	.9155
.1880	.2150	1.351	.8743	1.801	2.060	2.702	7.421	7.488	.9157
.1890	.2159	1.356	.8755	1.812	2.070	2.712	7.500	7.566	.9159
.1900	.2167	1.362	.8767	1.823	2.079	2.723	7.581	7.647	.9161
.1910	.2176	1.367	.8779	1.834	2.089	2.734	7.663	7.728	.9163
.1920	.2184	1.372	.8791	1.845	2.099	2.744	7.746	7.810	.9165
.1930	.2192	1.377	.8803	1.856	2.108	2.755	7.827	7.891	.9167
.1940	.2201	1.383	.8815	1.867	2.118	2.765	7.911	7.974	.9169
.1950	.2209	1.388	.8827	1.879	2.128	2.776	7.996	8.059	.9170
.1960	.2218	1.393	.8839	1.890	2.138	2.787	8.083	8.145	.9172
.1970	.2226	1.399	.8850	1.901	2.148	2.797	8.167	8.228	.9174
.1980	.2234	1.404	.8862	1.913	2.158	2.808	8.256	8.316	.9176
.1990	.2243	1.409	.8873	1.924	2.169	2.819	8.346	8.406	.9179
.2000	.2251	1.414	.8884	1.935	2.178	2.829	8.436	8.495	.9181
.2010	.2260	1.420	.8895	1.947	2.189	2.840	8.524	8.583	.9183
.2020	.2268	1.425	.8906	1.959	2.199	2.850	8.616	8.674	.9186
.2030	.2277	1.430	.8917	1.970	2.210	2.861	8.708	8.766	.9188
.2040	.2285	1.436	.8928	1.982	2.220	2.872	8.803	8.860	.9190
.2050	.2293	1.441	.8939	1.994	2.231	2.882	8.897	8.953	.9193
.2060	.2302	1.446	.8950	2.006	2.242	2.893	8.994	9.050	.9195
.2070	.2310	1.451	.8960	2.017	2.252	2.903	9.090	9.144	.9197
.2080	.2319	1.457	.8971	2.030	2.263	2.914	9.187	9.240	.9200
.2090	.2328	1.462	.8981	2.042	2.274	2.925	9.288	9.342	.9202
.2100	.2336	1.468	.8991	2.055	2.285	2.936	9.389	9.442	.9205
.2110	.2344	1.473	.9001	2.066	2.295	2.946	9.490	9.542	.9207
.2120	.2353	1.479	.9011	2.079	2.307	2.957	9.590	9.642	.9210
.2130	.2361	1.484	.9021	2.091	2.318	2.967	9.693	9.744	.9213
.2140	.2370	1.489	.9031	2.103	2.329	2.978	9.796	9.847	.9215
.2150	.2378	1.494	.9041	2.115	2.340	2.989	9.902	9.952	.9218
.2160	.2387	1.500	.9051	2.128	2.351	2.999	10.01	10.06	.9221
.2170	.2395	1.506	.9061	2.142	2.364	3.010	10.12	10.17	.9223
.2180	.2404	1.511	.9070	2.154	2.375	3.021	10.23	10.28	.9226
.2190	.2412	1.516	.9079	2.166	2.386	3.031	10.34	10.38	.9228
.2200	.2421	1.521	.9088	2.178	2.397	3.042	10.45	10.50	.9231
.2210	.2429	1.526	.9097	2.192	2.409	3.052	10.56	10.61	.9234
.2220	.2438	1.532	.9107	2.204	2.421	3.063	10.68	10.72	.9236
.2230	.2446	1.537	.9116	2.218	2.433	3.074	10.79	10.84	.9239
.2240	.2455	1.542	.9125	2.230	2.444	3.085	10.91	10.95	.9242
.2250	.2463	1.548	.9134	1.244	2.457	3.095	11.02	11.07	.9245
.2260	.2472	1.553	.9143	2.257	2.469	3.106	11.15	11.19	.9248
.2270	.2481	1.559	.9152	2.271	2.481	3.117	11.27	11.31	.9251
.2280	.2489	1.564	.9161	2.284	2.493	3.128	11.39	11.44	.9254
.2290	.2498	1.569	.9170	2.297	2.506	3.138	11.51	11.56	.9258

Table of functions of d/L_0

APPENDIX (Continued)

d/L_o	d/L	$2\pi d/L$	tanh $2\pi d/L$	sinh $2\pi d/L$	cosh $2\pi d/L$	$4\pi d/L$	sinh $4\pi d/L$	cosh $4\pi d/L$	H/H_o'
.2300	.2506	1.575	.9178	2.311	2.518	3.149	11.64	11.68	.9261
.2310	.2515	1.580	.9186	2.325	2.531	3.160	11.77	11.81	.9264
.2320	.2523	1.585	.9194	2.338	2.543	3.171	11.90	11.93	.9267
.2330	.2532	1.591	.9203	2.352	2.556	3.182	12.03	12.07	.9270
.2340	.2540	1.596	.9211	2.366	2.569	3.192	12.15	12.19	.9273
.2350	.2549	1.602	.9219	2.380	2.581	3.203	12.29	12.33	.9276
.2360	.2558	1.607	.9227	2.393	2.594	3.214	12.43	12.47	.9279
.2370	.2566	1.612	.9235	2.408	2.607	3.225	12.55	12.59	.9282
.2380	.2575	1.618	.9243	2.422	2.620	3.236	12.69	12.73	.9285
.2390	.2584	1.623	.9251	2.436	2.634	3.247	12.83	12.87	.9288
.2400	.2592	1.629	.9259	2.450	2.647	3.257	12.97	13.01	.9291
.2410	.2601	1.634	.9267	2.464	2.660	3.268	13.11	13.15	.9294
.2420	.2610	1.640	.9275	2.480	2.674	3.279	13.26	13.30	.9298
.2430	.2618	1.645	.9282	2.494	2.687	3.290	13.40	13.44	.9301
.2440	.2627	1.650	.9289	2.508	2.700	3.301	13.55	13.59	.9304
.2450	.2635	1.656	.9296	2.523	2.714	3.312	13.70	13.73	.9307
.2460	.2644	1.661	.9304	2.538	2.728	3.323	13.85	13.88	.9310
.2470	.2653	1.667	.9311	2.553	2.742	3.334	14.00	14.04	.9314
.2480	.2661	1.672	.9318	2.568	2.755	3.344	14.15	14.19	.9317
.2490	.2670	1.678	.9325	2.583	2.770	3.355	14.31	14.35	.9320
.2500	.2679	1.683	.9332	2.599	2.784	3.367	14.47	14.51	.9323
.2510	.2687	1.689	.9339	2.614	2.798	3.377	14.62	14.66	.9327
.2520	.2696	1.694	.9346	2.629	2.813	3.388	14.79	14.82	.9330
.2530	.2705	1.700	.9353	2.645	2.828	3.399	14.95	14.99	.9333
.2540	.2714	1.705	.9360	2.660	2.842	3.410	15.12	15.15	.9336
.2550	.2722	1.711	.9367	2.676	2.856	3.421	15.29	15.32	.9340
.2560	.2731	1.716	.9374	2.691	2.871	3.432	15.45	15.49	.9343
.2570	.2740	1.722	.9381	2.707	2.886	3.443	15.63	15.66	.9346
.2580	.2749	1.727	.9388	2.723	2.901	3.454	15.80	15.83	.9349
.2590	.2757	1.732	.9394	2.739	2.916	3.465	15.97	16.00	.9353
.2600	.2766	1.738	.9400	2.755	2.931	3.476	16.15	16.18	.9356
.2610	.2775	1.744	.9406	2.772	2.946	3.487	16.33	16.36	.9360
.2620	.2784	1.749	.9412	2.788	2.962	3.498	16.51	16.54	.9363
.2630	.2792	1.755	.9418	2.804	2.977	3.509	16.69	16.73	.9367
.2640	.2801	1.760	.9425	2.820	2.992	3.520	16.88	16.91	.9370
.2650	.2810	1.766	.9431	2.837	3.008	3.531	17.07	17.10	.9373
.2660	.2819	1.771	.9437	2.853	3.023	3.542	17.26	17.28	.9377
.2670	.2827	1.776	.9443	2.870	3.039	3.553	17.45	17.45	.9380
.2680	.2836	1.782	.9449	2.886	3.055	3.564	17.64	17.67	.9383
.2690	.2845	1.788	.9455	2.904	3.071	3.575	17.84	17.87	.9386
.2700	.2854	1.793	.9461	2.921	3.088	3.587	18.04	18.07	.9390
.2710	.2863	1.799	.9467	2.938	3.104	3.598	18.24	18.27	.9393
.2720	.2872	1.804	.9473	2.956	3.120	3.610	18.46	18.49	.9396
.2730	.2880	1.810	.9478	2.973	3.136	3.620	18.65	18.67	.9400
.2740	.2889	1.815	.9484	2.990	3.153	3.631	18.86	18.89	.9403
.2750	.2898	1.821	.9490	3.008	3.170	3.642	19.07	19.10	.9406
.2760	.2907	1.826	.9495	3.025	3.186	3.653	19.28	19.30	.9410
.2770	.2916	1.832	.9500	3.043	3.203	3.664	19.49	19.51	.9413
.2780	.2924	1.837	.9505	3.061	3.220	3.675	19.71	19.74	.9416
.2790	.2933	1.843	.9511	3.079	3.237	3.686	19.93	19.96	.9420
.2800	.2942	1.849	.9516	3.097	3.254	3.697	20.16	20.18	.9423
.2810	.2951	1.854	.9521	3.115	3.272	3.709	20.39	20.41	.9426
.2820	.2960	1.860	.9526	3.133	3.289	3.720	20.62	20.64	.9430
.2830	.2969	1.866	.9532	3.152	3.307	3.731	20.85	20.87	.9433
.2840	.2978	1.871	.9537	3.171	3.325	3.742	21.09	21.11	.9436

APPENDIX (Continued)

d/L_o	d/L	$2\pi d/L$	tanh $2\pi d/L$	sinh $2\pi d/L$	cosh $2\pi d/L$	$4\pi d/L$	sinh $4\pi d/L$	cosh $4\pi d/L$	H/H_o'
.2850	.2987	1.877	.9542	3.190	3.343	3.754	21.33	21.35	.9440
.2860	.2996	1.882	.9547	3.209	3.361	3.765	21.57	21.59	.9443
.2870	.3005	1.888	.9552	3.228	3.379	3.776	21.82	21.84	.9446
.2880	.3014	1.893	.9557	3.246	3.396	3.787	22.05	22.07	.9449
.2890	.3022	1.899	.9562	3.264	3.414	3.798	22.30	22.32	.9452
.2900	.3031	1.905	.9567	3.284	3.433	3.809	22.54	22.57	.9456
.2910	.3040	1.910	.9572	3.303	3.451	3.821	22.81	22.83	.9459
.2920	.3049	1.916	.9577	3.323	3.471	3.832	23.07	23.09	.9463
.2930	.3058	1.922	.9581	3.343	3.490	3.843	23.33	23.35	.9466
.2940	.3067	1.927	.9585	3.362	3.508	3.855	23.60	23.62	.9469
.2950	.3076	1.933	.9590	3.382	3.527	3.866	23.86	23.88	.9473
.2960	.3085	1.938	.9594	3.402	3.546	3.877	24.12	24.15	.9476
.2970	.3094	1.944	.9599	3.422	3.565	3.888	24.40	24.42	.9480
.2980	.3103	1.950	.9603	3.442	3.585	3.900	24.68	24.70	.9483
.2990	.3112	1.955	.9607	3.462	3.604	3.911	24.96	24.98	.9486
.3000	.3121	1.961	.9611	3.483	3.624	3.922	25.24	25.26	.9490
.3010	.3130	1.967	.9616	3.503	3.643	3.933	25.53	25.55	.9493
.3020	.3139	1.972	.9620	3.524	3.663	3.945	25.82	25.83	.9496
.3030	.3148	1.978	.9624	3.545	3.683	3.956	26.12	26.14	.9499
.3040	.3157	1.984	.9629	3.566	3.703	3.968	26.42	26.44	.9502
.3050	.3166	1.989	.9633	3.587	3.724	3.979	26.72	26.74	.9505
.3060	.3175	1.995	.9637	3.609	3.745	3.990	27.02	27.04	.9509
.3070	.3184	2.001	.9641	3.630	3.765	4.002	27.33	27.35	.9512
.3080	.3193	2.007	.9645	3.651	3.786	4.013	27.65	27.66	.9515
.3090	.3202	2.012	.9649	3.673	3.806	4.024	27.96	27.98	.9518
.3100	.3211	2.018	.9653	3.694	3.827	4.036	28.28	28.30	.9522
.3110	.3220	2.023	.9656	3.716	3.848	4.047	28.60	28.62	.9525
.3120	.3230	2.029	.9660	3.738	3.870	4.058	28.93	28.95	.9528
.3130	.3239	2.035	.9664	3.760	3.891	4.070	29.27	29.28	.9531
.3140	.3248	2.041	.9668	3.782	3.912	4.081	29.60	29.62	.9535
.3150	.3257	2.046	.9672	3.805	3.934	4.093	29.94	29.96	.9538
.3160	.3266	2.052	.9676	3.828	3.956	4.104	30.29	30.31	.9541
.3170	.3275	2.058	.9679	3.851	3.978	4.116	30.64	30.65	.9544
.3180	.3284	2.063	.9682	3.873	4.000	4.127	30.99	31.00	.9547
.3190	.3294	2.069	.9686	3.896	4.022	4.139	31.35	31.37	.9550
.3200	.3302	2.075	.9690	3.919	4.045	4.150	31.71	31.72	.9553
.3210	.3311	2.081	.9693	3.943	4.068	4.161	32.07	32.08	.9556
.3220	.3321	2.086	.9696	3.966	4.090	4.173	32.44	32.46	.9559
.3230	.3330	2.092	.9700	3.990	4.114	4.185	32.83	32.84	.9562
.3240	.3339	2.098	.9703	4.014	4.136	4.196	33.20	33.22	.9565
.3250	.3349	2.104	.9707	4.038	4.160	4.208	33.60	33.61	.9568
.3260	.3357	2.110	.9710	4.061	4.183	4.219	33.97	33.99	.9571
.3270	.3367	2.115	.9713	4.085	4.206	4.231	34.37	34.38	.9574
.3280	.3376	2.121	.9717	4.110	4.230	4.242	34.77	34.79	.9577
.3290	.3385	2.127	.9720	4.135	4.254	4.254	35.18	35.19	.9580
.3300	.3394	2.133	.9723	4.159	4.277	4.265	35.58	35.59	.9583
.3310	.3403	2.138	.9726	4.184	4.301	4.277	35.99	36.00	.9586
.3320	.3413	2.144	.9729	4.209	4.326	4.288	36.42	36.43	.9589
.3330	.3422	2.150	.9732	4.234	4.350	4.300	36.84	36.85	.9592
.3340	.3431	2.156	.9735	4.259	4.375	4.311	37.25	37.27	.9595
.3350	.3440	2.161	.9738	4.284	4.399	4.323	37.70	37.72	.9598
.3360	.3449	2.167	.9741	4.310	4.424	4.335	38.14	38.15	.9601
.3370	.3459	2.173	.9744	4.336	4.450	4.346	38.59	38.60	.9604
.3380	.3468	2.179	.9747	4.361	4.474	4.358	39.02	39.04	.9607
.3390	.3477	2.185	.9750	4.388	4.500	4.369	39.48	39.49	.9610

Table of functions of d/L_0

APPENDIX (Continued)

d/L_o	d/L	$2\pi d/L$	tanh $2\pi d/L$	sinh $2\pi d/L$	cosh $2\pi d/L$	$4\pi d/L$	sinh $4\pi d/L$	cosh $4\pi d/L$	H/H_o'
.3400	.3468	2.190	.9753	4.413	4.525	4.381	39.95	39.96	.9613
.3410	.3495	2.196	.9756	4.439	4.550	4.392	40.40	40.41	.9615
.3420	.3504	2.202	.9758	4.466	4.576	4.404	40.87	40.89	.9618
.3430	.3514	2.208	.9761	4.492	4.602	4.416	41.36	41.37	.9621
.3440	.3523	2.214	.9764	4.521	4.630	4.427	41.85	41.84	.9623
.3450	.3532	2.220	.9767	4.547	4.656	4.439	42.33	42.34	.9626
.3460	.3542	2.225	.9769	4.575	4.682	4.451	42.83	42.84	.9629
.3470	.3551	2.231	.9772	4.602	4.709	4.462	43.34	43.35	.9632
.3480	.3560	2.237	.9775	4.629	4.736	4.474	43.85	43.86	.9635
.3490	.3570	2.243	.9777	4.657	4.763	4.486	44.37	44.40	.9638
.3500	.3579	2.249	.9780	4.685	4.791	4.498	44.89	44.80	.9640
.3510	.3588	2.255	.9782	4.713	4.818	4.509	45.42	45.43	.9643
.3520	.3598	2.260	.9785	4.741	4.845	4.521	45.95	45.96	.9646
.3530	.3607	2.266	.9787	4.770	4.873	4.533	46.50	46.51	.9648
.3540	.3616	2.272	.9790	4.798	4.901	4.544	47.03	47.04	.9651
.3550	.3625	2.278	.9792	4.827	4.929	4.556	47.59	47.60	.9654
.3560	.3635	2.284	.9795	4.856	4.957	4.568	48.15	48.16	.9657
.3570	.3644	2.290	.9797	4.885	4.987	4.579	48.72	48.73	.9659
.3580	.3653	2.296	.9799	4.914	5.015	4.591	49.29	49.30	.9662
.3590	.3663	2.301	.9801	4.944	5.044	4.603	49.88	49.89	.9665
.3600	.3672	2.307	.9804	4.974	5.072	4.615	50.47	50.48	.9667
.3610	.3682	2.313	.9806	5.004	5.103	4.627	51.08	51.09	.9670
.3620	.3691	2.319	.9808	5.034	5.132	4.638	51.67	51.67	.9673
.3630	.3700	2.325	.9811	5.063	5.161	4.650	52.27	52.28	.9675
.3640	.3709	2.331	.9813	5.094	5.191	4.661	52.89	52.90	.9677
.3650	.3719	2.337	.9815	5.124	5.221	4.673	53.52	53.53	.9680
.3660	.3728	2.342	.9817	5.155	5.251	4.685	54.15	54.16	.9683
.3670	.3737	2.348	.9819	5.186	5.281	4.697	54.78	54.79	.9686
.3680	.3747	2.354	.9821	5.217	5.312	4.708	55.42	55.43	.9688
.3690	.3756	2.360	.9823	5.248	5.343	4.720	56.09	56.10	.9690
.3700	.3766	2.366	.9825	5.280	5.374	4.732	56.76	56.77	.9693
.3710	.3775	2.372	.9827	5.312	5.406	4.744	57.43	57.44	.9696
.3720	.3785	2.378	.9830	5.345	5.438	4.756	58.13	58.14	.9698
.3730	.3794	2.384	.9832	5.377	5.469	4.768	58.82	58.83	.9700
.3740	.3804	2.390	.9834	5.410	5.502	4.780	59.52	59.53	.9702
.3750	.3813	2.396	.9835	5.443	5.534	4.792	60.24	60.25	.9705
.3760	.3822	2.402	.9837	5.475	5.566	4.803	60.95	60.95	.9707
.3770	.3832	2.408	.9839	5.508	5.598	4.815	61.68	61.68	.9709
.3780	.3841	2.413	.9841	5.541	5.631	4.827	62.41	62.42	.9712
.3790	.3850	2.419	.9843	5.572	5.661	4.838	63.13	63.14	.9714
.3800	.3860	2.425	.9845	5.609	5.697	4.851	63.90	63.91	.9717
.3810	.3869	2.431	.9847	5.643	5.731	4.862	64.66	64.67	.9719
.3820	.3879	2.437	.9848	5.677	5.765	4.875	65.45	65.46	.9721
.3830	.3888	2.443	.9850	5.712	5.798	4.885	66.20	66.21	.9724
.3840	.3898	2.449	.9852	5.746	5.833	4.898	67.00	67.01	.9726
.3850	.3907	2.455	.9854	5.780	5.866	4.910	67.80	67.81	.9728
.3860	.3917	2.461	.9855	5.814	5.900	4.922	68.61	68.62	.9730
.3870	.3926	2.467	.9857	5.850	5.935	4.934	69.45	69.46	.9732
.3880	.3936	2.473	.9859	5.886	5.970	4.946	70.28	70.29	.9735
.3890	.3945	2.479	.9860	5.921	6.005	4.958	71.12	71.13	.9737
.3900	.3955	2.485	.9862	5.957	6.040	4.970	71.97	71.98	.9739
.3910	.3964	2.491	.9864	5.993	6.076	4.982	72.85	72.86	.9741
.3920	.3974	2.497	.9865	6.029	6.112	4.993	73.72	73.72	.9743
.3930	.3983	2.503	.9867	6.066	6.148	5.005	74.59	74.59	.9745
.3940	.3993	2.509	.9869	6.103	6.185	5.017	75.48	75.48	.9748

APPENDIX (Continued)

d/L_o	d/L	$2\pi d/L$	tanh $2\pi d/L$	sinh $2\pi d/L$	cosh $2\pi d/L$	$4\pi d/L$	sinh $4\pi d/L$	cosh $4\pi d/L$	H/H_o'
.3950	.4002	2.515	.9870	6.140	6.221	5.029	76.40	76.40	.9750
.3960	.4012	2.521	.9872	6.177	6.258	5.041	77.32	77.32	.9752
.3970	.4021	2.527	.9873	6.215	6.295	5.053	78.24	78.24	.9754
.3980	.4031	2.532	.9874	6.252	6.332	5.065	79.19	79.19	.9756
.3990	.4040	2.538	.9876	6.290	6.369	5.077	80.13	80.13	.9758
.4000	.4050	2.544	.9877	6.329	6.407	5.089	81.12	81.12	.9761
.4010	.4059	2.550	.9879	6.367	6.445	5.101	82.08	82.08	.9763
.4020	.4069	2.556	.9880	6.406	6.483	5.113	83.06	83.06	.9765
.4030	.4078	2.562	.9882	6.444	6.521	5.125	84.07	84.07	.9766
.4040	.4088	2.568	.9883	6.484	6.561	5.137	85.11	85.11	.9768
.4050	.4098	2.575	.9885	6.525	6.601	5.149	86.14	86.14	.9777
.4060	.4107	2.581	.9886	6.564	6.640	5.161	87.17	87.17	.9772
.4070	.4116	2.586	.9887	6.603	6.679	5.173	88.20	88.20	.9774
.4080	.4126	2.592	.9889	6.644	6.718	5.185	89.28	89.28	.9776
.4090	.4136	2.598	.9890	6.684	6.758	5.197	90.39	90.39	.9778
.4100	.4145	2.604	.9891	6.725	6.799	5.209	91.44	91.44	.9780
.4110	.4155	2.610	.9892	6.766	6.839	5.221	92.55	92.55	.9782
.4120	.4164	2.616	.9894	6.806	6.879	5.233	93.67	93.67	.9784
.4130	.4174	2.623	.9895	6.849	6.921	5.245	94.83	94.83	.9786
.4140	.4183	2.629	.9896	6.890	6.963	5.257	95.96	95.96	.9788
.4150	.4193	2.635	.9898	6.932	7.004	5.269	97.13	97.13	.9790
.4160	.4203	2.641	.9899	6.974	7.046	5.281	98.30	98.30	.9792
.4170	.4212	2.647	.9900	7.018	7.088	5.294	99.52	99.52	.9794
.4180	.4222	2.653	.9901	7.060	7.130	5.305	100.7	100.7	.9795
.4190	.4231	2.659	.9902	7.102	7.173	5.317	101.9	101.9	.9797
.4200	.4241	2.665	.9904	7.146	7.215	5.329	103.1	103.1	.9798
.4210	.4251	2.671	.9905	7.190	7.259	5.341	104.4	104.4	.9800
.4220	.4260	2.677	.9906	7.234	7.303	5.353	105.7	105.7	.9802
.4230	.4270	2.683	.9907	7.279	7.349	5.366	107.0	107.0	.9804
.4240	.4280	2.689	.9908	7.325	7.392	5.378	108.3	108.3	.9806
.4250	.4289	2.695	.9909	7.371	7.438	5.390	109.7	109.7	.9808
.4260	.4298	2.701	.9910	7.412	7.479	5.402	110.9	110.9	.9810
.4270	.4308	2.707	.9911	7.457	7.524	5.414	112.2	112.2	.9811
.4280	.4318	2.713	.9912	7.503	7.570	5.426	113.6	113.6	.9812
.4290	.4328	2.719	.9913	7.550	7.616	5.438	115.0	115.0	.9814
.4300	.4337	2.725	.9914	7.595	7.661	5.450	116.4	116.4	.9816
.4310	.4347	2.731	.9915	7.642	7.707	5.462	117.8	117.8	.9818
.4320	.4356	2.737	.9916	7.688	7.753	5.474	119.2	119.2	.9819
.4330	.4366	2.743	.9917	7.735	7.800	5.486	120.7	120.7	.9821
.4340	.4376	2.749	.9918	7.783	7.847	5.499	122.2	122.2	.9823
.4350	.4385	2.755	.9919	7.831	7.895	5.511	123.7	123.7	.9824
.4360	.4395	2.762	.9920	7.880	7.943	5.523	125.2	125.2	.9826
.4370	.4405	2.768	.9921	7.922	7.991	5.535	126.7	126.7	.9828
.4380	.4414	2.774	.9922	7.975	8.035	5.547	128.3	128.3	.9829
.4390	.4424	2.780	.9923	8.026	8.088	5.560	129.9	129.9	.9830
.4400	.4434	2.786	.9924	8.075	8.136	5.572	131.4	131.4	.9832
.4410	.4443	2.792	.9925	8.124	8.185	5.584	133.0	133.0	.9833
.4420	.4453	2.798	.9926	8.175	8.236	5.596	134.7	134.7	.9835
.4430	.4463	2.804	.9927	8.228	8.285	5.608	136.3	136.3	.9836
.4440	.4472	2.810	.9928	8.274	8.334	5.620	137.9	137.9	.9838
.4450	.4482	2.816	.9929	8.326	8.387	5.632	139.6	139.6	.9839
.4460	.4492	2.822	.9930	8.379	8.438	5.644	141.4	141.4	.9841
.4470	.4501	2.828	.9930	8.427	8.486	5.657	143.1	143.1	.9843
.4480	.4511	2.834	.9931	8.481	8.540	5.669	144.8	144.8	.9844
.4490	.4521	2.840	.9932	8.532	8.590	5.681	146.6	146.6	.9846

Table of functions of d/L_0

APPENDIX (Continued)

d/L_o	d/L	$2\pi d/L$	tanh $2\pi d/L$	sinh $2\pi d/L$	cosh $2\pi d/L$	$4\pi d/L$	sinh $4\pi d/L$	cosh $4\pi d/L$	H/H_o'
.4500	.4531	2.847	.9933	8.585	8.643	5.693	148.4	148.4	.9847
.4510	.4540	2.853	.9934	8.638	8.695	5.705	150.2	150.2	.9848
.4520	.4550	2.859	.9935	8.693	8.750	5.717	152.1	152.1	.9849
.4530	.4560	2.865	.9935	8.747	8.804	5.730	154.0	154.0	.9851
.4540	.4569	2.871	.9936	8.797	8.854	5.742	155.9	155.9	.9852
.4550	.4579	2.877	.9937	8.853	8.910	5.754	157.7	157.7	.9853
.4560	.4589	2.883	.9938	8.910	8.965	5.766	159.7	159.7	.9855
.4570	.4599	2.890	.9938	8.965	9.021	5.779	161.7	161.7	.9857
.4580	.4608	2.896	.9939	9.016	9.072	5.791	163.6	163.6	.9858
.4590	.4618	2.902	.9940	9.074	9.129	5.803	165.6	165.6	.9859
.4600	.4628	2.908	.9941	9.132	9.186	5.815	167.7	167.7	.9860
.4610	.4637	2.914	.9941	9.183	9.238	5.827	169.7	169.7	.9862
.4620	.4647	2.920	.9942	9.242	9.296	5.840	171.8	171.8	.9863
.4630	.4657	2.926	.9943	9.301	9.354	5.852	173.9	173.9	.9864
.4640	.4666	2.932	.9944	9.353	9.406	5.864	176.0	176.0	.9865
.4650	.4676	2.938	.9944	9.413	9.466	5.876	178.2	178.2	.9867
.4660	.4686	2.944	.9945	9.472	9.525	5.888	180.4	180.4	.9868
.4670	.4695	2.951	.9946	9.533	9.585	5.900	182.6	182.6	.9869
.4680	.4705	2.957	.9946	9.586	9.638	5.912	184.8	184.8	.9871
.4690	.4715	2.963	.9947	9.647	9.699	5.925	187.2	187.2	.9872
.4700	.4725	2.969	.9947	9.709	9.760	5.937	189.5	189.5	.9873
.4710	.4735	2.975	.9948	9.770	9.821	5.949	191.8	191.8	.9874
.4720	.4744	2.981	.9949	9.826	9.877	5.962	194.2	194.2	.9875
.4730	.4754	2.987	.9949	9.888	9.938	5.974	196.5	196.5	.9876
.4740	.4764	2.993	.9950	9.951	10.00	5.986	199.0	199.0	.9877
.4750	.4774	2.999	.9951	10.01	10.07	5.999	201.4	201.4	.9878
.4760	.4783	3.005	.9951	10.07	10.12	6.011	203.9	203.9	.9880
.4770	.4793	3.012	.9952	10.13	10.18	6.023	206.5	206.5	.9881
.4780	.4803	3.018	.9952	10.20	10.25	6.036	209.0	209.0	.9882
.4790	.4813	3.024	.9953	10.26	10.31	6.048	211.7	211.7	.9883
.4800	.4822	3.030	.9953	10.32	10.37	6.060	214.2	214.2	.9885
.4810	.4832	3.036	.9954	10.39	10.43	6.072	216.8	216.8	.9886
.4820	.4842	3.042	.9955	10.45	10.50	6.085	219.5	219.5	.9887
.4830	.4852	3.049	.9955	10.52	10.57	6.097	222.2	222.2	.9888
.4840	.4862	3.055	.9956	10.59	10.63	6.109	225.0	225.0	.9889
.4850	.4871	3.061	.9956	10.65	10.69	6.121	228.3	228.3	.9890
.4860	.4881	3.067	.9957	10.71	10.76	6.134	230.6	230.6	.9891
.4870	.4891	3.073	.9957	10.78	10.83	6.146	233.5	233.5	.9892
.4880	.4901	3.079	.9958	10.85	10.90	6.159	236.4	236.4	.9893
.4890	.4911	3.086	.9958	10.92	10.96	6.171	239.6	239.6	.9895
.4900	.4920	3.092	.9959	10.99	11.03	6.183	242.3	242.3	.9896
.4910	.4930	3.098	.9959	11.05	11.09	6.195	245.2	245.2	.9897
.4920	.4940	3.104	.9960	11.12	11.16	6.208	248.3	248.3	.9898
.4930	.4950	3.110	.9960	11.19	11.24	6.220	251.3	251.3	.9899
.4940	.4960	3.117	.9961	11.26	11.31	6.232	254.5	254.5	.9899
.4950	.4969	3.122	.9961	11.32	11.37	6.245	257.6	257.6	.9900
.4960	.4979	3.128	.9962	11.40	11.44	6.257	260.8	260.8	.9901
.4970	.4989	3.135	.9962	11.47	11.51	6.269	264.0	264.0	.9902
.4980	.4999	3.141	.9963	11.54	11.59	5.282	267.3	267.3	.9903
.4990	.5009	3.147	.9963	11.61	11.65	6.294	270.6	270.6	.9904
.5000	.5018	3.153	.9964	11.68	11.72	6.306	274.0	274.0	.9905
.5010	.5028	3.159	.9964	11.75	11.80	6.319	277.5	277.5	.9906
.5020	.5038	3.166	.9964	11.83	11.87	6.331	280.8	280.8	.9907
.5030	.5048	3.172	.9965	11.91	11.95	6.343	284.3	284.3	.9908
.5040	.5058	3.178	.9965	11.98	12.02	5.356	287.9	287.9	.9909

APPENDIX (Continued)

d/L_o	d/L	$2\pi d/L$	tanh $2\pi d/L$	sinh $2\pi d/L$	cosh $2\pi d/L$	$4\pi d/L$	sinh $4\pi d/L$	cosh $4\pi d/L$	H/H_o'
.5050	.5067	3.184	.9966	12.05	12.09	6.368	291.4	291.4	.9909
.5060	.5077	3.190	.9966	12.12	12.16	6.380	295.0	295.0	.9910
.5070	.5087	3.196	.9967	12.20	12.24	6.393	298.7	298.7	.9911
.5080	.5097	3.203	.9967	12.28	12.32	6.405	302.4	302.4	.9912
.5090	.5107	3.209	.9968	12.35	12.39	6.417	306.2	306.2	.9913
.5100	.5117	3.215	.9968	12.43	12.47	6.430	310.0	310.0	.9914
.5110	.5126	3.221	.9968	12.50	12.54	6.442	313.8	313.8	.9915
.5120	.5136	3.227	.9969	12.58	12.62	6.454	317.7	317.7	.9915
.5130	.5146	3.233	.9969	12.66	12.70	6.467	321.7	321.7	.9916
.5140	.5156	3.240	.9970	12.74	12.78	6.479	325.7	325.7	.9917
.5150	.5166	3.246	.9970	12.82	12.86	6.491	329.7	329.7	.9918
.5160	.5176	3.252	.9970	12.90	12.94	6.504	333.8	333.8	.9919
.5170	.5185	3.258	.9971	12.98	13.02	6.516	337.9	337.9	.9919
.5180	.5195	3.264	.9971	13.06	13.10	6.529	342.2	342.2	.9920
.5190	.5205	3.270	.9971	13.14	13.18	6.541	346.4	346.4	.9921
.5200	.5215	3.277	.9972	13.22	13.26	6.553	350.7	350.7	.9922
.5210	.5225	3.283	.9972	13.31	13.35	6.566	355.1	355.1	.9923
.5220	.5235	3.289	.9972	13.39	13.43	6.578	359.6	359.6	.9924
.5230	.5244	3.295	.9973	13.47	13.51	6.590	364.0	364.0	.9924
.5240	.5254	3.301	.9973	13.55	13.59	6.603	368.5	368.5	.9925
.5250	.5264	3.308	.9973	13.64	13.68	6.615	373.1	373.1	.9926
.5260	.5274	3.314	.9974	13.73	13.76	6.628	377.8	377.8	.9927
.5270	.5284	3.320	.9974	13.81	13.85	6.640	382.5	382.5	.9927
.5280	.5294	3.326	.9974	13.90	13.94	6.652	387.3	387.3	.9928
.5290	.5304	3.333	.9975	13.99	14.02	6.665	392.2	392.2	.9929
.5300	.5314	3.339	.9975	14.07	14.10	6.677	397.0	397.0	.9930
.5310	.5323	3.345	.9975	14.16	14.19	6.690	402.0	402.0	.9931
.5320	.5333	3.351	.9976	14.25	14.28	6.702	406.9	406.9	.9931
.5330	.5343	3.357	.9976	14.34	14.37	6.714	412.0	412.0	.9932
.5340	.5353	3.363	.9976	14.43	14.46	6.727	417.2	417.2	.9933
.5350	.5363	3.370	.9976	14.52	14.55	7.639	422.4	422.4	.9933
.5360	.5373	3.376	.9977	14.61	14.64	6.752	427.7	427.7	.9934
.5370	.5383	3.382	.9977	14.70	14.73	6.764	433.1	433.1	.9935
.5380	.5393	3.388	.9977	14.79	14.82	6.776	438.5	438.5	.9935
.5390	.5402	3.394	.9977	14.88	14.91	6.789	444.0	444.0	.9936
.5400	.5412	3.401	.9978	14.97	15.01	6.801	449.5	449.5	.9936
.5410	.5422	3.407	.9978	15.07	15.18	6.814	455.1	455.1	.9937
.5420	.5432	3.413	.9978	15.16	15.19	6.826	460.7	460.7	.9938
.5430	.5442	3.419	.9979	15.25	15.29	6.838	466.4	466.4	.9938
.5440	.5452	3.426	.9979	15.35	15.38	6.851	472.2	472.2	.9939
.5450	.5461	3.432	.9979	15.45	15.48	6.863	478.1	478.1	.9940
.5460	.5471	3.438	.9979	15.54	15.58	6.876	484.3	484.3	.9941
.5470	.5481	3.444	.9980	15.64	15.67	6.888	490.3	490.3	.9941
.5480	.5491	3.450	.9980	15.74	15.77	6.901	496.4	496.4	.9942
.5490	.5501	3.456	.9980	15.84	15.87	6.913	502.5	502.5	.9942
.5500	.5511	3.463	.9980	15.94	15.97	6.925	508.7	508.7	.9942
.5510	.5521	3.469	.9981	16.04	16.07	6.937	515.0	515.0	.9942
.5520	.5531	3.475	.9981	16.14	16.17	6.950	521.6	521.6	.9943
.5530	.5541	3.481	.9981	16.24	16.27	6.962	528.1	528.1	.9944
.5540	.5551	3.488	.9981	16.34	16.37	6.975	534.8	534.8	.9944
.5550	.5560	3.494	.9982	16.44	16.47	6.987	541.4	541.4	.9945
.5560	.5570	3.500	.9982	16.54	16.57	7.000	548.1	548.1	.9945
.5570	.5580	3.506	.9982	16.65	16.68	7.012	554.9	554.9	.9946
.5580	.5590	3.512	.9982	16.75	16.78	7.025	562.0	562.0	.9947
.5590	.5600	3.519	.9982	16.85	16.88	7.037	569.1	569.1	.9947

Table of functions of d/L_0

APPENDIX (Continued)

d/L_0	d/L	$2\pi d/L$	tanh $2\pi d/L$	sinh $2\pi d/L$	cosh $2\pi d/L$	$4\pi d/L$	sinh $4\pi d/L$	cosh $4\pi d/L$	H/H_0'
.5600	.5610	3.525	.9983	16.96	16.99	7.050	576.1	576.1	.9947
.5610	.5620	3.531	.9983	17.06	17.09	7.062	583.3	583.3	.9948
.5620	.5630	3.537	.9983	17.17	17.20	7.074	590.7	590.7	.9949
.5630	.5640	3.543	.9983	17.28	17.31	7.087	598.0	598.0	.9949
.5640	.5649	3.550	.9984	17.38	17.41	7.099	605.0	605.0	.9950
.5650	.5659	3.556	.9984	17.49	17.52	7.112	613.2	613.2	.9950
.5660	.5669	3.562	.9984	17.60	17.63	7.124	620.8	620.8	.9951
.5670	.5679	3.568	.9984	17.71	17.74	7.136	628.5	628.5	.9951
.5680	.5689	3.575	.9984	17.82	17.85	7.149	636.4	636.4	.9952
.5690	.5699	3.581	.9985	17.94	17.97	7.161	644.3	644.3	.9952
.5700	.5709	3.587	.9985	18.05	18.08	7.174	652.4	652.4	.9953
.5710	.5719	3.593	.9985	18.16	18.19	7.186	660.5	660.5	.9953
.5720	.5729	3.600	.9985	18.28	18.31	7.199	668.8	668.8	.9954
.5730	.5738	3.606	.9985	18.39	18.42	7.211	677.2	677.2	.9954
.5740	.5748	3.612	.9985	18.50	18.53	7.224	685.6	685.6	.9955
.5750	.5758	3.618	.9986	18.62	18.64	7.236	694.3	694.3	.9955
.5760	.5768	3.624	.9986	18.73	18.76	7.249	703.2	703.2	.9956
.5770	.5778	3.630	.9986	18.85	18.88	7.261	711.9	711.9	.9956
.5780	.5788	3.637	.9986	18.97	19.00	7.274	720.8	720.8	.9957
.5790	.5798	3.643	.9986	19.09	19.12	7.286	729.9	729.9	.9957
.5800	.5808	3.649	.9987	19.21	19.24	7.298	739.0	739.0	.9957
.5810	.5818	3.656	.9987	19.33	19.36	7.311	748.1	748.1	.9955
.5820	.5828	3.662	.9987	19.45	19.48	7.323	757.5	757.5	.9958
.5830	.5838	3.668	.9987	19.58	19.60	7.336	767.0	767.0	.9959
.5840	.5848	3.674	.9987	19.70	19.73	7.348	776.7	776.7	.9959
.5850	.5858	3.680	.9987	19.81	19.84	7.361	786.5	786.5	.9960
.5860	.5867	3.686	.9987	19.94	19.96	7.373	796.4	796.4	.9960
.5870	.5877	3.693	.9988	20.06	20.09	7.386	806.5	806.5	.9960
.5880	.5887	3.699	.9988	20.19	20.21	7.398	816.5	816.5	.9961
.5890	.5897	3.705	.9988	20.32	20.34	7.411	826.7	826.7	.9961
.5900	.5907	3.712	.9988	20.45	20.47	7.423	837.1	837.1	.9962
.5910	.5917	3.718	.9988	20.57	20.60	7.436	847.6	847.6	.9962
.5920	.5927	3.724	.9988	20.70	20.73	7.448	858.2	858.2	.9963
.5930	.5937	3.730	.9989	20.83	20.86	7.460	868.9	868.9	.9963
.5940	.5947	3.737	.9989	20.97	20.99	7.473	879.8	879.8	.9963
.5950	.5957	3.743	.9989	21.10	21.12	7.485	890.8	890.8	.9964
.5960	.5967	3.749	.9989	21.23	21.25	7.498	901.9	901.9	.9964
.5970	.5977	3.755	.9989	21.35	21.37	7.510	913.4	913.4	.9964
.5980	.5987	3.761	.9989	21.49	21.51	7.523	925.0	925.0	.9965
.5990	.5996	3.767	.9989	21.62	21.64	7.535	936.5	936.5	.9965
.6000	.6006	3.774	.9990	21.76	21.78	7.548	948.1	948.1	.9965
.6100	.6106	3.836	.9991	23.17	23.19	7.673	1,074	1,074	.9969
.6200	.6205	3.899	.9992	24.66	24.68	7.798	1,217	1,217	.9972
.6300	.6305	3.961	.9993	26.25	26.27	7.923	1,379	1,379	.9975
.6400	.6404	4.024	.9994	27.95	27.97	8.048	1,527	1,527	.9977
.6500	.6504	4.086	.9994	29.75	29.77	8.173	1,771	1,771	.9980
.6600	.6603	4.149	.9995	31.68	31.69	8.298	2,008	2,008	.9982
.6700	.6703	4.212	.9996	33.73	33.74	8.423	2,275	2,275	.9983
.6800	.6803	4.274	.9996	35.90	35.92	8.548	2,579	2,579	.9985
.6900	.6902	4.337	.9997	38.23	38.24	8.674	2,923	2,923	.9987
.7000	.7002	4.400	.9997	40.71	40.72	8.799	3,314	3,314	.9988
.7100	.7102	4.462	.9997	43.34	43.35	8.925	3,757	3,757	.9989
.7200	.7202	4.525	.9998	46.14	46.15	9.050	4,258	4,258	.9990
.7300	.7302	4.588	.9998	49.13	49.14	9.175	4,828	4,828	.9991
.7400	.7401	4.650	.9998	52.31	52.32	9.301	5,473	5,473	.9992

APPENDIX (Continued)

d/L_o	d/L	$2\pi d/L$	tanh $2\pi d/L$	sinh $2\pi d/L$	cosh $2\pi d/L$	$4\pi d/L$	sinh $4\pi d/L$	cosh $4\pi d/L$	H/H_o'
.7500	.7501	4.713	.9998	55.70	55.71	9.426	6,204	6,204	.9993
.7600	.7601	4.776	.9999	59.30	59.31	9.552	7,034	7,034	.9994
.7700	.7701	4.839	.9999	63.15	63.16	9.677	7,976	7,976	.9995
.7800	.7801	4.902	.9999	67.24	67.25	9.803	9,042	9,042	.9996
.7900	.7901	4.964	.9999	71.60	71.60	9.929	10,250	10,250	.9996
.8000	.8001	5.027	.9999	76.24	76.24	10.05	11,620	11,620	.9996
.8100	.8101	5.090	.9999	81.19	81.19	10.18	13,180	13,180	.9996
.8200	.8201	5.153	.9999	86.44	86.44	10.31	14,940	14,940	.9997
.8300	.8301	5.215	.9999	92.05	92.05	10.43	17,340	17,340	.9997
.8400	.8400	5.278	1.000	98.01	98.01	10.56	19,210	19,210	.9997
.8500	.8500	5.341	1.000	104.4	104.4	10.68	21,780	21,780	.9998
.8600	.8600	5.404	1.000	111.1	111.1	10.81	24,690	24,690	.9998
.8700	.8700	5.467	1.000	118.3	118.3	10.93	28,000	28,000	.9998
.8800	.8800	5.529	1.000	126.0	126.0	11.06	31,750	31,750	.9998
.8900	.8900	5.592	1.000	134.2	134.2	11.18	36,000	36,000	.9998
.9000	.9000	5.655	1.000	142.9	142.9	11.31	40,810	40,810	.9999
.9100	.9100	5.718	1.000	152.1	152.1	11.44	46,280	46,280	.9999
.9200	.9200	5.781	1.000	162.0	162.0	11.56	52,470	52,470	.9999
.9300	.9300	5.844	1.000	172.5	172.5	11.69	59,500	59,500	.9999
.9400	.9400	5.906	1.000	183.7	183.7	11.81	67,470	67,470	.9999
.9500	.9500	5.969	1.000	195.6	195.6	11.94	76,490	76,490	.9999
.9600	.9600	6.032	1.000	203.5	203.5	12.06	86,740	86,740	.9999
.9700	.9700	6.095	1.000	222.8	222.8	12.19	98,350	98,350	.9999
.9800	.9800	6.158	1.000	236.1	236.1	12.32	111,500	111,500	.9999
.9900	.9900	6.220	1.000	251.4	251.4	12.44	126,500	126,500	1.000
1.000	1.000	6.283	1.000	267.7	267.7	12.57	143,400	143,400	1.000

INDEX

Abyssal depths, 6, 163
Accretion (*see also* Siltation), 10, 18, 42, 85, 86, 93, 94, 98
Added mass, force, 36, 37, 286
Adriatic Sea, 18
Africa, 75, 91, 96
Alaska, 103
Algoa Bay, 83, 84
Amphidromes, 161–163
Anaerobic conditions, 220
Angle, breaking, 288, 291
–, internal friction, 36
Angled wave train, 4, 5, 11, 33, 40, 60, 134, 135, 141–145, 298–302, 307
Angular dispersion, wave, 3
Arctic coasts, 90
Arrested saline wedge, 206–208, 230
Asian coasts, 91
Asymmetry, wave, 12
Atlantic Ocean, 177, 188
Atmospheric pressure, 187, 188
Australia, 86, 96

Backwash, 11
Baltic Sea, 17
Bar, offshore, *see* Offshore bar
Barometric pressure, 187
– surge, 187
Basin tests (*see also* Model tests), 299–303
Bay, crenulate-shaped, *see* Crenulate-shaped bay
–, head, 101
Bays, 83, 84, 227, 272
Beach degradation (*see also* Erosion), 12–15, 72, 85
– encroachment, 25
– face, 10–12, 28
– model tests (*see also* Model tests), 293, 298
– nourishment, 145, 146
– plan, equation, (*see also* Crenulate-shaped bay) 77
– processes, 6–24
– profile, 1, 10–13, 16–19, 24–28, 39, 71, 72, 119, 128, 139, 295–298, 308, 309
– –, calculation of, 24

– –, storm, 12, 13, 24–28, 139, 308, 309
– –, swell, 10, 19, 24–28, 297, 308
– ridges, 99
– rock, 15
– slope, 296, 307
– zone, 128
Beaufort number, 179
Bed contours, 10, 83, 85, 93
– dunes, *see* Dunes, bed
– load, 227, 300
– roughness, 36, 300
– shear, 42
Bering Sea, 18, 103
Berm, beach, 11, 17, 27, 135
– height, 27
Bitumen sealing, 143
Black Sea, 17
Block, water *see* Water block
Borderland, continental, 101, 106–111
Bore, wave, 6
Bottom percolation, 29
Boundary layer, 21, 35–39, 44, 45, 54, 61, 299, 305
Breaker zone, 27, 33, 34
Breaking angle, 286, 289
–, wave, 2, 18–21, 59, 241, 287, 288, 296
Breakwaters, 130–134, 288
–, floating, 132–134
–, offshore, 130–132
–, rubble-mound, 288
Breezes, effect of, 272–274
Brick pattern, 50
By-passing, 139, 140

Cagliari, Gulf of, 88
Calcium carbonate, 23, 121, 123, 140, 228
Calculation of littoral drift (*see also* Littoral drift), 28, 30, 35
Cambay, Gulf of, 191, 192
Canal, 180
Cape Kennedy, 87
Carbonate, calcium, *see* Calcium carbonate
Caribbean Sea, 18
Carstens number, 53, 54
Caspian Sea, 17

Catchment, 114–118, 123
Cauchy number, 287
Caustic, wave, 144, 145
Cavitation, 152
Celerity, wave, 235
Centre, cyclonic, 2, 17, 128, 188, 274
Channel, dredged, *see* Dredged channel
Chart, cyclonicity, 17
–, hydrographic (*see also* Hydrographic survey), 21, 24, 115
Chemical plants (*see also* Power plants), 266
Chezy coefficient, 293
Chinese coast, 103
Circle, ecliptic, 161
Circular jet (*see also* Jet), 243–249, 251, 254–259
Clapotis, complete, *see* Standing wave
Clapotis gaufré, *see* Angled wave train
Clay, 228
Cliffs (*see also* Rocky coast), 84–87, 143
Climate (*see also* Wave climate), 100, 101
Clyde estuary, 229
Cnoidal wave, 29
Coastal defense, 127–153
Coastal Engineering I, 2–6, 40, 43, 58, 94, 127, 159, 166, 185, 188, 195, 196, 230, 232, 261, 298, 300
Coastal plain, 88, 89, 98–101, 108–111
Coasts (*see also* Bays, Continental margins, Gulfs), 17, 75, 86, 87, 90–96, 100–114, 145, 192, 270, 273
Coefficients, 34–38, 120, 167, 168, 179, 211–214, 218–222, 230, 232, 241, 268, 293
Colloids, 228
Concentration, sediment, 39
Continental borderland, 101, 106–111
– margin, 15, 84, 85, 94, 96
– patterns (*see also* Coasts), 90–94
– Shelf (*see also* Continental borderland, Continental margin), 1, 5–10, 18, 19, 24, 31, 34, 35, 53, 59–61, 82, 86, 89, 93, 94, 97, 99, 101, 119–123, 140, 148, 159–163, 174, 178, 183–190, 203, 215, 298, 299, 302
– slope, 6, 163, 177, 191–195
Contour, bed, 10, 83, 85, 93
– spreading, 85
Control, estuary, 228, 229
–, headland, 131–134, 136, 147
Convection, 261
Cooling water intake, 5, 239, 265–274, 278
– – coefficient, 268
– –, offshore, 269–272
– – outlet, 239, 262–266, 308
– – structures, 219, 239, 240, 262–279, 308

Coral, 97
Co-range lines (*see also* Tide), 161, 162
Coriolis force, 163, 288
Correlation coefficient, 120
Co-tidal lines (*see also* Tide), 161, 162
Crenulate-shaped bay (*see also* Equilibrium shape, bay), 73–75, 84–86, 90, 100, 108, 130–132, 136, 141, 146, 295
Cross-current, 260, 261
Current, 13, 18, 41, 42, 45, 120, 129, 136, 143, 147, 150, 163, 170, 171, 260, 261, 296
–, cross-, 260, 261
–, ocean, 1, 20
–, rip, 13, 129, 143
–, tidal, 136, 147, 150, 151, 163, 170, 171, 296
Curtain wall (*see also* Cooling water intake), 267, 278
Cycle, 23, 223, 292, 293, 303
–, tidal, 23, 223, 292, 293
–, wave, 303
Cyclones, 19, 139, 183–190, 195
–, extra-tropical, 96, 159, 183, 195
–, tropical, 19, 96, 139, 159, 184–190
Cyclonic centres, 17, 102, 105, 178, 188, 274
– –, polar, 102, 105
Cyclonicity chart, 17

Darcy's coefficient, 167, 168
Datum, 161
Decay zone, wave, 3
Declination, 160
Defense, coastal, 127–153
Degradation, beach (*see also* Erosion), 72, 85
Delta, 111–123
Deltaic plain, 107
Densimetric Froude number, 204, 205, 225, 226, 233, 243–253, 256, 257, 265, 268, 274–276, 292, 308
– velocity, 204
Density current (*see also* Stratification), 229
– differences (*see also* Stratification), 22, 224, 227
Depths, abyssal, 6, 163
Diffraction, wave, *see* Wave diffraction
Diffuser, multiport, 224, 256
Diffusion, 211–214, 218–222, 230, 232, 293
– coefficient, 211–214, 218–222, 230, 232, 293
–, molecular, 211
Dilution (*see also* Marine hydraulics works), 242–261
Dingwall harbour, 142, 145
Directional energy, wave, 3, 295

Discharge, freshwater, 216, 262
–, mean annual, 115
–, river, 203, 215, 223, 296
Dispersal zone, wave, 3
Dispersion model (see also Estuarine problems), 220, 221
– of pollutant, 220, 229, 231, 241
–, wave, 3, 4
Disposal, sewage, 240–261
Distortion (see also Model tests), 297, 308
Distributary, 117–122
Distribution, 38, 220, 242, 253
–, Gaussian, 220, 242, 253
–, Rayleigh, 38
Diurnal tide (see also Tide), 160
–, inequality, 160
Downcoast, 7
Drag, sediment, 34, 36–38, 40
– coefficient, 34, 36, 38
Drainage basin, 114–118, 123
Drakes Bay, 83
Dredge components, 149
Dredged channel, 16, 19, 127, 128, 137–140, 148, 150, 175
Dredging, 134, 139, 147–152, 229, 295
– spoil, 139, 147–150
Drift, see Littoral drift, Longshore drift
Dumping grounds (see also Dredged channel, Dredging), 229
Dune height, 35
Dunes, bed, 8, 9, 33, 34, 41, 43, 46–54, 169, 299, 302–305
–, vortex, 46
–, sand, 100, 101, 123
Duration, 6, 118
– swell, 118
– wind, 6
Dye, 44, 48, 49
Dyke, 151
Dynamic similitude, 286
– viscosity, 291

East China Sea, 18
Ebb tide (see also Tide), 21, 22, 137, 203, 210, 211, 216, 219, 222, 225, 228, 231
Ecliptic circle, 161
Ecological nuisance (see also Pollution), 239
Eddies (see also Turbulence, Turbulons), 21, 30, 33, 40, 45, 47, 48, 51
Effect of breezes, 272–274
Effects of permeability, 52
Effluents (see also Marine hydraulics works), 219, 228, 240, 241, 256, 259, 266, 274

Elasticity, force of, 284, 287, 291
Enclosed bodies of water (see also Enclosed seas), 178, 180
Enclosed seas (see also Seas), 15, 17, 21, 25, 28, 98
Energy dissipation, 29, 127
–, wave, 3, 17, 295
Equation of beach plan (see also Crenulate-shaped bay), 77
Equilibrium shape, bay (see also Crenulate-shaped bay), 72, 75–77, 81, 82, 90, 131, 132, 141, 146
Equinoctial spring tide (see also Tide), 161
Erosion, 1, 16–19, 26, 42, 57, 76, 81, 98, 113, 127–133, 141–145
–, beach, 12–15, 72, 85
Estuaries (see also Estuarine problems), 1, 18, 22, 203, 209
Estuarine problems, 146, 147, 203–233
– shoaling (see also Siltation), 23, 24, 121, 224–228
Estuary control, 228, 229
– mouth, 227, 294
– pollution, 218–223
Euler number, 287
European coasts, 90
Eustatism, 96
Evaporation, 262
Experimental coefficients, 241

Factor, safety, 256
Fauna, marine, 23, 60, 121, 123, 220, 229, 259
Feeder current, rip, 13
Fetch, wind, 6
Fickian law, 211
Fiords, 209
Fixed headlands, 72–74, 81–83, 90, 100, 146, 295, 296
Floating breakwater, 132–134
– debris (see also Pollution), 266
– groin, 133
Flocculation, 228, 229
Flood tide (see also Tide), 21–23, 137, 203, 210, 211, 216, 219, 222, 225, 228, 230
Flooding, 113, 114
Flora, marine, 219, 229, 259, 266
Florida, 87
Flow establishment (see also Jet), 242–246
– predominance (see also Estuarine problems), 210
Fluctuations, pressure, 40–43, 299, 304, 307
Flume tests (see also Model tests), 43, 143, 288, 296, 299, 300
Fluvial hydraulics, 294, 295

Force, 163, 168, 286–288, 291
– Coriolis, 163, 168, 288
–, elasticity, 286, 287, 291
–, gravity, 286, 287
–, inertia, 286
–, pressure, 286, 287, 291
–, surface tension 286, 287, 291
–, viscosity, 286, 287, 291
Fremantle harbour, 99
Freshwater discharge, 216, 262
Friction factor, 166–168, 293, 300
– loss, 152, 166–168
Froude number, 287, 293, 299
Froudian law, 290–293
Fully arisen sea, 5, 6, 43, 58
Funnel-shaped bay, 18, 22, 24, 146, 147, 163–165, 175

Ganges river (*see also* Delta), 114
Gaussian distribution, 220, 242, 253
Generation vortex, 33, 34, 43, 44
Geologic period, 84, 89, 96, 98, 101
– recent, 84, 98
Geometric distortion (*see also* Model tests), 286, 292, 293
– similitude, 286
Grain size, 35
Grains, incipient movement, *see* Incipient movement of grains
Grand Canyon, Colorado, 98
Gravel (*see also* Shingle), 97
Gravity force, 284, 287
Great Lakes of U.S., 17
Greens law, 166, 172
Groins, 128–133, 140, 146
–, floating, 133
–, spur, 140
Ground water, 19
Gulfs, 18, 88, 175, 190–192

Half heart (*see also* Crenulate-shaped bay), 72
Half Moon Bay, 83
Halocline, 209
Harbour, 140–143, 159, 307
– mouth, 140, 307
– seiche, 159
–shoreline, 140–143
Harbours, 99, 142, 145, 188, 189
Head, bay, 101
Headland control, 131–136, 147
– fixed (*see also* Rocky coast), 72–74, 81–83, 90, 100, 146, 295, 296
Height, dune, 35
–, wave, 3, 32, 33, 36, 40

High water (*see also* Tide), 217, 218
– – slack, 240, 241
Highly stratified (*see also* Stratified fluids), 208, 209, 230
Hokkaido, 101, 104, 105, 109, 110
Hong Kong (harbour) 188, 189
Honshu, 101, 113
Humps, 16, 25
Hurricane (*see also* Tropical cyclone), 185, 189, 190
–, standard project, 190
Hydraulic jump, 10–12, 265, 296
– interval, 265
– modelling (*see also* Model tests), 285–309
Hydraulics, fluvial, 294, 295
– works, marine, 239–279
Hydrographic chart, 21, 24, 115
– survey, 72, 84, 85, 90, 119, 161, 174, 295

Ice, 119, 123
Incipient movement of grains, 42–45, 48, 54–57, 60, 61, 299, 302–305
India, 192
Indian ocean, 177
Inertia force, 36, 37, 286
– coefficient, 36, 37
Inland Sea, 101, 108, 112
– seas, *see* Enclosed seas
– waterways, 117, 159
Inlet entrances, 134–138
Intake, cooling water, *see* Cooling water intake
Interaction, wave, 60
Interface (*see also* Stratified fluids), 22, 209, 230, 272, 273
Internal friction, sand, 36
– hydraulic jump, 265
– waves, 60
Intrusion length (*see also* Saline wedge), 217, 218, 231, 293
Island, offshore, 106
Isthmus (*see also* Sand spits), 109

Japan, 100–113, 270, 273
Jet, 241, 243–258
–, circular, 243–248, 251, 255–258
–, slotted, 249–254
–, experimental verification, 241, 254–256
– pump, water (*see also* Mixing tube), 151–153, 257, 258
Jump, hydraulic, 10–12, 265, 296
–, internal, 265

Kinematic similitude (*see also* Similitude), 286
– viscosity, 208, 291

INDEX

Kutch, Gulf of, 175, 190–192
Kyushu, 101–106

Lagoon, 87–89, 109–112, 118, 134
Lake (*see also* Enclosed seas), 111, 118, 178–183
Laminar conditions, 44, 45
Land reclamation, 148, 229
Landmass (*see also* Coasts), 71, 89
Laws, 42, 166, 172, 211, 290–293
Layer, boundary, *see* Boundary layer
Length, river, 115
Lift, sediment, 34, 36, 38, 40
– coefficient, 34, 36, 38
Limestone, 140, 150
Littoral current, 21, 23, 29, 57, 59, 74, 76, 141, 239, 269, 291
Littoral drift (*see also* Longshore drift), 3, 7, 15–17, 20, 21, 24, 29–32, 57, 227, 295
Logarithmic spiral (*see also* Crenulate-shaped bay), 74–84
Long-period waves, 159–196
– transition, 163–173
Longshore drift (*see also* Littoral drift), 24, 71, 75, 84–89, 94, 96, 100, 110, 113, 123, 127–136, 139–141, 147
Loss friction, 152, 166–168
Low water, 217
– – slack, 215, 231, 240, 241

Mach number, 287
Mach-stem effect, 145
Macroscopic view, 1, 42, 302
Macro-turbulence, 5, 33, 35, 40, 212, 256, 261, 299, 305–307
Magnus effect, 39, 303, 304
Malaysian peninsular, 75
Manchuria, 103
Margin, 15, 28, 29, 84, 85, 94, 96
–, continental, 15, 84, 85, 94, 96
–, oceanic, 28, 29
Marine fauna, 23, 60, 121, 123, 220, 229, 259
– flora, 219, 229, 259, 266
– hydraulics works, 239–279
– outfalls, 219, 239, 263, 274–276
Marshland, 118
Mass transport, 9, 10, 16, 21, 23, 27, 39, 42–45, 58–60, 71, 143, 211, 227, 231, 261, 291, 298–300, 305–307
Mean annual discharge, 115
– – sediment yield, 115
– highwater springs (*see also* Tide), 161
– spring tide (*see also* Tidal range), 18
Mediterranean Sea, 17, 230

Median diameter, sediment, 27, 28
Meteorological cycle, 1, 15
– tide, 178
Mexico, Gulf of, 18
Microscopic approach, 1, 39
Millibar, 188
Mindano deep, 97
Mineral-fired steam plants, 219
Mississippi river, 205
Mixing processes, 208–213, 230
Mixing tube (*see also* Jet), 239, 256–259, 277
Mizushima Bay, 272
Model tests (*see also* Flume tests), 41, 42, 72–83, 135, 219, 292–308
– basin, 292, 299–303
– beach, 293, 298
– wave, 287–290
– movable bed, 41, 294–308
Molecular diffusion, 211
Momentum, 291
Monsoon (*see also* Tropical cyclone), 102
Moon, 175
Mouths, 21–24, 88, 89, 108, 122, 134–140, 146, 227, 307
–, estuary, 227, 296
–, harbour, 140, 307
–, inlet, 134–138
–, river, 21–24, 88, 89, 108, 122, 134–136, 146, 296, 307
Movable bed model (*see also* Model tests), 41, 294–308
Movement, sediment (*see also* Incipient movement of grains), 3, 106
–, tectonic, 7, 53
Moving wind field, 183, 186, 187
Mud, *see* Silts
Multi-directional waves (*see also* Storm waves), 2, 3, 15, 41
Multiport diffusers, 224, 256

Natural period, basin, 161
Navigation (*see also* Dredged channel, Dredging), 113, 114, 146–148, 210
Neap tide (*see also* Tidal range), 161
Net sediment movement (*see also* Sediment transport), 84, 88, 90, 94–96
Netherlands, 114
Node, 175
Non-uniform density medium (*see also* Jet), 246–249, 253, 254
Norfolk coast, 145
North America, 75, 92
North Sea, 18
Norway, 205

Nourishment, beach, 145, 146
Nuclear power plants, 219, 266
Null point (*see also* Estuarine shoaling), 224–227, 233
Number, 53, 54, 179, 242, 287, 293, 299
–, Beaufort, 179
–, Carstens, 53, 54
–, Cauchy, 287
–, densimetric Froude, *see* Densimetric Froude number
–, Euler, 287
–, Froude, 287, 293, 299
–, Mach, 287
–, Reynolds, *see* Reynolds number
–, Schmidt, 242
–, Weber, 287
Nylon mesh, 134

Ocean currents, 1, 20
Oceanic margins, 28, 29
– spectrum, 61
Oceans (*see also* Storm zones of oceans), 161, 175, 177, 188
Offshore bar, 12–19, 71, 74, 129, 133, 139, 270, 296–298
– breakwater, 130–132
– intake, *see* Cooling water intake
– islands, 106
– wave action, 298–307
– zone, 1, 6–8, 11, 16, 18, 24–27, 33, 34, 41, 57, 61, 71, 90, 98, 127, 128, 131, 146, 174, 295, 296
Oil slick, 266
Onshore wave action, 296
Optimum surge, 187
Orbital motions, *see* Water orbital motions
–velocity, *see* Water orbital motions
Oristano, Gulf of, 88
Orthogonal, wave, 82, 83, 210
Oscillating sand bed (*see also* Model tests), 43, 303–305
– water mass (*see also* Model tests), 34, 44, 305, 306
Outfall, marine, 219, 239, 263, 274–276
Outlet, cooling water (*see also* Cooling water structures), 239, 262–266, 308
Oxygen uptake, 220, 221

Pacific coast of North America, 75
– coasts, 92, 101–106
– Ocean, 161, 175, 177
Panama zone, 175
Paradox, marine sedimentation, 96
Partial standing wave, 11, 175

Partially mixed (*see also* Estuarine problems), 209, 210, 230
Past geological period, 89, 96, 98, 101
Pattern, brick, 50
–, continental, 90
–-forming forces (*see also* Deltas), 114–119
–, world-wide wind, 94, 95
Peninsular (*see also* Promontories, Rocky coast), 100, 109
Percolation, 11, 14, 15, 29, 38
–, beach, 11, 14, 15
–, bottom, 29, 38
Period, tidal (*see also* Tide), 210, 215, 277
–, wave, 27, 36
Permeability, effects of, 52
Persistent swell, (*see also* Swell), 10, 13–16, 25, 29, 33, 71, 74, 75, 82, 89, 104, 128–130, 135, 136, 139–143, 146
Phase shift, 219
Physiographic feature, 22, 94, 100, 147
– unit, 18, 84–86, 89, 90

Quick sand, 12

Radial dispersion, 4
Radiation, heat, 261
Radioactive tracers, 57, 58
Rainfall, 23, 101, 114, 118, 121, 228
Range, tidal, *see* Tidal range
Rayleigh distribution, 38
Reach, wave, 53–57
Recent geologic period, 84, 98
– sediment, 71
Reclamation, land, 148, 229
Refineries (*see also* Power plants), 266
Reflection, *see* Wave reflection
Refraction, wave, 5, 7, 10, 20, 25, 27, 35, 85, 100, 107–109, 135, 144, 147, 288, 291
Regime, in (*see also* Sedimentation), 16, 71
Regression analysis, 82, 119–123
Resonant surge, 159, 163
Revetments (*see also* Seawalls), 143–145
Reynolds number, 37, 44, 168, 169, 204, 205, 213, 214, 221, 241, 287, 293, 299
Ridge, pitch, 45, 50
–, rolling grain, *see* Rolling grain ridge
Rip current, 13, 129, 143
– –, feeder to, 13
Ripple, sand, 33, 34, 43–45, 56, 60, 76, 89, 169, 298–303
–, pitch, 60
–, vortex, 46
– water, 287
Riser pipe, 269

River bar (*see also* River mouth), 134
– catchment, 114–118, 123
– discharge, 203, 215, 223, 296
– length, 115
– Mississippi, 205
– mouth, 21–24, 88, 89, 108, 122, 134–136, 146, 296, 307
– sand supply, 295
– system, 1, 14
– Thames, 211
Rocky coast (*see also* Cliffs), 94, 97
Rolling grain ridges, 44–47, 54
Root mean square wave height, 32
Rose, wind, 17
Rotterdam waterway, 215
Roughness, bed, 36, 300
Rubble-mound breakwater, 288

Safety factor, 256
Saint Francis Bay (*see also* Crenulate-shaped bay), 83
Sakhalin Island, 105
Saline wedge, 22, 23, 121–123, 203–213, 227, 230, 231, 272
– – arrested, 206–208, 230
Salinity, 119, 122
– distribution (*see also* Stratified fluid), 216, 217
Salt action (*see also* Incipient movement of grains), 52
– water diffusion, 40
– – intrusion (*see also* Stratified fluid), 114, 224, 230, 233
Sand, 12, 94, 97, 262
–, quick, 12
– dunes, 100, 101, 123
– particle motion (*see also* Incipient movement of grains), 43, 52
– roughness, 169
– spit (*see also* Sedimentation), 17, 89, 107, 110, 111, 131, 134
– supply, 295
Sandy Hook (*see also* Crenulate-shaped bay), 83
– protuberance (*see also* Siltation), 295
Sardinia, 88
Saturation, beach, (*see also* Water table), 143
Savannah estuary, 226, 227
Schmidt number, 242
Scouring, 177, 210
Sea, fully arisen, 5, 6, 43, 58
– state, 179
Sealing, bitumen, 143

Seas (*see also* Enclosed seas, Semi-enclosed seas), 17, 18, 101–104, 112, 230
Seasonal storm, 28
Seawalls (*see also* Revetment), 15, 19, 143–145, 308
Seaweed, 5, 262, 267, 269
Sediment movement (*see also* Incipient movement of grains), 3, 33–42, 84, 88, 90, 94–96, 106
– –, net, 84, 88, 90, 94–96
– –, theory, 33–42
– size, 27, 28, 36, 86
– sorting, 35
– supply, 71, 98
– transport, 43, 227, 300
– –, bed load, 227, 300
– –, wash load, 227, 300
– yield, mean annual, 115
Sedimentation, marine, 7
Seiche, 159, 178, 288
– harbour, 159
Semi-diurnal tide (*see also* Tide), 160, 171, 190
Semi-enclosed seas, 17, 18
Set-up, wave, 159
Sewage disposal, 240–261
– effluent (*see also* Effluents), 266, 274
Shadow zone (*see also* Wave diffraction), 74
Shape, crenulate, *see* Crenulate-shaped bay
–, equilibrium, *see* Equilibrium shape, bay
–, funnel, *see* Funnel-shaped bay
Shear, 52, 53
– stress, 35, 42, 45, 300
– – approach, 39–41, 51
– velocity, 33, 40, 293
Shelf, continental, *see* Continental shelf
– edge, 94, 174
Shell debris (*see also* Marine fauna), 23, 140, 150, 228
Shikoku, 101–106
Shingle (*see also* Gravel), 94, 97, 146
Shoals (*see also* Siltation) 87–89
Shoreline harbours, 140–143
– processes, 1–61
Short-crested wave, *see* Angled wave train
Significant height, wave, 3, 33
Siltation (*see also* Accretion), 1, 18, 23, 24, 57, 113, 127–134, 139–147, 177, 210, 211, 224, 227–229, 266, 307
–, estuary (*see also* Siltation), 23, 24, 121, 224–228
Silts, 93, 94, 97, 133, 134, 150, 211, 228, 229
Similitude, 286–297
–, dynamic, 286

Similitude, geometric, 286
—, kinematic, 286
—, laws of, 42
Sinusoidal, 4, 29, 40, 299, 305
Skimmer wall, 268, 269, 278
Slack, high water, 240, 241
Slope, continental, see Continental slope
Slotted jet, 249–254
Solar component (see also Tide), 163
Solitary wave, 29, 30
Sorting, sediment, 35
South Africa, 75, 83
South America, 93–96
South China Sea, 18
Spectrum, turbulence (see also Turbulence), 39, 42, 208, 212
Spectrum, wave, 21, 25, 32, 33, 43, 58
— power, 32, 33
Spit, sand, see Sand spit
Spoil, dredge (see also Dredging) 139, 147, 148, 150
Spring tide (see also Tide), 161, 176, 192
— — equinoctial, 161
Spur groin (see also Groins), 140
Stability, test of (see also Crenulate-shaped bay), 81–84
Stage, tidal (see also Tidal range), 209
Standard project hurricane (see also Tropical cyclone), 190
Standing wave, 11, 60, 134, 140–143, 175
— —, partial, 11, 175
Starvation, jet (see also Jet), 246, 259
Static wind field, 183–186
Steel works (see also Estuarine problems), 266
Steepness, wave, 27, 287, 288, 298
St. Francis Bay, 83
Stochastic process, 33, 43
Stone pitching (see also Seawalls), 143
Storm beach profile (see also Beach profile), 12, 13, 24–28, 139, 308, 309
— centre (see also Cyclones), 2, 17, 178, 188, 274
— sequence, 60, 72, 128, 130, 134
— surge (see also Wind stress), 13, 18–20, 25, 26, 99, 114, 123, 159, 160, 178–190, 195, 196
— waves, 2, 6, 8, 11, 12, 17–21, 32, 40–42, 71, 72, 94, 99, 105, 106, 118–121, 127–131, 134, 139–143, 146, 150, 151, 178
— zones, ocean (see also Cyclones), 71, 84, 86, 94, 96, 103
Stratification (see also Estuarine problems), 22, 224, 227

Stratified fluids, 22, 204–211, 214
Stress shear, 35, 42, 45, 300
Submarine canyon (see also Continental shelf), 97, 107, 108, 112
— feature, 85, 89, 106
— shelf (see also Continental shelf), 107
Submerged pipeline, 25
— objects, 290
Supply, sand, 295
Surf beat, 159
— zone (see also Beach profile), 1, 2, 6, 7, 15–19, 24–28, 33, 34, 57–60, 71, 85, 127, 139, 296, 298
Surface layer (see also Stratified fluids, Thermocline), 240, 241, 264, 265, 272, 273
— —, warm, 272, 273
— tension, force, 286, 287, 291
Surge, 159, 163, 187, 272, 273, 288
—, barometric, 187
—, optimum, 187
—, resonant, 159, 163
— height, 180–190
—, storm, see Storm surge
Survey, hydrographic, see Hydrographic survey
Swamp land (see also Estuarine shoaling), 110
Swash zone (see also Surf zone), 27
Swell (see also Persistent swell), 2, 4, 6, 10, 11, 20, 24, 28, 35, 42, 71, 72, 84, 90, 94, 99, 107, 118–121, 127, 130–135, 139, 141, 150
— beach profile (see also Beach profile), 10, 19, 24–28, 297, 308
— duration, polygon (see also Multi-directional waves), 118
Swellistence (see also Persistent swell), 16

Tables, tide, 177, 192
Tangent section (see also Crenulate-shaped bay), 72, 77, 82
Tectonic movements, 7, 53
Temperature (see also Cooling water structures), 114, 115, 118, 120, 123
Tephra, volcanic, 101
Terminal depth, 261
Terrigenous material, 24, 35, 295
Tests, flume, see Flume tests
—, model, see Model tests
—, tracer, see Tracer tests
Thames, 211
Theory, sediment movement (see also Sediment movement), 33–42
Thermal pollution (see also Pollution), 219, 225, 262

INDEX

Thermocline (*see also* Surface layer), 262, 263, 268–270, 278, 279
Tidal constituents (*see also* Tide), 170
– current, 22, 136, 147, 150, 151, 163, 170, 171, 296
– cycle, 23, 223, 292, 293
– excursion (*see also* Saline wedge), 217, 222
– height, 25, 26
– inlet (*see also* Littoral drift), 136, 137
– oscillations (*see also* Tide), 203, 208, 211–214, 222
– period, 210, 215, 277
– prism, 137, 215, 222, 223, 231
– range, 18, 99, 101, 143, 150, 162, 173–177, 191–195
– shoals (*see also* Estuarine shoaling), 106, 175
– waves, *see* Tsunami
Tide (*see also* Ebb tide, Flood tide), 1, 13, 17–19, 60, 118–123, 159–164, 171, 176, 178, 190–195, 229, 232, 288, 295
–, diurnal, 160
–, meteorological, 178
–, semi-diurnal, 160, 171, 190
– generator (*see also* Model tests), 294
– tables, 177, 192
Tombolo, 17, 89, 107, 110, 111, 131, 134
Topography, 115, 117
Tracer tests, 57, 58, 84, 136
– –, radioactive, 57, 58
Trade winds, 102
Train, wave, 297
Transition, long wave, 163–173
Transport, mass, *see* Mass transport
–, sediment, 43, 57, 227, 293, 300
– rate, 57, 293
Tray, model (*see also* Model tests), 47, 300, 304, 305
Triangular wind field (*see also* Storm surge), 186, 187, 195
Tributary, 116, 211
Tropical cyclone (*see also* Hurricane, Typhoon), 19, 96, 139, 159, 180–186, 189, 190
Tropics, 19, 24
Tsunami (*see also* Long-period wave), 159
Tunnel, water (*see also* Model tests), 43, 44, 305, 306
Turbulence (*see also* Macro-turbulence), 9, 39, 40, 42, 51, 61, 208, 212, 300, 304
– spectrum, 39, 42, 208, 212
Turbulons (*see also* Macro-turbulence), 208, 212, 258, 303, 306
Typhoon (*see also* Tropical cyclones), 41, 102, 103, 185, 189, 195, 196, 270

Typical tide (*see also* Tide), 174, 176

Underwater contours, 10, 83, 85, 93
Uni-directional flow (*see also* Current), 33, 41, 45, 49
Uniform density medium (*see also* Jet), 243–245, 249–252, 255
– wind field (*see also* Storm surge), 184, 187
Upcoast (*see also* Longshore drift), 7, 90
Uprush (*see also* Beach profile), 10, 11

Variance (*see also* Pattern forming forces), 120
Vector resultant, wave, 84, 118
Vegetation (*see also* Deltas), 114, 118, 123
Vehicle, submersible, 10
Velocity cap (*see also* Cooling water intake), 269–272, 278
–, densimetric, 204
– fluctuations (*see also* Water orbital motions), 40, 42, 43
– predominance curve (*see also* Estuarine shoaling), 224, 225
Viscosity, 208, 286–288, 291
–, dynamic, 291
– force of, 286–288
–, kinematic, 208, 291
Viscous effects (*see also* Mass transport), 21
– friction (*see also* Model tests), 296
– sublayer (*see also* Shear stress approach), 41, 45, 293
Volcanic tephra, 101
Volume I, *Coastal Engineering (see Coastal Engineering, I)*
Vortex dunes (*see also* Dunes, sand), 46
– generation (*see also* Vortices), 33, 34, 43, 44
Vortices (*see also* Macro-turbulence), 40, 46–50, 60, 175, 269, 270, 302, 303, 306

Warm surface layer (*see also* Surface layer), 272, 273
– water (*see also* Thermal pollution), 219
– – effluent (*see also* Effluents), 266
Wash load, 227, 300
Water block (*see also* Model tests), 306, 307
– jet pump (*see also* Mixing tube), 152, 153, 257, 258
– orbital motions, 4, 6, 35, 36, 40, 41, 44–50, 56, 60, 162, 163, 169, 291
– – velocity, 46–49, 291
– ripples (*see also* Model tests), 287
Watershed (*see also* Deltas), 100, 114–118, 123
Waterways inland (*see also* Estuarine problems), 117, 159, 215
– –, Rotterdam, 215

Wave action, 296–307
– –, offshore, 298–307
– –, onshore, 296
– asymmetry (*see also* Wave steepness), 12
– base, 53–57, 85, 101
– breaking, 2, 18–21, 59, 241, 287, 288, 296
–, caustic, 144, 145
– celerity, 287
– characteristics, 1–5
– climate, 10, 25, 28, 33, 40, 71, 98, 99, 127, 128, 139, 146, 151
–, cnoidal, 29
– cycle, 303
– diffraction, 25, 35, 74, 75, 85, 100, 104–112, 135, 141, 144, 147, 291, 295
– dispersion, 3
– dissipation, 29, 127
– energy, 3, 17, 295
– –, directional, 3, 295
– flumes, *see* Flume tests
– generator (*see also* Model tests), 289, 290, 307, 309
– height, 3, 32, 33, 36, 40
– –, root mean square, 32, 40
– –, significant, 3, 33
– interaction, 60
–, internal, 60
– orthogonal, 82, 83, 210
– period, 27, 36
–, progressive, 33, 87, 166
– reach, 53–57, 85, 101
– reflection, 11, 29, 85, 87, 127, 128, 134–136, 141–147, 172, 269, 288–291, 301, 307
– refraction, *see* Refraction wave
– set-up, 159
–, sinusoidal, 4, 29, 40, 299, 305
–, solitary, 29, 30

– spectrum, 21, 25, 43, 58
–, standing, *see* Standing wave
– steepness, 27, 287, 288, 298
– train, 299
Waves, 2, 3, 15, 41, 159–196, 261
– long-period, 159–196
– multi-directional, 2, 3, 15, 41
–, storm, *see* Storm waves
Weathering, 35, 96–98
Weber number, 287
Wedge, saline, *see* Saline wedge
Well mixed (*see also* Estuarine problems), 208–213, 230
Western Australia, 99
Wind, 6, 43, 102, 114
–, duration, 6
–, trade, 102
– field, 183–187, 195
– –, moving, 183, 186, 187
– –, static, 183–186
– –, triangular, 186, 187, 195
– –, uniform, 184, 187
– rose, 17
– stress, 159, 178–190, 239, 241, 262, 272, 273
– – coefficient, 179
– systems, 94–96, 102
World-wide wind pattern, 94, 95

Yanquina Bay, 227

Zone, 18, 27, 33, 34, 57, 59–60, 84–86, 89, 90
–, beach, 128
–, breaker, 27, 33, 34
–, physiographic, 18, 84–86, 89, 90
–, surf, 27, 33, 34, 57, 60
–, offshore, *see* Offshore zone

G-7